Introduction to Chemical Product Design

Whereas chemical engineering has traditionally focused on the design of chemical processes to produce commodity chemicals, there has been increasing attention in the discipline on the engineering of high-performance chemical products. Chemical product design principles are now being incorporated into university curricula to ensure that new chemical engineers can help meet industry demands. Bridging the gap between teaching and practice, *Introduction to Chemical Product Design* provides a comprehensive overview of chemical product design, explains the role of chemical engineers in different stages of chemical product design and covers stepwise approaches used to design different types of chemical products.

- Includes chemical product design steps with examples and case studies.
- Features mini projects to illustrate the connection between product design stages.
- Focuses on the most common types of problems that engineers encounter in their career.
- Offers solved examples and exercise problems on manufacturing steps.
- Considers the role of quality assurance and the tools to quantify environmental, health, and safety aspects.

Aimed at undergraduate- and graduate-level students in chemical engineering, this textbook provides readers with practical know-how to apply chemical product design principles to develop and manufacture high-value products. It will also be of interest to practising chemical engineers seeking to expand their knowledge of chemical product design.

The book features a Solutions Manual for adopting professors.

Green Chemistry and Chemical Engineering

Series Editor
Dominic C.Y. Foo
University of Nottingham, Malaysia

Hemicelluloses and Lignin in Biorefineries
Jean-Luc Wertz, Magali Deleu, Séverine Coppée, and Aurore Richel

Materials in Biology and Medicine
Sunggyu Lee and David Henthorn

Resource Recovery to Approach Zero Municipal Waste
Mohammad J. Taherzadeh and Tobias Richards

Hydrogen Safety
Fotis Rigas and Paul Amyotte

Nuclear Hydrogen Production Handbook
Xing L. Yan and Ryutaro Hino

Water Management: Social and Technological Perspectives
Iqbal Mohammed Mujtaba, Thokozani Majozi and Mutiu Kolade Amosa

Efficiency and Sustainability in the Energy and Chemical Industries: Scientific Principles and Case Studies, Third Edition
Krishnan Sankaranarayanan

Process Integration for Resource Conservation, Second Edition
Dominic C.Y. Foo

Industrial Decarbonization: Materials, Methods, and Developments
Qingxu "Bill" Jin, Ida Chen, Fan Shi and Dushyant Shekhawat

Valorisation of Food Waste: Process and Product Design
Chien Hwa Chong

Managing Biogas Plants: A Practical Guide
Mario Alejandro Rosato

Introduction to Chemical Product Design
Nishanth G. Chemmangattuvalappil, Jecksin Ooi and Mario R. Eden

For more information about this series, please visit: https://www.routledge.com/Green-Chemistry-and-Chemical-Engineering/book-series/CRCGRECHECHE

Introduction to Chemical Product Design

Nishanth G. Chemmangattuvalappil,
Jecksin Ooi and Mario R. Eden

CRC Press
Taylor & Francis Group
Boca Raton London New York

CRC Press is an imprint of the
Taylor & Francis Group, an **informa** business

Designed cover image: © 2026 Shutterstock

First edition published 2026
by CRC Press
2385 NW Executive Center Drive, Suite 320, Boca Raton FL 33431

and by CRC Press
4 Park Square, Milton Park, Abingdon, Oxon, OX14 4RN

CRC Press is an imprint of Taylor & Francis Group, LLC

ISBN: 978-1-032-70889-8 (hbk)
ISBN: 978-1-032-70890-4 (pbk)
ISBN: 978-1-032-70891-1 (ebk)

DOI: 10.1201/9781032708911

Typeset in Times LT Std
by Apex CoVantage, LLC

Access the Instructor and Student Resources: https://routledge.com/9781032708898

Contents

Preface

Chemical product design has emerged as an exhilarating subject that has captured the attention of the chemical engineering community in recent times. Historically, the primary focus of chemical engineers lay in the efficient production of bulk chemicals through optimal processes. However, the rapid expansion of the product industry has brought forth numerous fresh opportunities and challenges for chemical engineers to tackle.

A product-centric approach to design introduces a distinct set of challenges, setting it apart from traditional process design. While engineers working in the product industry still leverage their fundamental knowledge acquired during their chemical engineering education, they must also apply several additional concepts within this unique environment. For example, the societal aspects exert a profound influence on product design decisions. Identifying, quantifying and comparing these societal attributes with the technical requirements pose significant and intricate challenges for the designers. Hence, a comprehensive understanding of these issues and the strategies employed to address them undoubtedly represents an indispensable skill set for a 21st-century chemical engineer.

Due to the changing landscape of the industry, the significance of chemical product design has prompted its inclusion in both graduate and undergraduate chemical engineering programs, although dedicated product design courses are only offered in a limited number of programs. Teaching chemical product design at the undergraduate level poses certain challenges, such as determining the most relevant topics for practising chemical engineers while ensuring a natural progression from the fundamental chemical engineering courses typically covered in such programs. In this book, we have placed special emphasis on the topics that require special attention when working in the chemical product industry. Additionally, we have explored several subjects that are typically only superficially covered in a standard undergraduate-level program but extensively employed in the product industry. We have also contextualized various processes engineers may encounter within the framework of product design. Lastly, we have presented case studies that demonstrate the application of product design tools in real-world problems, providing a comprehensive description of the various steps involved, from problem formulation to manufacturing.

This book is well-suited for undergraduate students who have received comprehensive training in fundamental chemical engineering topics and design methodologies. As a result, it can serve as a textbook for both graduate and undergraduate courses focusing on chemical product design or related subjects. The material presented in this book has been successfully utilized in teaching students within our organizations over the past several years. The inclusion of examples and practice problems throughout the book facilitates a solid understanding of the subject matter with minimal external assistance. Moreover, this book serves as an introductory resource on the principles of chemical product design for professionals working in this field. The systematic approaches outlined in this book for tackling chemical product design problems have the potential to enhance the efficiency of existing processes.

This book has been organized in the following manner. The initial chapters focus on distinguishing between commodity products and chemical products. These chapters also underscore the role of engineers in product design and their close collaboration with other team members involved in the design process. The subsequent stage delves into an exploration of some of the most powerful technical and non-technical tools available for product design. The concept of risk and uncertainty, as well as their impact on economic decision-making, is also presented. Moving forward, the book provides detailed coverage of engineering techniques that are not adequately emphasized in process design courses. These techniques are elucidated through the use of illustrative examples. Finally, the book extensively discusses the processes that find widespread application in the chemical product industry, accompanied by illustrative examples to facilitate understanding.

Lastly, we extend our sincere appreciation to all who have contributed to the successful completion of this book. We are particularly grateful to the undergraduate chemical engineering students at the University of Nottingham Malaysia Campus over the years, especially during the period 2012–2024. Special acknowledgement is due to Ms. Jaqlyn Chong, Ms. Palisha Shakya and Dr. Joon Yoon Ten, who were responsible for developing the solutions manual and case studies accompanying this book. We also wish to recognize our colleagues for their constructive feedback, which has been invaluable in improving the quality of this textbook.

Authors

Nishanth G. Chemmangattuvalappil is Professor of Chemical Engineering at Sunway University, Selangor, Malaysia. He earned his bachelor's degree in Chemical Engineering from the University of Calicut, Kerala, India, and his PhD in Chemical Engineering from Auburn University, Alabama, USA. He has over 13 years of teaching experience in chemical process and product design-related courses, along with five years of industrial experience. His primary areas of expertise include sustainable chemical product design, the application of machine learning tools in product development, carbon capture and storage using ionic liquids, and the integration of computer-aided molecular design techniques into biorefinery processes.

Jecksin Ooi serves as Assistant Professor in the Department of Chemical Engineering at Heriot-Watt University Malaysia. She received her MEng (Hons) in Chemical Engineering from the University of Nottingham Malaysia and PhD in Engineering from the same institution. Her main area of expertise covers process and molecular design, particularly in computer-aided molecular design with the integration of environmental aspects. Her research interests also include multi-criteria decision making, application of machine learning tools on chemical process and molecular design and sustainable design of chemical products.

Mario R. Eden is the Dean of the Samuel Ginn College of Engineering and Joe T. & Billie Carole McMillan Endowed Chair at Auburn University, where he has been on the faculty since 2004 and has served as Dean since 2023. Dr. Eden previously served as Chair of the Department of Chemical Engineering for almost 11 years (2012–2023), served for two years as the Acting Director of the Alabama Center for Paper and Bioresource Engineering (2014–2016) and directed the first-ever NSF-IGERT on Integrated Biorefining (2011–2018) at Auburn University. His main areas of expertise include process design, integration and optimization, as well as molecular synthesis and product design. His group focuses on the development of systematic methodologies for process and product synthesis, design, integration and optimization. He has mentored eight postdoctoral fellows and graduated 16 PhD students, along with seven MS students.

1 Introduction

The role of chemical engineers in the process industry has traditionally focused on the design of processes. Most chemical engineering textbooks put a heavy emphasis on the design of processes suitable for the production of bulk chemicals through continuous processes, whereas the production of speciality chemicals and fine chemicals is only superficially covered in most of the chemical engineering programmes. Chemical product design has become a specialized topic only since the early 2000s (Cussler and Moggridge, 2011). In the past, this has always been considered a sub-topic of the broader topic of process design. However, there are significant differences between the design philosophies involved in process and product designs. In process design, the final product is fixed in the beginning, and the focus is on the efficient manufacture of this product from raw materials. In product design, the main focus is the identification of the product to be manufactured that can generate high income for the company.

In recent years, there has been an increasing interest in product-centric design in the chemical industry (Cussler and Wei, 2003), mainly due to the increase in the number of companies that produce high-value-added products. Moreover, the field of chemical product design has been recognized as a specialization by the scientific community. While the formulation of products is often developed by chemists (Hill, 2009), chemical engineers contribute unique skill sets to the design process by addressing the processing steps required to produce these products on a commercial scale. Chemical engineers also possess unique backgrounds that allow the integration of expertise from a variety of different, yet interrelated disciplines.

Chemical product design involves creating new chemical-based products or enhancing existing ones, with an emphasis on their functionality, cost-efficiency, safety and sustainability. It combines expertise from diverse disciplines, such as various branches of engineering, chemistry, biology and material science, to develop products that fulfil specific consumer demands. Many of these topics are already covered in core chemical engineering courses, albeit in different contexts. Consequently, chemical engineering graduates can relatively easily extend their expertise into chemical product design. This book aims to provide an introduction to the tools used in chemical product design, facilitating this transition.

In this introductory chapter, the significance of chemical product design and its areas of application will be discussed. Unique attributes of chemical products will be highlighted from a chemical engineer's perspective. The value of chemical products is presented, along with a comparison of process and product design approaches. Finally, the chemical product design template, which will be used in subsequent chapters, is introduced.

1.1 CHEMICAL PRODUCTS

Before exploring the topic of chemical product design, it is important to have a clear understanding of what constitutes a chemical product. A chemical product is

DOI: 10.1201/9781032708911-1

a substance or mixture that has been manufactured or processed through chemical processes to serve specific purposes in industrial, commercial or household applications. There are different types of chemical products, such as commodity chemicals, consumer products and speciality chemicals. Traditionally, the majority of chemical engineering practitioners work in the commodity chemical industry. The commodity chemical industry is characterized by large production volumes, continuous operation, a constant production rate and the use of highly specialized equipment. These chemicals are typically composed of a single molecule or a simple mixture of a small number of molecules. Generally, chemical plants producing commodity chemicals are highly capital-intensive, with equipment optimized for specific operations and production volumes. The selling price of commodity chemicals is uniform across companies, as there is little differentiation in the product. Consequently, the profit margin for the commodity chemical industry is narrow. Opportunities to increase profit margins include making the process more cost-effective through automation, energy integration, manpower reduction and local production of raw materials and energy.

Other classes of chemical products, such as speciality chemicals, are generally produced in smaller quantities. Unlike commodity chemicals, these products are differentiated based on their performance. For example, a new medicine for headaches may replace existing headache tablets if it performs better. Performance can be evaluated using various criteria, such as how quickly it alleviates the headache, potential side effects, comfort, duration of relief and other factors. Similarly, consumers will be willing to pay a premium price for a paint of superior quality. In this case, the quality may be defined in terms of aesthetic factors, environmental factors, durability and special features. The target of such products in many cases is hard to predict as it is often related to the customer behaviour. In addition, several chemical products are not simple molecules. Instead, many of these products are complex mixtures of several chemicals where the microstructure of the final product determines the product performance (Cussler and Moggridge, 2011). In addition, most chemical products have a limited lifespan due to several factors. Customer preferences change over time, superior products emerge that render existing products obsolete, and pharmaceutical chemicals or agrochemicals become ineffective as bacteria or insects develop resistance. For these reasons, the introduction of new products is essential to remain competitive in this market. Consequently, significant investment in research and development is required in this field. This also necessitates a workforce composed of highly skilled employees. In fact, the quality of a company's scientific team can be the determining factor between success and failure.

High-value-added materials that are differentiated based on performance are categorized as chemical products (Cussler and Moggridge, 2011). Typical examples of chemical products are pharmaceuticals, rubber chemicals, personal care products, semiconductors, food products, special inks, special solvents and paint (Favre et al., 2002). The nature of the design of such products is rather unique as it requires a totally different approach to develop acceptable solutions. The most fundamental difference of such problems is the shift in focus from the cost to the quality. Conventional chemical engineering design focuses on improvements in the process because the prices of raw materials and final products are determined by the market and are generally inflexible. Therefore, the focus is to reduce the production cost. In product-centric design, the value lies in the quality and uniqueness of the product. These qualities are complex functions of market science, customer behaviour, chemical structure, product microstructure and numerous other known and unknown phenomena. Solving such problems, therefore, requires a fundamentally different approach, making this a unique topic. In fact, chemical product design has been regarded as a new paradigm in chemical engineering because it demands a completely different mindset to address such challenges (Hill, 2009).

1.2 SIGNIFICANCE OF PRODUCT-CENTRIC DESIGN

Chemical product-centric design is becoming increasingly important due to shifting market demands for products that are marketed based on their performance (Seider et al., 2009). In modern industries, there are additional requirements such as the need for sustainable solutions and environmentally friendly alternatives to existing products. Customers are increasingly demanding customized, high-performance and eco-friendly products, prompting manufacturers to design chemical products tailored to specific applications while minimizing environmental impact.

Many commonly used chemicals and products from the past have been found to pose significant safety, health or environmental risks. Furthermore, ongoing research has raised concerns about the potential adverse effects of several widely used products. For instance, asbestos is a group of minerals known for its remarkable heat and fire resistance, as well as its exceptional fibre strength and low cost. Due to these attributes, asbestos was widely used in building materials such as roofing, pipes and insulation. However, in the 1970s, asbestos was discovered to cause severe lung diseases, including cancer. As a result, many countries have banned the use of asbestos in construction materials. The challenge in this context is to develop a replacement for asbestos that retains its desirable qualities while eliminating its harmful effects. Similar stories can be found across all facets of the consumer product industry, as various once-popular products were replaced due to changing regulations and increased public awareness.

There has also been an increase in the 'technology-pushed' industries due to the tremendous advances in technology in the last century. These types of companies focus on developing newer technologies and finding applications of those technologies on consumer products (Wei, 2007). Advances in material science, nanotechnology and biotechnology have enabled innovative product development, while stricter global regulations necessitate compliance with safety, health and environmental standards. Because of this, there is a growing need for engineers who can understand these fields and effectively extract useful information to address specific problems. One advantage for chemical engineers in this context is that their curricula provide exposure to most of these topics, although not always in the most relevant context. However, the ability to extract information from these closely related fields is implicitly ingrained in chemical engineering graduates. As a result, it is easier for them to extend their skills into product design and engineering with the support of guidelines and templates.

Companies also face pressure to differentiate their products in competitive markets, making product-centric design crucial for creating value-added offerings. Since the product design targets are not just restricted to engineering or science, cross-disciplinary collaboration between chemists, engineers, managers, marketing and business professionals is essential to ensure that the products are designed holistically to meet performance and production needs. Companies tend to follow an approach known as project organization for the design of new products (Cussler and Moggridge, 2011). In project organization, a core team comprised of members with different backgrounds and expertise works in parallel towards efficient solutions.

It is to be noted that the traditional chemical industry that produces bulk chemicals has reached saturation, and the required innovation for further development is rather limited. Furthermore, the rising demand for speciality chemicals with precise formulations underscores the importance of aligning design with functionality, sustainability and economic competitiveness. It should be noted that several speciality chemicals may become a commodity if they are found to have bulk demand (Turton et al., 2012). Many polymers, such as nylon, had limited applications when they were invented but soon found applications in various fields, which were not anticipated at the time of their invention. Another possibility is the increase in demand for a speciality chemical due to its decreasing effectiveness. Several agrochemicals, such as pesticides, were initially required in small quantities. But, over a period of time, as the pests improve their immunity towards these chemicals, the dosage significantly increases to meet the desired outcome. In order to stay in competition, the companies producing these chemicals will need to produce them in larger quantities at a lower price. In such situations, there will be a significant compromise on the profit unless the company develops a new chemical with improved functionalities that justify the higher cost.

1.3 VALUE OF CHEMICAL PRODUCTS

The value of a chemical product is often significantly higher than the raw materials from which it is produced. In addition, the cost of such products is often orders of magnitude higher than commodity chemicals. For example, several pharmaceutical products are produced through fermentation in a dilute form with several impurities. Unlike commodity chemicals, the limit of impurities in such products should not be more than a few ppm. In addition, the nature of the impurity may also impact the performance of the final product. Because of the level of innovation needed to identify the product, the cost of these products may even exceed $1000 per gram, whereas the price range of commodity chemicals is often in the order of a few dollars per kilogram. This difference in the prices for such products makes innovations/improvements in the related processes insignificant as it is usually a small fraction of the profit margin of the final product. The reward for identifying the right product exceeds the potential savings in producing it through an optimized process.

Another important consideration is the volume of the products produced. A typical commodity chemical plant/manufacturer produces thousands of tonnes of products each year because of the large demand. At the same time, the price of the product remains constant across different manufacturers as there is little to differentiate the products from various producers other than the price. The only product-specific target is purity, which needs to be maintained by all the suppliers. Once the purity requirement is met, there is no major factor to be considered that differentiates the quality of the products. However, for a chemical product, quality often determines its selling price and differentiates it from similar products offered by competitors (Figure 1.1).

The major differences between commodity chemicals and chemical products are shown in Table 1.1. Please note that there can be exceptions to these general rules.

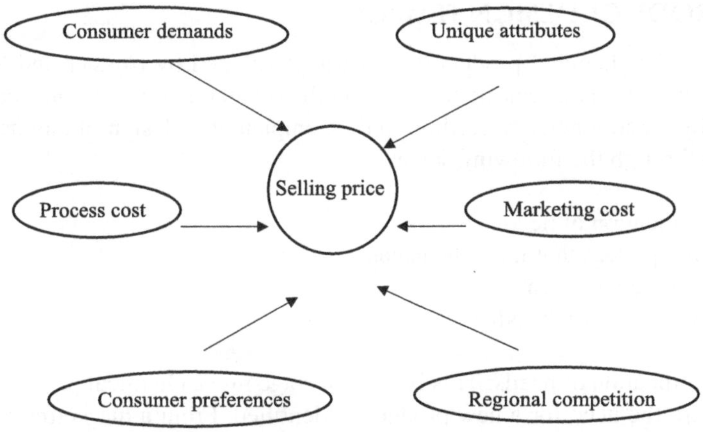

FIGURE 1.1 Price of chemical products.

TABLE 1.1
Commodity Chemicals versus Chemical Products

	Commodity Chemicals	Chemical Products
Quantity	>1000 ton/year, more often more than 10,000 ton/year	< 1 ton/year
Differentiation in the Market	Price	Performance and quality
Value	Low added value to the raw materials	High added value
Profit Margin	Low per unit product mass	Very high
Competition	Less	Intense
Speed of Design	Irrelevant	Need to be fast
Research Activities	Limited	Intense
Manufacturing	Continuous, large-scale, & purpose-built plants	Batch, generic equipment
Design team	Only specialized in plant design, commissioning, & optimization	An integrated team from inception to marketing
Process Efficiency	Very important	Not the primary focus
Energy Integration	Important	Secondary value
Market lifetime	Long lifetime	Short lifetime, which makes it necessary to reach the market fast
Examples	Ethylene, propylene, acetic acid, PVC, etc.	Pharmaceutical & healthcare products, agrochemicals, special solvent, personal care products, etc.

1.4 PRODUCT DESIGN TEMPLATE

A template for chemical product design was developed by Cussler and Moggridge (2011) in their groundbreaking textbook on the subject, and it has since been widely adopted by practitioners. According to this template, the design of any new product has to go through the following stages:

1. Identification of needs
2. Develop ideas that meet the needs
3. Select the best idea
4. Manufacture the product

1. Identification of needs: This is the first stage in the chemical product design where the need for a new product is identified. From a marketing perspective, this is the most important step in product design as targeting the wrong needs will result in rejection by customers. From an engineering perspective, the key challenge lies in the translation of qualitative needs into quantitative properties and understanding the underlying principles.
2. Development of ideas: Ideas are generated to meet the targets identified in the first stage by a team of experts. A screening is performed on the short-listed ideas to eliminate obviously inferior/non-practical ideas. Ideas can be obtained from a variety of sources, such as expert opinion, published and unpublished literature, company logbooks and surveys.
3. Selection of the best idea: From the short-listed set of ideas, select the best idea based on a more thorough investigation, risk assessment and economic analysis. This is the stage at which the designers use more detailed and accurate analysis based on published work, complex calculations and pilot plant studies.
4. Manufacture of the product: Proceed to the commercial production of the new product based on the selected idea. The majority of the tools mentioned here are commonly used in conventional process design as well, but there are some significant differences. Considering the short lifetime of the products, it is not economical to use specialized and optimized equipment in many designs. We also need to consider the challenges involved in fast scale-up.

In this book, the first five chapters provide an overview of the steps followed in the chemical product design template. One of the most challenging steps in product design is identifying the most relevant needs and quantifying those needs in terms of measurable attributes. Furthermore, not all needs hold the same importance, and during the design process, it is essential to recognize the relative significance of each need to prioritize the targets effectively. An overview of different types of chemical products is covered in Chapter 2. Common strategies to approach this issue are covered in Chapter 3. In Chapter 4, various strategies for obtaining the most promising ideas are discussed. Chapter 5 introduces some of the quantitative tools used in the final selection of ideas. While this chapter is nowhere near an exhaustive coverage, it provides a glimpse into the approaches used at this stage.

The second half of this book focuses on the tools used in the manufacture of chemical products. One advantage for chemical engineers in addressing these problems is that the fundamentals underlying most of these tools are covered in either traditional chemical engineering topics or closely related subdomains. In Chapter 6, the economic analysis for chemical products is explained with a special emphasis on uncertainties and short payback periods. The concept of intellectual properties has also been covered. Chapter 7 discusses the scale-up and scale-down of processes across different scales. Since most chemical products are produced via batch processes, a chapter is dedicated to covering this topic in detail. In the following chapters, common unit operations used in the chemical product industry are discussed. Finally, three complete case studies on chemical product design are presented in the last chapter. The topics covered in these chapters provide sufficient background to explore more advanced chemical product design challenges that future engineers may face, allowing them to make specific adjustments to address their unique problems.

1.5 CONCLUSIONS

Chemical product design is the procedure to translate consumer needs into commercial products. Unlike commodity chemicals, the decision to make the right chemical product is the most critical step in chemical product design. The need for a new product is not always obvious and, in many situations, ill-defined. Therefore, formulation of the problem statement is a key step in product design, whereas this step is generally obvious in the design of commodity chemicals. Based on the identified needs, the design involves the generation of ideas to meet the needs, selecting the best idea and eventually manufacturing the product. The cost of product development gets higher and higher as the design passes through these stages, and it is essential to have a thorough review at each stage to prevent major financial losses at a later stage. Overall, it can be seen that the involvement of management and finance is much higher than in the commodity chemical industry, where such reviews take place only towards the end of the process development stage.

REFERENCES

Cussler, E. L., & Moggridge, G. D. (2011). Chemical Product Design. In *Cambridge Series in Chemical Engineering* (2nd ed.). Cambridge University Press. https://doi.org/10.1017/CBO9781139035132

Cussler, E. L., & Wei, J. (2003). Chemical Product Engineering. *AIChE Journal, 49*, 1072.

Favre, E., Marchal-Heusler, L., & Kind, M. (2002). Chemical Product Engineering: Research and Educational Challenges. *Chemical Engineering Research & Design, 80*(1), 65–74.

Hill, M. (2009). Chemical Product Engineering – The Third Paradigm. *Computers & Chemical Engineering, 33*(5), 947–953.

Seider, W. D., Seader, J. D., Lewin, D. R., & Widagdo, S. (2009). *Product and Process Design Principles: Synthesis, Analysis, and Evaluation* (3rd ed.). Hoboken: Wiley.

Turton, R., Bailie, R. C., Whiting, W. B., Shaeiwitz, J. A., & Bhattacharyya, D. (2012). *Analysis, Synthesis, and Design of Chemical Processes* (4th ed.). Upper Saddle River, NJ: Pearson Education International.

Wei, J. (2007). *Product Engineering: Molecular Structure and Properties*. Oxford; New York: Oxford University Press.

2 Chemical Products

2.1 TYPES OF CHEMICAL PRODUCTS

In the previous chapter, we introduced the definition of a chemical product and discussed how chemical products are categorized within the chemical industry. There are three main sectors, namely commodity chemicals, speciality chemicals and consumer products. They differ in terms of production volume, value and end-user applications. Commodity chemicals are manufactured in bulk quantities, but their profit margins are low because of competitive selling prices. Speciality chemicals are normally produced in smaller amounts as they are high-value chemicals targeted for specific applications. Consumer products such as detergents, shampoo and cleaning agents are directly sold to customers. In traditional process design, these classifications help in developing a suitable production strategy and economic analysis.

However, the chemical industry is experiencing a shift in paradigm, that is, from valuing materials for their purity, such as commodity chemicals, to focusing on materials sold based on their performance. Emphasis has been shifted from reducing the cost of production to enhancing the product functionality. From the perspective of chemical product design, the aforementioned product classification provides limited guidance for chemical engineers in the identification or design of potential high-value chemicals. This is because the functionality or performance of a chemical product often depends on its molecular structure, microstructure and composition. As suggested by Cussler and Moggridge (2011), chemical products can be more effectively classified into commodity chemicals, molecular products and microstructured products. In this chapter, we will provide an overview of these three different types of chemical products.

2.2 COMMODITY CHEMICALS

Traditionally, the focus of chemical engineering education was on the design of process plants that produce commodity chemicals. These are chemical products produced in large quantities to meet the global needs. These chemicals are basic raw materials that are used in the production of other chemicals with higher value. Some examples include ammonia, sulfuric acid, benzene and ethylene.

Let's take ammonia as an example. Ammonia serves as a key chemical used in the agrochemical industry to produce fertilizers such as urea and ammonium nitrate. Besides its application in the agriculture industry, ammonia is also used to manufacture textiles, pharmaceuticals, plastics and other chemicals. Ammonia is mainly produced through the Haber-Bosch process, the first large-scale industrial process developed for ammonia synthesis. This process converts nitrogen from atmospheric air and hydrogen from fossil fuels into ammonia using a metal catalyst. It is an energy-intensive process and depends heavily on non-renewable fossil fuels. Efforts have

DOI: 10.1201/9781032708911-2

been made to explore alternative processes that are less costly while also complying with safety, health and environmental regulations. For instance, there are works that focus on identifying new catalysts for the Haber-Bosch process to enhance reaction conversion to ammonia and allow the reaction to take place under milder operating conditions (Lin et al., 2023; Minteer et al., 2019). There are also studies (Ishaq and Dincer, 2020; Lee et al., 2022) that look for alternative renewable energy sources to produce hydrogen for ammonia synthesis instead of relying on fossil fuels. We can see that most patents related to ammonia production are focused on the manufacturing process rather than the specific function of ammonia. This is because ammonia remains chemically the same (NH_3) regardless of the process used to manufacture it. For example, there was an invention that proposed an ammonia synthesis system and an ammonia production method that reduces the energy required for producing ammonia (Yagi and Yamaguchi, 2025). There is little to no difference in terms of its functionality or performance, regardless of how it is produced. In chemical product design, we consider this kind of chemical as an undifferentiated product.

Commodity chemicals are usually considered as undifferentiated products as their properties and functionality are similar across different manufacturers. Using sulphuric acid and gasoline as additional examples, they are similar to ammonia in that they are all produced on a large scale and have similar applications regardless of the manufacturing process. To date, sulphuric acid remains the largest volume commodity chemical produced in the world. Sulphuric acid is mainly produced through the contact process (Carl, 1937) and has its largest application in producing phosphate fertilizers. Its chemical formula is fixed as H_2SO_4, and its purity is standardized across different suppliers at 95–98%. Thus, it makes no difference whether you purchase it from supplier A or supplier B, except for the cost and logistics.

Gasoline is a mixture of simple hydrocarbon molecules produced in large volume, which is primarily used as a transportation fuel to power internal combustion engines. There are different grades of gasoline, which depend on the Research Octane Rating (RON). RON is an indicator that measures how well the fuel resists knocking in the engine during combustion. Different countries may have their own octane rating systems. For instance, the two common grades of gasoline sold in Malaysia are RON 95, known as regular unleaded fuel, and RON 97, which is considered a premium fuel. The higher the octane rating, the better the performance, but it often comes at a higher cost. While RON 95 is sufficient for most car models, some consumers may opt for RON 97 if they are concerned about performance and fuel efficiency. RON 95 and RON 97 gasoline is produced by various companies such as Shell, Petronas and Caltex. Nonetheless, consumers do not usually differentiate gasoline from one brand to another when it comes to refuelling their car. This is because gasoline of the same grade, such as RON 95, is almost identical in performance. Moreover, in the context of Malaysia, RON 95 is sold at the same price regardless of the manufacturer. Though RON 97 is sold at a higher price, the price remains consistent across different companies and petrol stations throughout Malaysia.

From the previous examples, it is obvious that commodity chemical manufacturers are not able to compete with each other based on product uniqueness. They can only remain competitive by offering attractive product prices, which come with a reduced profit margin. Thus, optimizing the manufacturing process is key for commodity chemicals (Cussler and Moggridge, 2011). The opportunities lie within the use of cheaper and/or renewable raw materials, enhancing efficiency as well as reducing capital and operation costs of equipment used in the plants. Manufacturers should look for innovative technologies that are able to enhance product yield and reduce energy consumption to improve profitability.

2.3 MOLECULAR PRODUCTS

These products consist of single molecules with molecular weights significantly larger than those of the chemicals used as commodities. These chemicals have unique properties and are utilized for specific applications such as pharmaceutical products, agrochemicals and more. Unlike commodity products, which are normally manufactured using continuous processes, molecular products are often manufactured in batch or semi-batch operations. They are produced in smaller quantities compared to commodity products. Despite its lower production volume, molecular products often have higher profit margins as they are being sold at higher unit prices. The focus is not on reducing capital or operational cost, nor on maximizing process efficiency, but on the discovery of the molecular product and how fast it can be introduced to the market. The first molecular product designed for a specific application often ends up making about two-thirds of the total sales in that respective market (Cussler and Moggridge, 2011).

Let's take one of the oldest pharmaceutical products as an example – acetylsalicylic acid, more commonly known as aspirin, or by its chemical name, 2-acetoxybenzoic acid. Aspirin has the chemical formula of $C_9H_8O_4$, and its molecular structure is shown in Figure 2.1. It is commonly used to relieve pain, fever and inflammation as well as to reduce the risk of cardiovascular disease. It was first synthesized in 1897 by Felix Hoffman at a Bayer laboratory in Germany. Two years later, it was introduced to the market in powder form under the trademark name 'Aspirin' (Jack, 1997). Aspirin is produced through a chemical reaction between salicylic acid and acetic anhydride, in the presence of a catalyst.

The functionality of aspirin arises directly from its molecular structure. Its anti-inflammatory and antithrombotic effects are attributed to the presence of acetyl group,

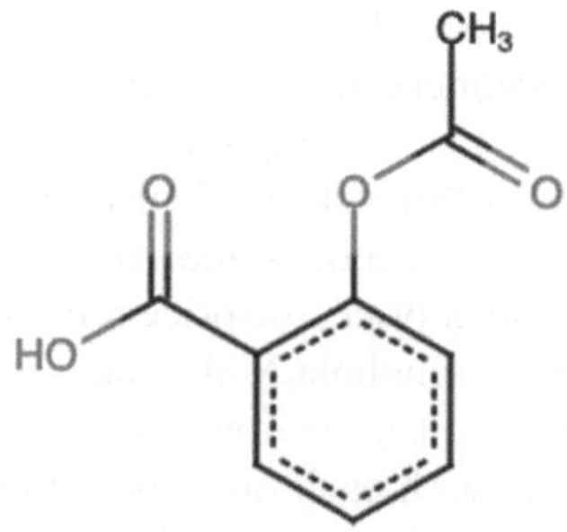

FIGURE 2.1 Molecular structure of aspirin.

which irreversibly inhibits cyclooxygenase (COX) enzymes, primarily COX-1 and COX-2 (Vane and Botting, 2003). This inhibition blocks the synthesis of prostaglandin and thromboxane, which are responsible for inflammation and platelet aggregation. From this example, we can see that the therapeutic effect of a pharmaceutical product is closely linked to its molecular structure. Although aspirin was first introduced in powder form, it is now available in different dosage forms, including regular tablets, effervescence, chewable tablets and coated tablets. Despite the presence of many aspirin manufacturers today, Bayer, being the first company to commercialize aspirin, continues to hold the largest share of the global market.

In the agrichemical sectors, herbicides, insecticides and fungicides are examples of molecular products. For instance, the first synthetic herbicide to be commercially available was 2,4-dichlorophenoxyacetic acid (2,4-D). It is widely used to control weeds on orchards and crops without harming grasses. 2,4-D is known as auxin-type herbicide as it kills the weeds by imitating the natural plant hormone auxin to disrupt the normal growth processes. Upon exposure to 2,4-D, cell division in the vascular tissue becomes uncontrollable, eventually leading to damage in the vascular tissue, causing the weed to collapse and die. It is structurally similar to the natural auxin, indole-3-acetic acid (IAA) and, thus, functions in a similar way. As shown in Figures 2.2 and 2.3, both molecules contain a carboxyl group and an aromatic ring. There are seven classes of synthetic auxin-type herbicides, classified based on the position of the carboxyl group and the type of aromatic group (Busi et al., 2018). They differ in terms of selectivity and spectrum of weed control.

From the examples provided earlier, it can be clearly seen that the effectiveness of molecular products is closely linked to their molecular structures. The functional

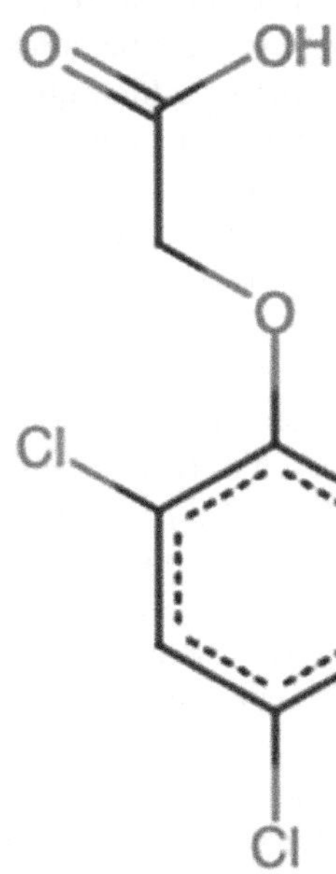

FIGURE 2.2 Molecular structure of 2,4-D.

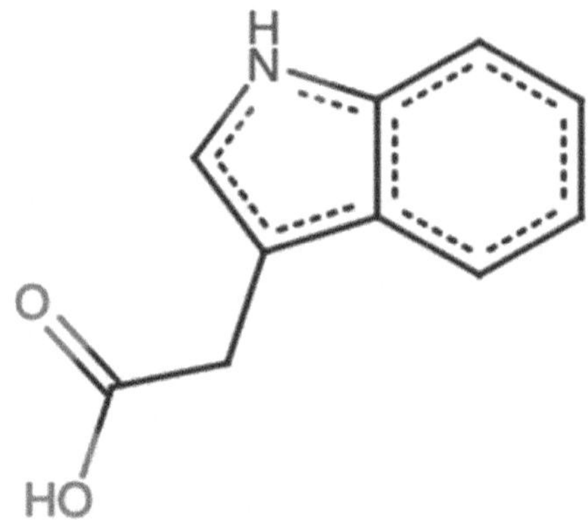

FIGURE 2.3 Molecular structure of auxin (IAA).

groups within a molecule play an important role in defining its properties and functions. The key to success in developing new products in this category lies in how quickly a suitable structure can be identified, tested and brought to market. At the same time, risk analysis is crucial due to the various uncertainties inherent at different stages of the development process.

2.4 MICROSTRUCTURED PRODUCTS

These are products whose value is strongly linked to their microstructure. Chemically similar products can exhibit significant differences in performance due to variations in their microstructure. For example, a low-quality ice cream may have a chemically similar structure to that of a high-quality ice cream, but the significant price difference is largely due to variations in their microstructure. Similarly, chemically dissimilar materials may be used for the same application if they possess similar microstructures. For instance, pigments used in paints can be made from different materials but still produce comparable results. The design focus here is driven by meeting customers' needs.

To illustrate this in the context of daily life, we consider laundry detergents as an example. Nowadays, laundry detergents are available in different physical forms such as powder, liquid and pods. Each of them exhibits different morphological and microstructural properties. While all of them often share similar basic ingredients and serve the same purpose, that is, cleaning our laundry, their performances may vary due to the differences in solubility, dissolution rate, stability and convenience for use. The dissolution rate of powder detergent is influenced by the powder size and granule porosity. Powder detergent with smaller size and greater porosity has a higher dissolution rate. The pourability and stability of liquid detergent are affected by the viscosity. The viscosity of liquid detergent should be carefully controlled to ensure the flowability of the product and, at the same time, maintain its stability.

These differences in physical properties directly affect their performance during use. Powder detergents require a longer time to dissolve in cold water compared to liquid detergents. As a result, powder detergents tend to have a higher potential for leaving residues on clothes when it is not rinsed properly. On the other hand, liquid detergents dissolve more easily at all water temperatures and are gentler on clothes. However, liquid detergents are slightly more costly compared to powder detergents. Detergent pods offer convenience to consumers in terms of storage and usage as they come in compact packaging and are pre-measured. Nonetheless, they are sold at a premium price, and the fixed dose also results in limited flexibility.

To qualify as a laundry detergent, the product must meet certain minimum requirements, commonly known as primary quality factors in the context of chemical product design (Fung and Wibowo, 2007). The opportunity to stand out from other products comes from secondary quality factors, which are designed to satisfy customers' needs. We refer to these as value-added benefits, which go beyond the basic functions of the product. For example, some detergents offer skin-sensitive formulations, biodegradable packaging and/or more pleasant fragrances. Some detergents are formulated to preserve the brightness of coloured fabrics. There are also detergents that contain anti-bacterial agents to make them suitable for indoor drying. They often come at different prices, and the choice depends heavily on customers' needs and personal preferences.

Another example of a microstructured product is toothpaste. Toothpaste often has complex physical microstructure properties. It behaves as a shear-thinning fluid where its viscosity decreases with increasing shear rate. This property allows toothpaste to flow easily from the tube when the tube is squeezed and also prevents it from leaking when at rest (Cussler and Moggridge, 2011). Like laundry detergents, toothpastes often share a similar basic formulation but can contain different active ingredients tailored to address different consumers' needs. The differences in the composition of ingredients among toothpastes lead to distinct microstructure and rheological properties. For example, children's toothpaste is either fluoride-free or contains a lesser amount of fluoride compared to adult toothpaste. They often come in kid-friendly flavours such as strawberry, orange, grape and apple. On the other hand, adult toothpaste formulation usually incorporates mint or herbal flavours. Moreover, adult toothpastes include different active ingredients tailored for different functions. For example, toothpaste designed for sensitive teeth normally contains potassium nitrate, whereas whitening toothpaste contains hydrated silica or hydrogen peroxide.

The examples given earlier clearly illustrate that the value and performance of microstructured products depend heavily on their microstructures. However, the main challenge is to link the performance of such products to the product attributes. Therefore, when designing these products, focus should be placed on identifying customers' needs and translating those needs into specifications.

2.5 CONCLUSIONS

This chapter provides an overview of various types of chemical products, which are classified into commodity products, molecular products and microstructured products. Commodity products are usually mass produced and often undifferentiated, and, thus, the strategy to improve their profit margin is either through the use of lower cost raw materials, improving process efficiency or minimizing the capital and operation costs of plant equipment. In contrast, molecular products are designed for specific applications and can be sold at a higher price, resulting in better profit margins. Microstructured products are another important category, where even a small change in the structure/formulation can lead to significant differences in their performance. Its selling price depends on how well their structures add value to the customers.

REFERENCES

Busi, R., Goggin, D. E., Heap, I. M., Horak, M. J., Jugulam, M., Masters, R. A., Napier, R. M., Riar, D. S., Satchivi, N. M., Torra, J., Westra, P., & Wright, T. R. (2018). Weed Resistance to Synthetic Auxin Herbicides. *Pest Management Science*, *74*(10), 2265–2276. https://doi.org/10.1002/ps.4823

Carl, F. (1937). *Manufacture of Sulphuric Acid by the Contact Process (Patent No. US2079760A)*. United States Patent Office.

Cussler, E. L., & Moggridge, G. D. (2011). Chemical Product Design. In *Cambridge Series in Chemical Engineering* (2nd ed.). Cambridge University Press. https://doi.org/10.1017/CBO9781139035132

Fung, H.-K., & Wibowo, C. (2007). Chemical Product Design: Toward a Perspective Through Case Studies. *Computer Aided Chemical Engineering*, *23*, 239–274. http://www.sciencedirect.com/science/article/pii/S1570794607800113

Ishaq, H., & Dincer, I. (2020). Design and Simulation of a New Cascaded Ammonia Synthesis System Driven by Renewables. *Sustainable Energy Technologies and Assessments*, *40*, 100725. https://doi.org/10.1016/j.seta.2020.100725

Jack, D. B. (1997). One Hundred Years of Aspirin. *The Lancet*, *350*(9075), 437–439. https://doi.org/10.1016/S0140-6736(97)07087-6

Lee, B., Winter, L. R., Lee, H., Lim, D., Lim, H., & Elimelech, M. (2022). Pathways to a Green Ammonia Future. *ACS Energy Letters*, *7*(9), 3032–3038. https://doi.org/10.1021/acsenergylett.2c01615

Lin, B., Nowrin, F. H., Rosenthal, J. J., Bhown, A. S., & Malmali, M. (2023). Perspective on Intensification of Haber–Bosch to Enable Ammonia Production Under Milder Conditions. *ACS Sustainable Chemistry & Engineering*, *11*(27), 9880–9899. https://doi.org/10.1021/acssuschemeng.2c06711

Minteer, S. D., Christopher, P., & Linic, S. (2019). Recent Developments in Nitrogen Reduction Catalysts: A Virtual Issue. *ACS Energy Letters*, *4*(1), 163–166. https://doi.org/10.1021/acsenergylett.8b02197

Vane, J. R., & Botting, R. M. (2003). The Mechanism of Action of Aspirin. *Thrombosis Research*, *110*(5), 255–258. https://doi.org/10.1016/S0049-3848(03)00379-7

Yagi, T., & Yamaguchi, K. (2025). *Ammonia Synthesis System and Ammonia Production Method (Patent No. US12209025B2)*. United States Patent Office.

3 Identification and Quantification of Product Needs

3.1 NEEDS FROM BUSINESS SIDE

In order for a new product to be successful, it should be able to meet some of the requirements of the customers not met by the existing product. In addition, those products must be able to meet the requirements met by the existing products as well. The additional requirements could be related to product performance or new government regulations. Generally, this is considered a marketing problem rather than a technical challenge. In the subsequent sections, we will see that the technical team's role is to ascertain that the needs identified by the marketing team are achievable and that there are no scientific hurdles to meeting those objectives. At the same time, the marketing team plays a central role in determining consumer needs.

Identifying the right customer needs is essential for the success of a new product. To achieve this, marketing teams typically employ several strategies. Traditionally, these strategies have included interviewing customers, collecting feedback through surveys and questionnaires, and conducting discussions in focus groups (Cussler and Moggridge, 2011). However, there is now a major focus on incorporating data analytics tools to identify underlying patterns in customer behaviour, using web analytics to track consumer interests, and monitoring social media platforms. Most companies use tools to track mentions, hashtags and sentiment around products or services. Additionally, companies routinely monitor competitors to understand which needs they are addressing and to identify gaps or opportunities in the market. However, predicting consumer attitudes towards a new product or in a new market has always been challenging, and focusing on the wrong needs has often resulted in major marketing mishaps.

One of the most frequently cited examples of a marketing misstep is Kellogg's entry into the Indian market in the early 1990s (Haig, 2011). The case is often used to illustrate how even well-established global brands can struggle when they fail to align their products with local consumer habits and preferences. Kellogg's launched its breakfast cereals in India with the idea of introducing a healthier alternative to traditional breakfasts. At the time, the company faced limited direct competition, and awareness around health and nutrition was beginning to grow among Indian consumers. Despite a high-profile launch and significant marketing investment, the products failed to resonate with the target audience. Analysts have since studied the case in detail, and many agree that the main issue was a misunderstanding of what Indian consumers actually wanted from a breakfast experience. Indian breakfasts

DOI: 10.1201/9781032708911-3

are typically rich in carbohydrates, full of flavour and served in generous portions. In contrast, a small bowl of cornflakes felt unsatisfying to many people. On top of that, the concept of eating cereal with cold milk didn't sit well with most Indian consumers, who were more accustomed to warm meals in the morning. Those who tried to adapt the product by using hot milk found that it turned the cornflakes soggy and unappealing. While some consumers were curious enough to try something new, few continued to buy the product regularly. The relatively high price of Kellogg's cereals, especially when compared to the cost of a typical homemade Indian breakfast, further limited its appeal to the broader market.

Kellogg's failed entry into the Indian market serves as a powerful reminder of how important it is to accurately identify and understand local consumer needs. Despite strong marketing efforts, the company's inability to align its product with the preferences and expectations of Indian consumers highlights that understanding customer needs is key to achieving success in any market.

3.2 NEEDS FROM TECHNICAL SIDE

The customer needs identified by marketing teams are often qualitative in nature and may not directly translate into practical inputs for product development. Therefore, an important step in the process is to derive quantitative insights from this qualitative information to guide design and development decisions. In the later stages, these needs are used to formulate product specifications, and thus the quantification is always a key consideration. At the same time, it is important not to impose overly restrictive constraints during the early stages of product design, as this can limit creativity and innovation. To support a structured yet flexible approach, the following guidelines are commonly followed during the need identification phase:

1. Present needs in terms of functionalities.
2. State customer needs as specifically as possible, in alignment with the raw data, to preserve the original intent and avoid loss of information.
3. Use positive phrasing when translating needs into quantitative terms, as this promotes clearer interpretation and actionable outcomes.
4. Frame each need as an attribute of the product being designed, to facilitate the conversion of qualitative requirements into measurable specifications.
5. Avoid using terms such as 'must' or 'should' at this stage, as they may suggest a premature prioritization of needs that has not yet been validated.

Example 3.1 illustrates the identification of needs from a product development team.

Example 3.1. Targets for a New Hair Spray

A hairspray formulation consists of a copolymer blend, solvents, a propellant and additives such as neutralizing agents, plasticizers and fragrances. The polymers serve as the active ingredients (AIs), responsible for holding curls. The solvent mixture is typically a water-based blend of organic chemicals. Along with the propellant, the solvent mixture forms the delivery system. Neutralizing agents, which are usually alkaline, are required to make the product water-soluble (rinseable with water). Plasticizers are included to enhance the flexibility of the hair. In this example, the AI and the neutralizing agent are known, while the solvent mixture has yet to be defined, though a short list of candidate solvents is available.

Suggest all the attributes that may be considered for the design of this product.

Solution

A list of important targets is given next:

- Good retention and holding power
- Shine and lustre
- Easy to rinse with water
- Shouldn't make frizzy hair – low static charge
- Non-flammable
- Non-toxic
- Low VOC emissions
- Long lasting
- Non-sticky
- Odourless
- Sprayable
- No white residue
- Should not decolourize coloured hair
- UV protection
- Quick drying
- Good smell
- Skincare on sensitive skin
- Waterproof
- Less quantity is needed

3.3 DIFFERENT TYPES OF NEEDS

At this stage, it is important to realize that it may not be practical to develop new products that outperform the existing products in all areas of competition at a reasonable cost. So, it is important to prioritize the targets before investigating the solutions. One of the most useful approaches for differentiating various types of needs was developed by Noriaki Kano in the 1980s and is known as the Kano Model or Kano Map (Figure 3.1). This model provides a framework for distinguishing product attributes based on different aspects of customer demands. According to the Kano model, customer preferences fall under one of the following types of needs:

1. Must be needs

   Must be needs are essentially basic needs. The product cannot be sold unless these fundamental needs are met. However, customers do not differentiate a product based on these essential requirements. For example, in the case of a new toothpaste, the ability to clean teeth is a fundamental necessity.

Regardless of any additional attributes the toothpaste may offer, if it fails to perform its primary function of cleaning teeth, customers will not be satisfied. However, merely fulfilling this basic requirement does not serve as a distinguishing feature for the product.

2. One-dimensional needs
   One-dimensional needs are performance needs. The satisfaction level of customers increases with the degree of performance of these needs. Unlike 'must-be' needs, failing to meet one-dimensional needs does not necessarily lead customers to reject the product. However, since customer satisfaction increases with higher levels of performance, companies prioritize enhancing these needs to attract customers. For example, a car's fuel efficiency is a one-dimensional need, as greater mileage leads to higher customer satisfaction.

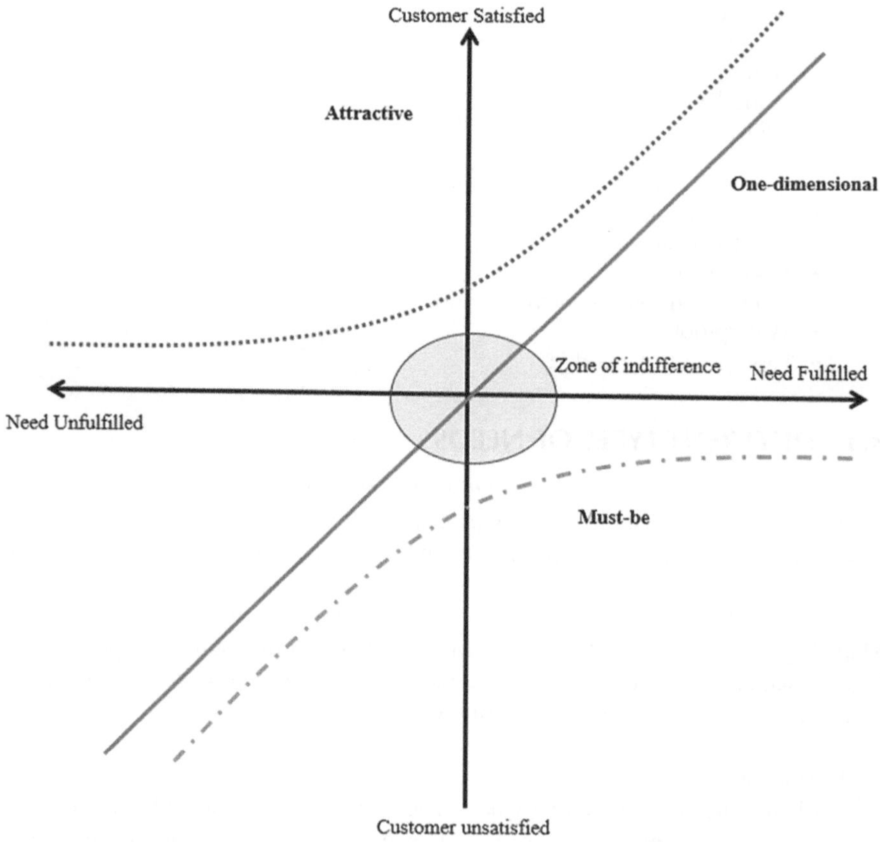

FIGURE 3.1 Kano map.

3. Attractive needs
   These needs add an element of excitement for customers when satisfied. At the same time, they are neither a primary requirement nor essential for fulfilling the core purpose for which the product was developed. However, having some of the attractive elements in the product can make a new product successful in a competitive market. For example, the flavour of a toothpaste does not affect the way it performs its primary function. However, having a nice flavour is an added advantage if that toothpaste also cleans the teeth well (primary need) and is reasonably priced (one-dimensional need).

Example 3.2. Kano's Map of Hair Spray Example

Classify the needs identified in Example 3.1 according to Kano's model. Suggest how these attributes can be quantified.

Solution

A possible classification is shown in Table 3.1, along with the quantification methods.

TABLE 3.1
Kano Map and Quantification of Needs

Attribute	Category	Quantification Approach
Good retention and holding power	M	Appearance
Shine and lustre	A	Active ingredients
Easy to rinse with water	A/M	Solubility in water
Shouldn't make frizzy hair – low static charge	1D/M	Dielectric constant
Non-flammable	A	Flash point
Non-toxic	M	LC50
Low VOC emissions	A	Concentration of VOC
Long lasting	1D	Viscosity
Non-sticky	A/M	Viscosity/density
Odourless	A	Solubility of co-polymers in alcohol
Sprayable	M	AI
No white residue	M/A	AI
Should not decolourize coloured hair	A/M	Evaporation time (T_{90})
UV protection	A	Smell test
Quick drying	1D/A	Clinical trials
Good smell	A	Solubility
Skincare on sensitive skin	M/A	Measurement
Waterproof	A	
Needed quantity	1D	

3.4 NEEDS TO SPECIFICATIONS

When working on a new chemical product, setting specifications can be particularly challenging, especially if the product is something entirely new or quite different from what's already available. In many cases, developers begin by benchmarking against existing products to get a sense of reasonable targets. But when no close comparison exists, it becomes harder to know what those targets should be. The fundamental scientific knowledge is helpful in making meaningful comparisons. Basic principles of chemistry, material science, thermodynamics and chemical kinetics can help form the basis for early estimates. Experiments and analysis also play a key role in shaping the product's specification, especially when there's little prior data to draw from. It's important to remember that in these early stages, many things will still be uncertain. There are no fixed rules to follow, and the process often involves a lot of iteration and learning as the product takes shape. Example 3.3 illustrates the thought process involved in setting specifications for a drug that releases the active ingredient at a controlled rate.

Example 3.3. Controlled Drug Release

For many disease`s, it is critical to deliver the medicines at a controlled rate. This is because the concentration of medicine in blood changes as a function of time (Cussler, 2009). The concentration is high when the medicine is taken/administered, and it drops with time. One disadvantage of the approach is that all the medicines have a safe limit and effective level for concentration. Below the effective level, the medicine may not be effective, and the values above the safe limit may have undesirable side effects on some patients. In addition, the medicine provides the best results if the concentration of the blood is within a recommended limit. The dosage and interval of medicines are suggested based on these limits. However, the limitation with this approach is that many patients do not strictly follow the instructions on dosage, which may affect the performance of the medicine or the patient's wellbeing. For the diseases for which the dosage is critical, it is recommended to develop a drug that controls the delivery of active ingredients at a constant rate.

Identify the needs and the methods to develop specifications for the controlled drug delivery.

Solution

Based on the description of the problem, a list of needs for the new drug delivery system is identified:

- The drug delivery should not depend on the patients' discipline/motivation/attitude
- Drug delivery should not be dependent on the patient's mood/memory, mental or physical state
- Medicine to be delivered uniformly
- Should not cause discomfort/pain
- The risk level must be low
- Cheap (should not be more expensive than the drug)

The most challenging parameter to quantify is the delivery rate. There are various possible approaches to achieving controlled drug delivery, such as osmotic pumps and membrane-based devices. However, at this stage, we are not committing to a specific approach, as doing so could rule out other potentially better innovations. Therefore, the specifications at this stage are drug delivery rate determined based on the daily dosage and the cost – typically suggested by the marketing team.

3.5 REPRESENTATION OF NEEDS

Up to this stage, we have compiled a list of needs and calculated several measurable product attributes or performance indices. This information needs to be presented to a general audience in a clear and structured format-typically using charts, tables or graphs. One effective way to present this information comprehensively is through the House of Quality (Temponi et al., 1999). The House of Quality is an interconnectivity matrix that links product needs with performance indices. It also serves as a tool to highlight where specifications may complement or conflict with one another. Additionally, it is used to display quantitative comparisons with benchmark products (Figure 3.2).

The House of Quality diagram is constructed through the following steps:

Left side of the diagram: Identified needs
Top: Performance indices
Middle: Interconnection between needs and the performance indices
Bottom: Unit of measurement, relative importance, benchmark and some target values

For each performance index, a + or – sign is added to indicate whether the objective is to maximize or minimize each index. For example, ease of carrying a device can be quantified using the weight of the device, and a lower weight is the desired attribute. Therefore, a '-' sign is put in the relevant column. In the last stage, the inter-relationship between different needs is considered. This is indicated on the roof of the diagram.

The application of the use of House of Quality is illustrated in example 3.4.

Example 3.4. Need Representation for a New Raincoat

In this example, we will discuss the process of identifying the right needs for a raincoat to be manufactured. The product design starts by outlining the needs required by the customer and then quantifying the needs with the market benchmark of currently pre-existing raincoats. The following needs are identified after the preliminary discussion:

1. Is waterproof
2. Is easy to carry along
3. Is fairly durable
4. Is easy to wear
5. Is safe
6. Convenient to store
7. Has reasonable price
8. Enables keeping user warm

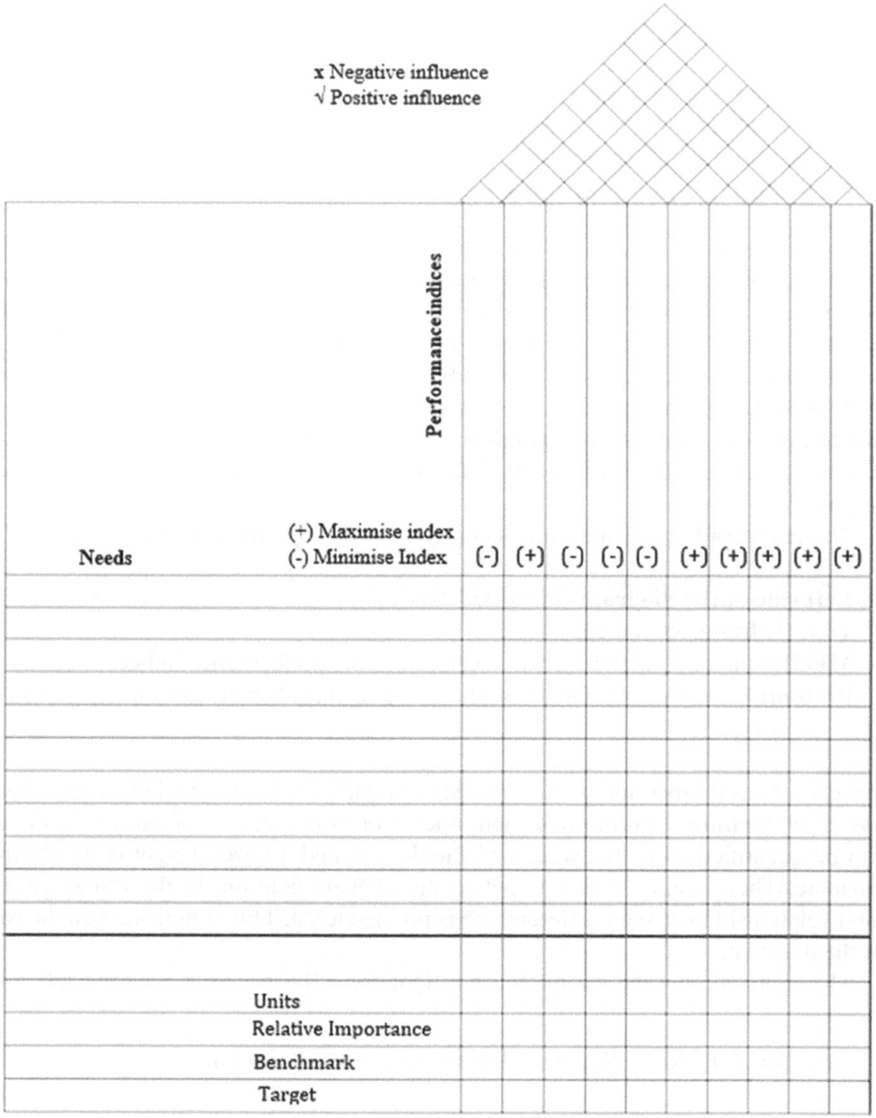

FIGURE 3.2 House of Quality diagram.

9. Is environmentally friendly
10. Is biodegradable

With many existing products in the market, the benchmark can be easily set. By considering the convenience and usage of raincoat, efficiency to repel water, weight, volume and even time taken to wear are defined subsequently. The product can keep the user warm as the material of product has high thermal conductivity. The performance indices and benchmarks are shown in Table 3.2.

TABLE 3.2
Performance Indices

Performance Index	Need Captured	Importance	Benchmark
1. Efficiency of water proofing (mm of water repelled)	1	1	5000–20,000
2. Total weight (g)	2,6	0.9	250–450
3. Durability (tensile strength nylon MPa)	3	0.7	78
4. Time taken to put on coat (s)	4	0.7	< 7
5. Safety (index)	5	1	–
6. Total volume (L)	2,6	0.8	1–3
7. Price (RM)	7	0.7	50–100
8. Thermal conductivity [W/(m*K)]	8	0.6	0.25
9. Environmentally friendly index (years)	9,10	0.7	30–40

Present this information using a House of Quality diagram (Figure 3.3).

Example 3.4. Alternative Refrigeration System

A refrigeration system plays an important role in modern human life. Many homes have refrigerators, which work by compressing gas into liquid and then decompressing and letting it evaporate. This evaporation removes energy from the system and creates a cold environment for storing foodstuffs. The product idea is based on the need to remove heat by evaporation, although convective heat transport is also considered. During hot days or undertaking strenuous activities, people heat up; this heat should be removed naturally, but in warmer equatorial climates the body's own self-cooling mechanism by perspiration may not be enough. Profuse sweating may also lead to dehydration. Therefore, a self-cooling water bottle which can both replace bodily fluids and reduce temperature by drinking a cold beverage would be beneficial.

Identify the relevant needs and quantify the identified needs in terms of measurable ideas. Categorize the needs and compare those against a suitable benchmark.

Solution

Step 1: Need identification

Scope of the project: Alternative refrigeration system to replace traditional cooling fridge
Main target market: Equatorial countries
Second target market: Other seasonal countries

The most relevant needs for the new product are presented in Table 3.3, and these needs are classified according to the Kano model in Table 3.4.

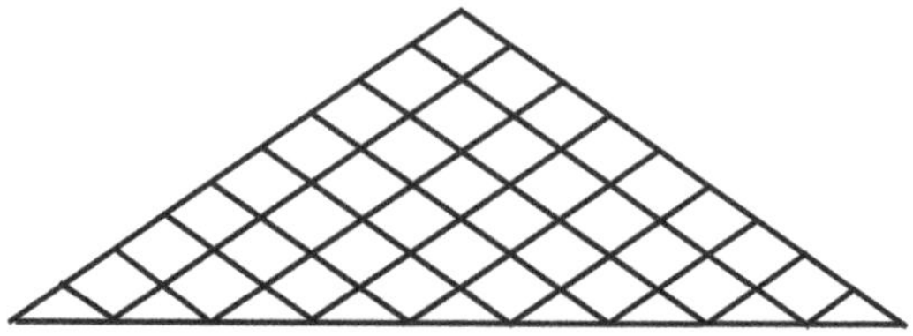

Needs \ Performance Indices	1. Efficiency of water proofing (+)	2. Total Weight (-)	3. Durability (+)	4. Time taken to put on coat (-)	5. Safety (+)	6. Total Volume (-)	7. Price (-)	8. Thermal Conductivity (-)	9. Environmentally friendly (+)
1. Waterproof	O								
2. Light-weighted and portable		O				O			
3. Durable			O						
4. Easy to wear				O					
5. Safety Index (Skin irritation)					O				
6. Ease of storage		O				O			
7. Affordable							O		
8. Keep Warm								O	
9. Environmentally Friendly									O
10.Biodegradable									O

Units	**mm**	**g**	**Mpa**	**s**		**L**	**$**	**W/ (mK)**	**Years**
Relative importance	1	0.9	0.7	0.7	1	0.8	0.7	0.6	0.7
Benchmark	5000 to 20000	250 to 450	78	Less than 7	-	1 to 3	20	0.25	30 to 40
Target	4000	250	+10	-3		1	60	-0.1	-10

FIGURE 3.3 Need representation for a new raincoat.

TABLE 3.3
Identified Needs

Needs	
1. Higher efficiency of cooling	2. Comfortable
3. Cool drinks	4. Long shelf life
5. Non-toxic	6. Long-lasting cooling effect
7. Operate quickly	8. No side effects
9. Environmentally friendly material	10. Reusable
11. High safety protection for application	12. Recyclable
13. Convenience	14. Not corrosive
15. Affordable price	16. Designed in a range of attractive colours
17. Non-combustible	18. Created in a range of attractive designs
19. Light weight	20. High mechanical resistance
21. User friendly	22. Produced with existing equipment and technology

Step 2: Need Classification

TABLE 3.4
Classified Needs Using Kano Model

Class	Need
Must be	Cool drinks
	Produced with existing equipment and technology
	Non-toxic
	High safety protection for application
	Non-combustible
	No side effects
	Not corrosive
One dimensional	Long shelf life
	Long-lasting cooling effect
	High mechanically resistant
	Higher efficiency of cooling
	Operate quickly
	Affordable price
	Light weight
Attractive	Designed in a range of attractive colours
	Created in a range of attractive designs
	Environmentally friendly material
	Reusable/recyclable
	User friendly
	Convenience
	Comfortable

Kano map for the classified needs is shown in Figure 3.4.

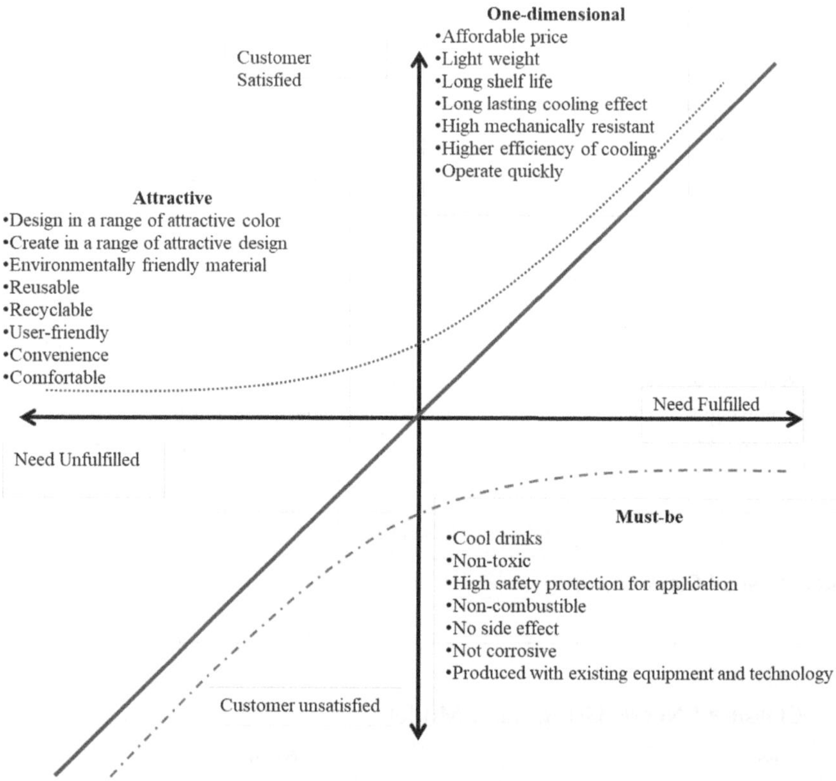

FIGURE 3.4 Need classification for alternate refrigeration system.

Step 3: Comparison with benchmark

To enhance the quality of product, comparison with existing product is necessary. A commercial self-chilling drink is chosen as a benchmark. The relative importance of the needs is identified as well. The results are shown in Table 3.5, and the complete product design targets are presented using the house of quality diagram in Figure 3.5.

House of Quality

(+) maximize index
(-) minimize index
(v) positive influence
(x) negative influence

TABLE 3.5
The Performance Indices for Alternate Refrigeration System

Performance Index	Need Captured	Importance	Existing Product as Benchmark
Temperature drop provided (°C)	1,2	1	30
Energy removed (kJ)	1,2	1	37.6
Total volume (mL)	7	0.9	300
Total weight (g)	7,10	0.9	50
Time to achieve temperature drop (min)	4	0.8	3
Safety (index)	3,6,9,15,18	1	9
Time drink stays cold (min)	14	0.8	25
Price (%)	8	0.7	$4
Versatility (index)	11,12,13,19,20,21,22	0.6	4
Number of times it can be reused (-)	8,16	0.7	0
Cost in reuse (%)	8,16,17	0.7	–
Environmental friendliness (index)	5,16,17	0.6	5

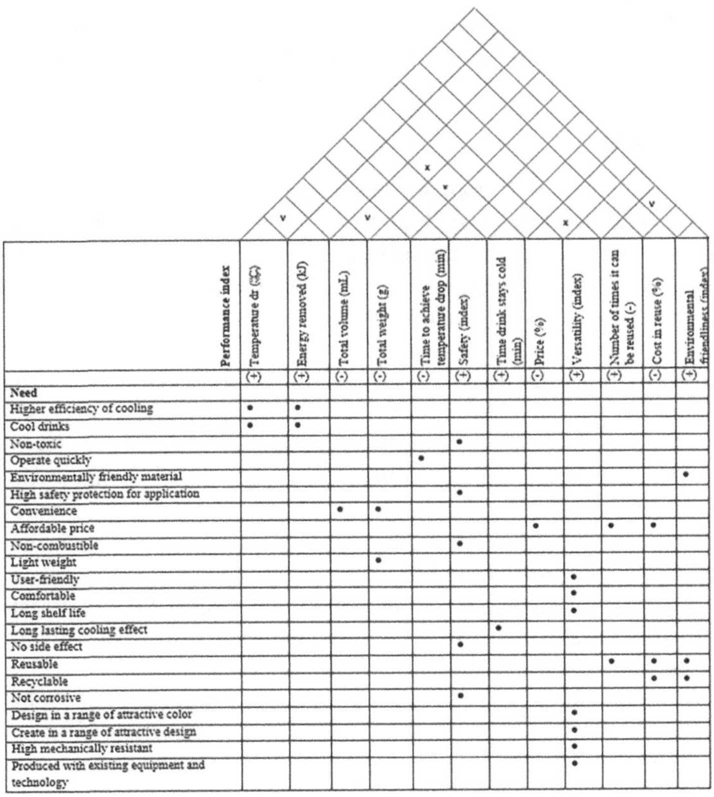

FIGURE 3.5 House of Quality for alternate refrigeration system.

3.6 CONCLUSIONS

This chapter has introduced the first stage of chemical product design, which involves the identification and quantification of needs. It has been demonstrated that customer needs may differ in interpretation from marketing and technical perspectives, and this distinction has been illustrated with relevant examples. A structured tool for categorizing needs in order to establish their relative priority has also been presented. For the technical team, it is crucial to translate these needs into measurable specifications. Finally, the House of Quality – a well-established framework for systematically representing targets, needs, specifications and their interrelationships – has been discussed and illustrated through an example. In the next chapter, various strategies for generating ideas that address the identified needs will be presented.

REFERENCES

Cussler, E. L. (2009). *Diffusion: Mass Transfer in Fluid Systems* (3rd ed.). Cambridge: Cambridge University Press.

Cussler, E. L., & Moggridge, G. D. (2011). Chemical Product Design. In *Cambridge Series in Chemical Engineering* (2nd ed.). Cambridge University Press. https://doi.org/10.1017/CBO9781139035132

Haig, M. (2011). *Brand Failures: The Truth about the 100 Biggest Branding Mistakes of All Time* (2nd ed.). London: Kogan Page.

Temponi, C., Yen, J., & Tiao, W. A. (1999). House of Quality: A Fuzzy Logic-Based Requirements Analysis. *European Journal of Operational Research, 117*(2), 340–354.

3.7 PROBLEMS

1. Beverages purchased from takeaway counters in small-scale restaurants have several drawbacks, such as being slippery and providing less insulation, which makes them uncomfortable for the user and causes the beverage to lose its optimal conditions. Identify some of the core needs for the new product and identify the measurable properties to quantify these needs.
2. For years, a cosmetic company has produced a massage cream, a type of cold cream normally used as a cleansing agent for facial skin. Recently, this product (*A*) begins to lose market share. Market research reveals that the customers have been switching to a competitor's product (*B*), which they claim has a stronger cleansing effect. It is decided to reformulate the cream to increase its cleaning power. The target of the new design is to retain the good existing properties of the cream while increasing the cleansing power. The following characteristics must be achieved for the new cream:
   i) The product should be sold in a plastic tube (similar to toothpaste tube). The cream should not flow by itself but can be squeezed easily out of the tube.
   ii) It is required to increase the proportion of the oily materials in the cream in order to dissolve the fats and oils accumulated on the face. At the same time, there should not be any greasy feeling on the face upon application.
   iii) Low toxicity

iv) Good skin feeling – must be soft and smooth when applied to skin
v) Easiness to rinse off

Based on all the customer needs, identify and list the required performance indices and plot all the properties into the 'House of Quality' diagram. Ignore the 'benchmark' and 'target values' in the House of Quality diagram. Explain how the chosen indices can fulfil the needs.

3. A cosmetic company wants to produce a hand lotion, the function of which is to restore and to maintain the skin in a fully moisturized condition. Based on market research on customers' expectations, the following targets have been set for the new product.
   i) The hand lotion should be easily applicable and rub in quickly
   ii) Hard bottles would be used as the product containers. The product would flow when poured from the bottle but should not appear runny
   iii) Nice odour
   iv) Low cost
   v) Low toxicity
   vi) Anti-bacterial properties
   vii) Maintain the skin in a fully moisturized condition
   viii) Good appearance
   ix) Good skin feeling
   x) Stability
   xi) Good adhesive power on skin
      a) Based on the customer demands listed in the problem definition, identify the desirable physical properties of the lotion to address the first five (i–v) targets mentioned (Hint: For the first two targets, provide the answer based on the expected non-Newtonian behaviour of the lotion).
      b) Based on all the aforementioned needs, identify the categories of needs built on the Kano model and plot all the properties into the Kano map. If any of the needs fall into multiple regions, briefly explain the reason.
4. Dried fruits are common snacks in many southeast Asian countries. As a manufacturer in the dried fruits business, your company has decided to expand the current business by introducing a new dried fruit snack, that is, dried durian, into the market. In revising product specification, your team were asked to compare the new product with a 'benchmark'. In your opinion, what will be a good benchmark to be used? Suggest three (3) specifications to be revised by comparing the new product with this benchmark.
5. Your company is planning to introduce a new insect repellent lotion to the market. Water should be one of the formulation ingredients, because of safety and cost concerns. Some well-known insect repellents in the spray form are based on water-alcohol mixtures.

   Problem definition: Consumers want a product with the following characteristics: high effectiveness against mosquitoes (the main function of the product); good sensorial factors and cosmetic properties, that is, nice odour, appearance and good skin feeling; a water-soluble product, for safety and

TABLE 3.6
Feedback of Customer *j* on Sample *i*

Customer	Sample 1	Sample 2	Sample 3
1	1	2	3
2	4	3	3
3	6	6	6
4	9	7	7

toxicology reasons; low price; long durability (it should not be needed to apply the product often during exposure to mosquitoes); low toxicity; high stability (no separation of phases); good user friendliness, such as a spray product; long shelf life.

a) Based on the customer demands listed in the problem definition, come out with the product specifications based on physical properties/quantitative parameters to select the components for the lotion. Address any five of the attributes listed in problem definition.
b) Based on all the aforementioned needs, identify the categories of needs based on the Kano model. If any of the needs are falling into multiple categories, briefly explain the reason.

6. A company has decided to produce a fruit juice with a new flavour. Four (4) professional taste testers were asked to rate three (3) samples of fruit juice with different sugar content, and the scores given by them are documented in Table 3.6. It is decided to use both the ratio of assessment and category assessment for the evaluation of fruit juice sweetness. Rank the samples in terms of sweetness using both these methods and reflect on the results. Comment on the reasons for the differences in results from these methods?
7. A company specializing in high-quality chocolates is planning to launch a new chocolate product that is enriched with cocoa polyphenols. Such a product is currently not available in the market. Polyphenols are best known for their antioxidant properties and are beneficial to human health, while also improving the shelf-life of the chocolate. However, the addition of polyphenols to the chocolate is typically associated with a bitter and astringent flavour during consumption. After some interviews and preliminary market research, the company's product team has drawn up a list of the following needs associated with this product.

   Need 1: Melting point of chocolate is at body temperature
   Need 2: The distinct aftertaste can be recognized easily
   Need 3: The price of chocolate is affordable
   Need 4: Health claim is indicated on the packaging material
   Need 5: The product has a good shelf-life

   Based on the aforementioned needs, identify the categories of needs based on the Kano model and briefly explain the choice of your selection.

4 Generation of Ideas

The likelihood of a single idea translated into a major success is remote. To increase our chances, a large number of ideas need to be generated in the preliminary stages. Experienced product developers suggest that every successful idea requires evaluating at least 100 ideas. The two major types of approaches for idea generation are human sources and from nature (Cussler and Moggridge, 2011). The human sources of ideas can stem from various sources, including customers, competitors, consultants and members of the product development team. This is the stage for unrestrained, creative thinking! However, very often, the number of ideas generated is insufficient, requiring additional methods to stimulate creativity. Nature is also a valuable source of ideas – especially for the development of pharmaceutical products. Several chemical structures with medicinal properties have been found in the traditional medicines popular in different cultures. In addition, observations of natural phenomena and wildlife have inspired several inventions by mimicking what has been observed. Depending on the type of target products, the ideas generated from various sources are compiled before proceeding to the final selection.

There could be different possible approaches to meet the identified needs. Because of that, the identification of ideas that meet the needs is done in a systematic way. In the first step, a large number of ideas are generated. After this, quick screening methods are used to identify the most promising ideas from the initial set. Finally, quantitative and rigorous analysis followed by experimental verifications are conducted to select the best idea. In this chapter, the generation and screening of the ideas are discussed in the following sections. The final selection of ideas will be discussed in the next chapter.

4.1 HUMAN SOURCES OF IDEAS

This is the most widely used approach for the generation of ideas for novel products. The product development team in organizations comprise members of different backgrounds with varying levels of experience. Therefore, the members of the team themselves can generate several promising ideas. The next source of idea generation is through an extensive literature review. There is a vast number of publications on almost every topic, and in the current age, access to the published literature is also easier than ever. In the last decade, there has also been an unprecedented growth in open-access journals, which also makes data availability easier. While much of the published literature may be academic in nature, an experienced practitioner can identify potential ideas from the technical papers. This is one of the reasons to have highly skilled practitioners in the product design team. In addition, there are engineering magazines that publish industrially relevant articles from practitioners and field experts. While the articles published in the magazines are not as rigorously reviewed as the journal articles, they often contain practical information and

DOI: 10.1201/9781032708911-4

observations from similar areas. Another source of ideas is patents, which can also highlight the practical side of the proposed concepts. It should be noted that the biggest challenge experts have faced in the last decade is not the availability of data but the overflow of information. The information is available in various formats, and extracting useful insights from these sources is akin to drinking water from a fire hose (Venkatasubramanian and Morris, 2008).

Another increasingly popular source of ideas is conferences and trade shows. In addition to the formal presentations of new developments, these platforms provide opportunities for direct interaction with the proposers, helping designers identify the practical aspects of new findings. Many conferences are thematic, and by carefully selecting the right ones, the chances of discovering the most relevant ideas can be maximized. There are also product consultants who can provide valuable service in identifying the ideas. However, it is unlikely to obtain groundbreaking ideas from this source, though their service can be utilized in analysing the ideas identified through other methods.

Finally, the once-popular method of surveying customers is still used by many organizations to gather promising ideas. While most of the proposed ideas may not be of much practical values, they may provide some tips to the core design team.

After collecting the raw ideas through the aforementioned sources, the core design teams generally conduct brainstorming sessions to generate more ideas (Cussler and Moggridge, 2011). The purpose of brainstorming is to generate as many ideas as possible without any kind of screening or making the proposer responsible for the quality of the ideas. This has been considered as teamwork, and the participants are encouraged to propose any idea, even if those ideas sound impractical or absurd. An example of a raw list of ideas has been presented in Table 4.1.

Example 4.1. New Toothpaste

A company wants to develop a new toothpaste and looks for fresh ideas for this project. A list of raw ideas generated in this project is listed in Table 4.1.

4.2 CHEMICAL SOURCES OF IDEAS

There are several areas in chemical product design where nature can provide ideas for new directions. These sources of ideas are especially valuable in the pharmaceutical field. There are three major types of product design approaches based on the inspirations from nature (Cussler and Moggridge, 2011):

1. Identify chemical in plant, extract or synthesize in laboratory (duplicate)
2. Extract precursor from plant, synthesize derivative in laboratory (use as building blocks).
3. Mimic properties of natural products (biomimetics)

Nature has a large supply of several valuable chemicals, and several commonly used chemicals today are initially extracted from plants or animals. Several common alkaloids and phenolic compounds are still primarily extracted from their natural

TABLE 4.1
Raw List of Ideas for New Toothpaste

1	Add salt
2	Make toothpaste sprayable
3	Toothpaste acts as acne solution as well
4	Fluoride free
5	Toothpaste which cures ulcers
6	Multi-active mouthwash
7	Toothpaste as de-appetizing
8	Toothpaste & tooth conditioner (formulated) 2 in 1
9	Actilux – light-activated toothpaste (whitening)
10	Double foaming action toothpaste
11	Iso-active (gel to foam technology)
12	Enzyme toothpaste (seaweed enzyme)
13	Smoke stain-removing toothpaste
14	2 in 1 toothbrush+toothpaste (pen-liked design)
15	Coffee stain-removing toothpaste
16	Toothpaste with remedy for scald
17	(SensiShield Technology) Recalcifying and whitening (iQ+ActiveOxi Tech) mouthwash
18	Dry mouth relief mouthwash
19	Tongue cleaning toothpaste
20	Anti-ageing mouthwash
21	Chewable toothbrush (mint flavour, no toothpaste needed)
22	Add 'Keep-32' molecule, which kills germs in 60 seconds, to toothpaste
23	Toothy tabs (solid) easy for storage
24	Origami tube toothpaste (to fully squeeze)
25	Biodegradable tube
26	Green tea flavoured mouthwash
27	Toothpaste is able to give dual (fire-freeze) sensation – long-lasting cooling effect
28	Add liquid calcium to toothpaste – recalcify teeth (by filling in microscopic holes in tooth enamel)
29	Toothpaste natural tea tree oil and neem with baking soda wintergreen
30	Titanium oxide toothbrush (plague-removing electrochemical reactions)
31	Blue LED lighting toothbrush (gingivitis treatment)
32	Silver child toothbrushes – antibacterial
33	Toothpowder
34	Natural charcoal toothbrush-cure bad breath, tooth whitening
35	Toothbrush that has bio-sensory effect
36	Light-activated ionic toothbrush
37	Toothbrush that has Bluetooth connectivity
38	Enzymatic mouthwash
39	Lighted-activated mouthwash
40	Mouthwash with active oxygen
41	Thermal-sensitive mouthwash

Continued

TABLE 4.1 (*Continued*)
Raw List of Ideas for New Toothpaste

42	Mouthwash with light timer cap
43	Toothswab – lift mucus
44	Chocolate-flavoured toothpaste
45	Strawberry-flavoured toothpaste
46	Vanilla-flavoured toothpaste
47	Cinnamon-flavoured toothpaste
48	Toothpaste tube with 2 heads
49	Tooth auto-cleaning machine which has visual and sensory effect
50	Air floss technology (use of rapid burst of air and water droplets to floss)
51	Mouthwash as smoking deterrent

sources. On many occasions, instead of extracting pure components, a plant extract may be directly used for its applications, especially in traditional medicines. Morphine is the first medically significant alkaloid separated in pure form. This is an alkaloid separated in pure form from the Opium poppy in 1805 (Heinrich, 2013). Several other alkaloids are used as analgesics, antiseptic agents, antimicrobial agents and also for their cardiovascular activities (Stary, 1996). One of the most important chemicals of this category is quinine, which is obtained from Cinchona trees (Saboon et al., 2019). It is used for the treatment of malaria for a long time until better molecules with fewer side effects have been found. Many of those molecules are also alkaloids extracted from various plants.

Polyphenols are the other major category of chemicals that are extracted from plants. Among a wide variety of polyphenols, flavonoids are the most abundant and well-studied class of plant phenols (Dai and Mumper, 2010). These chemicals have been used in a variety of medicinal products due to their antioxidant, antiviral and anti-inflammatory properties. Some of the common plant-based chemicals and their natural source are shown in Table 4.2.

In the past, the scientific community focused on extracting the valuable chemical directly from the source. Many of the commercially attractive chemicals are originally purified from their natural source. In fact, until the historic synthesis of urea in the laboratory in 1828 by a German scientist, Friedrich Wöhler, organic compounds were believed to be made only with the help of vital force present in living matter, such as animals or plants. This theory was known as vitalism, which restricted the scientists from developing synthetic pathways for the production of organic compounds, and the production of organic chemicals relied only on natural sources. The synthesis of urea in the laboratory proved this theory wrong and paved way to the birth of organic synthesis. Even though several plants, fungi and animals are sources of valuable chemicals, extracting pure chemicals from them is not always practical due to the extremely low concentrations in which they are present. Therefore, it is advantageous to identify the chemical with desirable attributes and develop a synthetic route for its commercial production. For example, vanillin is

TABLE 4.2
Common Plant-Based Chemicals and Their Source

Chemical	Source	Uses
Sorbicillactone A	Penicillium chrysogenum	HIV treatment
Codeine	Opium	Analgesic
Dragmacidin alkaloids	Marine sponge	Leukaemia treatment
Ephedrine	Ephedra	Antiasthma
Chelerythrine	Poppy flowers	Antibacterial
Caffeine	Coffee beans	Stimulant
Vincamine	Vinca minor	Vasorelaxant
Epicatechin	Lychee	Anti-breast cancer activity
Quercetin	Common grains	Against colorectal cancer
Salicin	Willow	Anti-inflammatory
Quinine	Cinchona	Anti-malarial
Pilocarpine	Rutaceae	Treatment of glaucoma
Paclitaxel	Pacific Yew	Cancer treatment

the primary molecule in vanilla bean extract that provides the pleasant aroma and unique flavour characteristic of the extract, which is primarily used in food products. However, vanillin can be synthesized in the laboratory at a fraction of the cost of real vanilla essence while retaining all its favourable attributes. Interestingly, several independent taste tests have shown that the majority of consumers cannot distinguish between natural and synthetic vanilla essence. In many cases, the synthetic product is even preferred over the natural one. In addition, several chemicals with similar physical and chemical properties are available, which makes it hard to separate them from one another. In many chemicals with therapeutic value, it is important to produce chemicals with high purity because even a minor quantity of certain impurities can have a detrimental effect on some patients. Moreover, even chemically similar compounds may have different performance as medicine. For example, aloe vera had been used extensively in traditional medicines to treat several ailments, most notably for treating jaundice. However, it has been found that the use of aloe vera can also cause liver damage in some patients because of the presence of some of the chemicals that get consumed along with the medicinal molecules (Yang et al., 2010). Therefore, the knowledge from nature is usually used to identify the chemicals that can be valuable for different applications.

In some cases, the compound primarily identified from a natural source can motivate the discovery of alternate pathways that are more practical or economical. For example, taxol (Paclitaxel) is an invaluable chemical used for the treatment of several types of cancers. In the second half of the 20th century, taxol was almost exclusively produced from the barks of Pacific yew, an evergreen conifer native to the Pacific coast ranges in North America. Since the tree gets killed during the extraction of its bark, there have been widespread concerns that it may soon be extinct. Because of this, there have been widespread efforts to develop an alternative method to produce taxol through alternative routes such as total synthesis, semi-total synthesis, biological synthesis and so on. Currently, alternative processes have been developed to produce taxol from European yew, which is far more common and abundant than Pacific yew. Efforts have also been made for a total synthesis method for the production of taxol though they are not yet commercialized.

If the target compound is not easy to synthesize or extract from natural sources, there are two approaches to developing practical solutions:

1. Obtain a precursor from a natural product, and use it as a foundation to construct a more complex molecule.
2. Analyse the active ingredient in a natural product and use it as a template for synthesizing a similar compound.

Another commonly followed approach is to mimic the properties of natural products. For example, the tensile strength of spider silks is even higher than that of steel. Studying the structure of spider silks helps researchers to develop bulletproof vests with low weight and better ability to resist bullets. Another example is the development of Velcro hook and loop fasteners commonly used in household products such as bags and shoes. This product was developed by George de Mestral, a Swiss engineer. The idea came to him when he noticed that the burrs of burdock stuck to his clothes and his dog's fur. He has noticed that the bur of burdock has several hooks that get caught on anything with a loop, such as clothes. This idea has been extended into the development of fasteners with two components: one with hooks and another with loops.

Engineers and scientists often take inspiration from nature, and biomimetics is a scientific field dedicated to studying and applying nature's designs and processes to solve human challenges. In addition to the examples given earlier, there are hundreds of scientific developments that used natural phenomena to make commercial products. The key is to study biological organisms, materials and ecosystems to understand how they function efficiently in their environments. When researchers find a useful feature in nature, they figure out its main ideas – like self-healing, saving energy or flying efficiently – and turn them into engineering or design ideas. After understanding these ideas, they use them to develop new products or technologies. Some additional examples include the development of less painful needles for injections inspired by the mosquito's proboscis, drag-reducing swimwear inspired by shark skin, and an underwater surveillance robotic vehicle inspired by manta rays.

Example 4.1. Chemical Sources of Ideas

A number of active ingredients that can be used to treat diseases are present in plants/fungi and can be extracted. However, most of these are synthesized in the laboratory instead of extracting from the plants/fungi. Explain the possible reasons for companies choosing synthetic chemistry route to produce pharmaceutical ingredients.

Solution

The concentration of active ingredients (AI) in the plants/fungi can be very low, and the cost of extracting these into pure components can be expensive. The cost of producing the same component from a known component can be significantly cheaper. It is also much faster to produce via synthetic routes. Extraction from plants requires further purification. Finally, a number of trees may be cut to produce the AI, whereas this can be avoided if synthesis route is used.

Example 4.2. Ideas from Fish Skin

It has been reported that a fish known as pirarucu, found in the rivers of the Amazon, infested with piranhas (known for razor-sharp teeth, incredible bite strength and speed), has a unique body skin that can withstand the bite of piranhas. Scientists believe that the knowledge of the skin of pirarucu can help to develop better quality bulletproof vests.

a) Identify the major attributes of the pirarucu skin that may be relevant for the design of bulletproof vests. Relate the required attribute of vests to the attributes of pirarucu skin.
b) How is the observation on pirarucu skin related to the chemical product design? How can this idea be further enhanced for the design of vests?
c) Comment on the possible challenges in developing a new bulletproof vest based on the observations on pirarucu skin.

Solution

a) The imperviousness of the pirarucu skin to withstand the sharp teeth of piranhas, the ability to absorb shock to withstand the strength and speed of piranhas' bite, flexibility to be a part of the body of the fish that allows it to swim swiftly and light weight – as light as the skin.
b) Mimic properties of natural product to develop a commercial product. Analyse the structure of the fish skin to identify the chemicals/structure/materials that give the unique attributes and use them in the design of vests.
c) Evolution takes long time to develop. Combining all attributes into a new material can be challenging.

4.3 SORTING OF IDEAS

The idea generation stage generally produces a large number (typically 100 or more) of crude ideas. From this list, the most suitable idea must be identified for manufacture. This process has often been compared with the hiring of new employees in a company. Once a position is advertised, the human resources department receives hundreds of applications. In the first step, all the applicants without the necessary credentials, experience and obvious red flags are screened out. From the reduced list, candidates with the most relevant experience and qualifications are identified.

Further screening on this list is performed to identify the candidates with the best match to the requirements of the specific tool. These candidates are usually invited for an interview to make the final selection. We can see that the first two levels of screening can be done very quickly, even by a person with no technical background or by a computer with a screening algorithm. The third level of screening often requires the involvement of a technical person, though this can also be conducted very quickly. Only the final stage of screening requires a detailed scrutiny; however, the effort needs to be focused on a small list of candidates.

The selection of the best idea from a crude list often follows the same pattern. Different levels of screening will be used in the early stage to eliminate the unsuitable ideas without spending much effort. A high level of analysis is often reserved for the most promising ideas during the selection stage.

The preliminary list of ideas often contains redundant ideas, repetition, different phrasing of the same concept, expensive ideas and obviously impractical ideas. At this stage, the generated list of ideas is scrutinized to remove any redundancies and obviously impractical ideas from the list. However, care must be taken to ensure that the ideas that sound impractical are not an 'outside the box' idea that offers an unexpected yet feasible solution. For example, disposable shaving razors were initially thought to be of no competition to the traditional safety razors. However, companies were able to market their portability and lack of required maintenance, which made them attractive to many customers. Currently, disposable razors dominate the market and significantly exceed the popularity of safety razors.

Once the redundancies and infeasible ideas are eliminated, the remaining ideas need to be classified into different categories. The main purpose of categorization is to group ideas of a similar type so they can be compared within the same category – it is always easier to compare apples to apples!

Example 4.3. New Carwash

A product design team was given a task to come up with an innovative product which is an effective and improved version of the current methods of washing cars. The main priority was to keep the use of water to a minimum in comparison to washing cars conventionally. To compete with conventional car washing, the following criteria must be met: limited or no water must be used, cleaning performance should be achieved with easy application and be friendly to both human health and the environment. The initial raw set of ideas are shown in Table 4.3.

From the unsorted list, the ideas 2, 10, 18, 22 and 36 have been removed without detailed scrutiny as they seemed impractical. The remaining ideas have been categorized into different categories and sub-categories:

i) New shampoo
   a) Chemistry (4, 13, 15, 19, 20, 23, 24, 29, 31, 35, 45, 48, 56)
   b) New paint (9, 33, 58)
   c) Biochemical (21, 28, 37, 48)
   d) Natural ingredients (25, 26, 27, 57)
ii) New cleaning technology
   a) Equipment design (3, 16, 17, 38, 47, 49, 52, 53, 54, 60)
   b) Ultrasonic and pressure waves (6, 46)

c) Thermal energy (5, 41)
d) Light (7)
iii) Waterless washing
a) Improved dry cleaning (1,14, 43)
b) New solvents (8, 39, 48)
c) Minor parts of car (25, 30, 50,51)
iv) New directions
a) Reusable materials (36)
b) High technology (9, 11, 12, 32, 34, 40, 42, 44, 55, 58, 59)

TABLE 4.3
Raw Ideas for Better Carwash

1. Use little/no water when washing
2. Remove paint
3. A tape which can stick up dirt and dust, and can remove them by friction
4. A dry shampoo which will adhere to the car surface when using; effect can only be seen after raining (effect with rainwater)
5. Freeze car, blow off dirt
6. Wash ultrasonically (by using the resonance effect to shake off the dirt)
7. A dry shampoo which will disperse the dirt particles under UV light
8. Use supercritical CO_2 for dry cleaning
9. Use a hydrophobic paint (only a layer of dust on the surface, therefore only cleaning by cloth is needed)
10. Wrap the car (change the wrapping)
11. Sprayable coating (after the coating dried, it will form a layer of tearable mask whereby the dirt particles will be adsorbed on it)
12. Nanoparticles carwash spraying system (Less volume of shampoo needed)
13. Advanced shampoo with rain repellent and waxing effect
14. Get a new dry cleaning solvent
15. Wash with Fuller's Earth
16. Portable vacuum cleaning with brush (brush the shampoo on the surface)
17. Portable water gun with refillable shampoo + blowing system + operates using battery
18. A dry shampoo which can change the colour of the car
19. A dry shampoo with anti-rust and scratch remover effect
20. A dry shampoo with anti-static effect (static might cause explosion when adding petrol without stopping engine)
21. Use enzyme to wash car
22. Change another car
23. Mop up dirt with particles
24. Use specific chemical interactions
25. Use vinegar to wash car (good for polishing glass)
26. Use baking soda + dishwashing liquid + a few drops of pure essential oil of lavender, tea tree oil or rosemary (optional) (effective car cleaning product)
27. Use lemon as scrubber instead of microfibre cloth (cutting a lemon in half)
28. Use denatured alcohol to wash car

Continued

TABLE 4.3 (*Continued*)
Raw Ideas for Better Carwash

29. A plant-based dry shampoo instead of petroleum based
30. Use toothpaste to clean the interior of the car (mild stains on leather or vinyl seats)
31. Use some peanut butter or solid shortening to wipe off tree sap or tar (then wipe it off with cloth)
32. Sandblasting technology (not harmful to the surface of the car)
33. Instant wash paint – removing the painting of the car, and replacing with the new layer again
34. Laser cleaning
35. All-purpose dry shampoo, which provides not only waterless carwash but also waterless home cleaning (kitchen)
36. Use newspaper to clean car windows (reused)
37. Use dish detergent
38. Use toothbrush (instead of microfibre cloth)
39. Use another supercritical solvent
40. Electrostatic (remove dirt and coat the shampoo evenly)
41. Car washing with steam + microfibre cloth
42. Double-action atomizer (less time consumption)
43. Foam/snow washing + microfibre cloth
44. Aerosol technology (gas stored in liquid form – spray out as gas phase, wiping not needed)
45. Sodium ash added inside the shampoo to prevent hard water from bonding
46. Use pressure waves
47. Create a microfibre towel that is thick and contains detergent injection (whenever you squeeze the cleaning liquid comes out of it) – just need to press on it and wipe on the surface
48. Powerful solvent which can remove hard stains (Alkyl polyglycosides)
49. Small kit to wash and dry the microfiber towels on the spot instantly
50. Use mineral oil in the solution (give shine to the glass)
51. Use polymer in the solution which gives tyres shine
52. Use tissue towel instead of microfibre cloth
53. Car wash system that uses high-pressure air jets
54. Car wash system that uses dry shampoo jets
55. Submit car in vacuum condition to remove all dirt on the surface
56. Different types of fragrances for dry shampoo
57. Use petroleum-based form to wash the car
58. A paint which can repel dirt and dust – frictionless surface
59. Shock wave technology to disturb air pressure system to force dirt out from surface
60. A device which can pump the cleaning agent (convenient for huge car)

4.4 SCREENING MATRIX

At this stage of the design process, we have identified a significant number of raw ideas and eliminated the obviously unsuitable options. Typically, this step removes anywhere between 10% and 20% of the initial list. However, sorting ideas can serve as a more effective method for preliminary screening. Even if some ideas are truly promising, if they fall outside the company's expertise or lack the potential to be profitable, they may not be worth further consideration. Therefore, a large number of initially chosen ideas could be eliminated at this stage. Our objective is to bring down the ideas to about 20–30 at this stage. Once we get to that number, a semi-quantitative approach known as a screening matrix can be applied to further reduce the number of ideas. The short-listed ideas at this stage will go to the selection stage for a more detailed and comprehensive scrutiny.

The screening matrix is a quick way to identify the most promising ideas from a large set without performing complex analysis. To do this, we need to identify the most important factors involved in deciding the suitability of the product. Since there are several factors involved in these decisions, we also need to determine the relative importance of these objectives. For example, if a new toothpaste is introduced into the market, the company may want to highlight some unique attribute of this product – such as the one-dimensional or attractive needs. As mentioned in the previous chapter, effectiveness may not be a key target even though the product may not be able to attract customers if it is not effective. Among these needs, the cost, flavour and additional gum protection could be the target attributes. However, not all these are equally important. If the unique selling point is gum protection, this attribute should be given a higher weightage. The weightage given to an attribute among *n* attributes is given by Equation 4.1.

$$\sum_{i=1}^{n} w_i = 1 \tag{4.1}$$

where w_i is the weighting factor for the attribute *i*. So, in the previous example, the gum protection may receive a score of 0.4 or above, depending on the discretion of the design team. It may be noted that the allocation of score may impact the final decision, and the scores must be assigned judiciously. Despite that, the effect of uncertainty in assigning the score will be analysed during the next step – which is sensitivity analysis.

In the next step, different ideas are compared within each category. Typically, a score range of 1 to 10 is used for comparison, where a higher score indicates better performance. To make the comparison easier, a benchmark is chosen at this stage. The benchmark is usually an established product or a potential competitor, which is assigned a score of 5 (Cussler and Moggridge, 2011). If an idea performs better than the benchmark in a certain category, it is assigned a score higher than 5. Conversely, if the benchmark outperforms the idea in a certain category, the assigned score must be lower than 5. For example, if a new idea has a better flavour but a higher cost, it may receive a score between 6 and 10 in the flavour category. However, if the new idea results in a more expensive product, a score of 4 or below must be assigned in the cost category. Scores are allocated to all ideas across all categories based on their relative performance in each category.

Once the scores for each idea have been assigned, the total score is calculated by determining the weighted average, as shown in Equation 4.2.

$$\text{Score}_j = \sum_{i=1}^{n} w_i s_{ij} \tag{4.2}$$

where j is the index for the idea, and s_{ij} is the score allocated to the idea j for the attribute i. The ideas receiving higher scores will be further analysed during the selection stage.

The following are the typical attributes that need to be considered during the screening stage:

1. Performance-based targets – especially if the new product has a few new selling points.
2. Feasibility targets – how mature the suggested idea is in terms of scientific and engineering capabilities.
3. Safety, health and environmental impacts or improvements.
4. Cost – this is subjective, and it is recommended not to put a high weightage on cost at this stage. We do not want to eliminate an otherwise promising idea too early based only on approximations.

A few points to be noted while performing the idea screening:

1. Generally, whole numbers are used in allocating the scores. Avoid using scores like 4.2. Similarly, avoid using too many decimal places in the weightage.
2. For allocating the scores, it is recommended to have an agreement between the design team members and also incorporate the suggestions from an expert group. When necessary, the opinion of customers may also be considered.
3. It is recommended to limit the number of attributes to the most relevant ones at this stage. While there may be several factors considered in the decision-making process, screening needs to be performed based on the most dominating factors – typically 3 or 4. Having more attributes at this stage will reduce the influence of the most important factors.
4. The main objective of idea screening is not the final selection – but elimination of non-promising ideas. Therefore, the ranks of shortlisted ideas are not so important.
5. Some ideas lead to groundbreaking inventions as a result of truly innovative thinking. In such cases, it may not be meaningful to compare with another set of ideas which may only provide incremental advantages from existing products.
6. The comparison is meaningful in the one-dimensional and attractive needs. However, it is important to eliminate ideas that do not fulfil the must-be needs (which also consist of government regulations) and are also prohibitively expensive. It is important to realize that the non-promising ideas (because of their inability to meet the must-be needs and cost) should not even be considered till the screening stage, regardless of their overall appeal.

Example 4.4. Insect Repellent Ideas

A product design team is investigating the possibility of developing a better and environmentally benign insect repellent. After exploring hundreds of ideas, 25 ideas are shortlisted for the idea screening stage. Table 4.4 shows the application

TABLE 4.4
Screening Matrix for Insect Repellent Ideas

Selection Criteria	Weighting Factor	Benchmark	Mosquito Coil	Ultrasonic Insect Repellent	Patch with Dissolving Repellent	Body Wash	Plant-Based Repellent Spray
Cost	0.2	5	8	3	8	4	5
Environmental sustainability	0.3	5	3	8	4	8	6
Performance	0.4	5	4	4	6	2	5
Convenience	0.1	5	4	3	6	8	5
Total	1	5	4.5	4.9	5.8	4.8	5.3

of idea screening technique on five of those shortlisted ideas. A commercial insect repellent spray has been chosen as the benchmark for this screening exercise.

From the screening matrix, it can be seen that two ideas: patch with dissolving repellent and plant-based repellent spray, and perform better than the benchmark. At this point, we do not have much information about product development. However, these two ideas seem promising, whereas the rest of the ideas underperform even with the limited information available at this time. Similar screening needs to be performed on the rest of the short-listed ideas to identify five to six most promising ideas to be taken to the selection stage.

4.5 SENSITIVITY ANALYSIS

As we have noticed during the previous section, the screening process depends heavily on the assigned scores, chosen criteria and selected weightage. At this stage, the scores and weightages are changed within reasonable limits. The effect of different screening criteria may also be investigated. It is to be noted that the purpose of sensitivity analysis is to see how sensitive the scores, weightages and additional criteria are, which may not be foreseen beforehand. This is not a random changing of scores or ideas; instead, make the changes only to see the effect of uncertainties in the decision-making process. For example, any additional screening criteria will be introduced if there are scenarios where the additional criteria are justified or if there is a difference in judgement whether a certain criterion is important. For scores, we make changes only to ensure that the difference in numbers does not have an impact on the decision. For example, if two ideas are given scores of 6 and 7, it is unlikely to swap those scores or allocate a score less than 5 as the design team already agreed that the second idea is better than the first and both ideas are better than the benchmark. So, the only uncertainty is about the scores. Instead of 6 and 7, we may allocate scores of 6 and 8 or 7 and 9, depending on the subjective judgement. By doing sensitivity analysis, the design team can see how robust the decisions are to the subjectivity involved. If there are major changes after the sensitivity analysis, we should re-examine our selection criteria. Otherwise, we may proceed with the results to the selection stage.

An example of the application of sensitivity analysis is shown in Example 4.5.

Example 4.5. Insect Repellent Ideas – Sensitivity Analysis

Conduct a sensitivity analysis on Example 4.4 by considering the following uncertainties.

i) The design team has some disagreement on whether the customers prefer convenience over cost and environmental sustainability.
ii) Patch with dissolving repellent is just as expensive as the benchmark, and the plant-based repellent spray is more expensive.
iii) Add comfort as an additional screening criterion as consumers may be concerned about their comfort when using the insect repellent, according to some expert team members.

Solution

Sensitivity analysis has been performed for the three cases, and the results are shown in Tables 4.5–4.7.

i) Weightage has been changed in the screening matrix to meet the requirements It can be seen that previously identified ideas are still potential candidates even after the change of preferences. However, body wash is also a potential candidate with the current preferences.
ii) New scores are assigned to represent the new scenario.
   Patch with dissolving repellent still performs better than the benchmark, whereas plant-based repellent spray performs just as good as the benchmark. Therefore, there will be less confidence to pursue this option in this scenario.
iii) Additional criterion has been introduced.

TABLE 4.5
Sensitivity Analysis with New Weightage

Selection Criteria	Weighting Factor	Benchmark	Mosquito Coil	Ultrasonic Insect Repellent	Patch with Dissolving Repellent	Body Wash	Plant-Based Repellent Spray
Cost	0.1	5	8	3	8	4	5
Environmental sustainability	0.2	5	3	8	4	8	6
Performance	0.4	5	4	4	6	2	5
Convenience	0.3	5	4	3	6	8	5
Total	1	5	4.2	4.4	5.8	5.2	5.2

TABLE 4.6
Sensitivity Analysis with New Scores

Selection Criteria	Weighting Factor	Benchmark	Mosquito Coil	Ultrasonic Insect Repellent	Patch with Dissolving Repellent	Body Wash	Plant-Based Repellent Spray
Cost	0.2	5	8	3	5	4	3
Environmental sustainability	0.3	5	3	8	4	8	6
Performance	0.4	5	4	4	6	2	5
Convenience	0.1	5	4	3	6	8	5
Total	1	5	4.5	4.9	5.2	4.8	5.0

TABLE 4.7
Screening Matrix for Insect Repellent Ideas

Selection Criteria	Weighting Factor	Benchmark	Mosquito Coil	Ultrasonic Insect Repellent	Patch with Dissolving Repellent	Body Wash	Plant-Based Repellent Spray
Cost	0.2	5	8	3	8	4	5
Environmental sustainability	0.2	5	3	8	4	8	6
Performance	0.3	5	4	4	6	2	5
Convenience	0.1	5	4	3	6	8	5
Comfort	0.2	5	3	6	6	6	5
Total	1	5	4.4	4.9	6.0	5.0	5.2

In this scenario, both ultrasonic insect repellent and plant-based repellent spray are still preferred options.

After analysing the results from all three scenarios covered in sensitivity analysis, it can be concluded that both ultrasonic insect repellent and plant-based repellent spray are not much sensitive to the minor variations in the scores and weightage. The addition of new scenarios also does not make these ideas less preferable. Therefore, it is safe to assume that these two are robust solutions and can be considered further as possible business ideas and need further scrutiny at the selection stage.

Example 4.6. Ideas for a New Perfume

In the previous sections, various aspects of idea generation and screening were presented. In this example, we will go over all the stages involved in a product design problem.

A company plans on developing a new perfume. However, this is a well-established field, and it is imperative that the new product fulfils customer demands through innovative ideas. The overall process of idea screening is illustrated in this example.

Step 1: Generation of ideas
A fraction of the ideas generated by the design team is shown in Table 4.8.

TABLE 4.8
Raw List of Ideas

1. Do not wear perfumes
2. Sprayable perfume
3. Biodegradable packaging
4. Essential oils (citrus, aromatic, floral, green, fruity, spicy, balsamic, woody)
5. Animal secretion
6. Lab-synthesized perfume
7. Recyclable packaging
8. More base notes scent (longer shelf life, need time to prepare formula)
9. Higher composition of fixative
10. Essential oils with lower volatility
11. Add fixatives
12. Application of Vaseline before applying perfume
13. Spray perfume on comb and comb through hair
14. Apply perfume after a bath
15. Spray at low-humidity areas (especially in humid countries)
16. Solvent with high volatility
17. Alcohol-free perfume
18. Plant-based ingredients
19. Non-toxic ingredients
20. Organic perfume
21. Nutritional ingredients, nutritionally packed perfume
22. Hijab perfume (perfect for Hijab)
23. Non or low flammable solvents
24. Safer and sustainable manufacturing process
25. Storage at wider temperature ranges
26. Animal cruelty-free
27. Sustainable sourcing (e.g. rose and jasmine do not cause harm to the land)
28. Sustainable manufacturing process
29. Essential oils with high purity (small volume with strong smell)
30. Baked goods scent
31. Chocolate scent
32. New money scent – I want to smell rich
36. Natural colouring for perfume
37. Colour-changing
38. Unisex perfume
39. Add ionic liquid (salt in liquid state) – for smelling pleasantly as you sweat

TABLE 4.8 (*Continued*)
Raw List of Ideas

40. Motion-sensing microcapsule (the more friction, the more smell)
41. Perfume bottle design that cannot be toppled (inspired by Johnnie Walker's Swing)
42. Solid perfume
43. Spraying perfume at constant intervals
44. Perfume shower (spray into the air and walk through it)
45. Spray perfume on your clothes

Step 2: Remove redundancy and classification
At this point, ideas that are obviously irrelevant, redundant or illogical have been removed. The remaining ideas have been classified under different titles so that the company can also make the shortlisting based on their expertise.

Design and manufacturing process

Basic

Sprayable perfume
Lab-synthesized perfume
Essential oils with lower volatility

Complex

More base notes scent (longer shelf life, need time to prepare formula)
Higher composition of fixative
Solvent with high volatility
Safer and sustainable manufacturing process
Storage at a wider temperature range
Essential oils with high purity (small volume with strong smell)

Alternative methods

Unconventional

Alcohol-free perfume
Add fixatives
Non-toxic ingredients
Non- or low-flammable solvents
Animal cruelty free
Sustainable sourcing (i.e. rose and jasmine do not cause harm to the land)
Sustainable manufacturing process
Scent of perfume evolve over time
Solid perfume

Natural Sources

Essential oils (citrus, aromatic, floral, green, fruity, spicy, balsamic, woody)
Animal secretion
Biodegradable packaging
Plant-based ingredients
Organic perfume
Natural colouring for perfume

Creative ideas

Logical

Nutritional ingredients, nutritionally packed perfume
Hijab perfume (perfect for Hijab)
Perfume bottle design that cannot be toppled (inspired by Johnnie Walker's Swing)
Baked goods scent
Chocolate scent

Unorthodox

Recyclable packaging
Unisex perfume
Add ionic liquid (salt in liquid state) – for smelling pleasantly as you sweat
Motion-sensing microcapsule (the more friction, the more smell)
New money scent – I want to smell rich
Magazine scent
Old books scent
Colour changing
Perfume shower (spray into the air and walk through it)

Miscellaneous

Easy

Application of Vaseline before application of perfume
Spray perfume on comb and comb perfume on hair
Put perfume after bath
Spraying perfume at constant intervals
Spray perfume on your clothes

Hard

Spray at low-humidity area (especially in humid countries)
Apply at pulse points of the body

Step 3: Screening matrix

At this point, a coarse screening has been applied to remove the ideas that (1) obviously have less potential, (2) fall outside the company's expertise, or (3) economically unattractive. Typically, 15–20 ideas will be taken to the next step. At this stage, three to five ideas will be compared at one time using a screening matrix. All ideas are compared using screening matrices to identify a list of the most promising ideas. One of such screening matrices is shown in Table 4.9.

Sensitivity Analysis

A sensitivity analysis has been carried out to analyze how sensitive the results are to the weightages, scores and new criteria. The results are presented in Tables 4.10–4.12.

Scenario 1: Engineering maturity has higher priority.

Scenario 2: Re-prioritizing due to Health & Safety Environment (HSE) rules & regulations.

TABLE 4.9
Screening Matrix

Selection Criteria	Weightage	Benchmark	Solid Perfume	Motion-Sensing Microcapsule Aerosol	Ionic Liquid
Cost	0.40	5	7	6	3
Engineering maturity	0.30	5	8	4	2
Environment	0.15	5	4	2	3
Safety	0.15	5	5	1	4
Total	1.00	5	6.55	4.05	2.85

TABLE 4.10
Sensitivity Analysis with Difference in Weightage

Selection Criteria	Weightage	Benchmark	Solid Perfume	Motion-Sensing Microcapsule Aerosol	Ionic Liquid
Cost	0.30	5	7	6	3
Engineering maturity	0.40	5	8	4	2
Environment	0.15	5	4	2	3
Safety	0.15	5	5	1	4
Total	1.00	5	6.65	3.85	2.75

TABLE 4.11
Sensitivity Analysis with Difference in Scores

Selection Criteria	Weightage	Benchmark	Solid Perfume	Motion-Sensing Microcapsule Aerosol	Ionic Liquid
Cost	0.2	5	7	6	3
Engineering maturity	0.1	5	8	4	2
Environment	0.3	5	4	2	3
Safety	0.4	5	5	1	4
Total	1.0	5	5.4	2.6	3.3

TABLE 4.12
Sensitivity Analysis with Additional Criterion

Selection Criteria	Weightage	Benchmark	Solid Perfume	Motion-Sensing Microcapsule Aerosol	Ionic Liquid
Cost	0.40	5	7	8	8
Engineering maturity	0.30	5	8	4	2
Environment	0.15	5	4	2	3
Safety	0.15	5	5	1	4
Total	1.00	5	6.55	4.85	4.85

Scenario 3: Funding support for new technology.
In this scenario, it is assumed that the new technology will receive additional funding from the authorities. In that case, both motion-sensing microcapsule aerosol and ionic liquid will be cheaper to produce, whereas there will be no impact on the cost of solid perfume.

It can be seen that the benefits of solid perfume in the rest of the areas outweigh the cost benefits of its competitors.

The solid perfume performs significantly better than the other options and the benchmark in different scenarios. Therefore, it can be chosen as the idea that needs further investigation. Similar screening will be conducted on other sets of ideas to identify four to six most promising ideas, which will be scrutinized further in the selection stage.

4.6 CONCLUSIONS

This chapter describes the second step in chemical product design – the generation and screening of ideas. Ideas are accepted from a variety of sources, including the dedicated core team in the organization, customers, competitors and consultants. Brainstorming sessions will play a key role in generating a broad range of ideas in an open and non-critical environment. Such sessions are targeted to generate approximately 100 ideas, though this number varies across organizations. The large number of fragmented ideas is organized into classifications, and redundant ideas are removed. Finally, concept screening methods are applied to judge the advantages and disadvantages of the remaining ideas to focus attention on the most promising ideas. These ideas will be analysed in more detail during the selection stage. Promising ideas are presented for management review, and if they are convinced of the potential and practicality of the proposed ideas, the project progresses to the next stage.

REFERENCES

Cussler, E. L., & Moggridge, G. D. (2011). Chemical Product Design. In *Cambridge Series in Chemical Engineering* (2nd ed.). Cambridge University Press. https://doi.org/10.1017/CBO9781139035132

Dai, J., & Mumper, R. (2010). Plant Phenolics: Extraction, Analysis and Their Antioxidant and Anticancer Properties. *Molecules*, *15*, 7313–7352.

Heinrich, M. (2013). Ethnopharmacology and Drug Discovery. In Reedijk, J. (ed.), *Elsevier Reference Module in Chemistry, Molecular Sciences and Chemical Engineering*. Oxford: Elsevier.

Saboon, S. K. C., Arshad, S., Amjad, M. S., & Akhtar, M. S. (2019). Natural Compounds Extracted from Medicinal Plants and Their Applications. In Akhtar, M., Swamy, M., & Sinniah, U. (eds.), *Natural Bio-Active Compounds*. Singapore: Springer. https://doi.org/10.1007/978-981-13-7154-7_7

Stary, F. (1996). *The Natural Guide to Medicinal Herbs and Plants*. New York: Barnes & Noble Inc.

Venkatasubramanian, V., & Morris, K. R. (2008). Drinking from a Fire Hose. *Pharmaceutical Technology Europe*, *20*, 40–46.

Yang, H. N., Kim, D. J., Kim, Y. M., Kim, B. H., Sohn, K. M., Choi, M. J., & Choi, Y. H. (2010). Aloe-Induced Toxic Hepatitis. *Journal of Korean Medical Science*, *25*(3), 492–495.

4.7 PROBLEMS

1. A chemical found in the roots of a rare plant species in the rainforests of Costa Rica is found to have a unique ability to control uric acid levels in blood, which leads to gout (gout is a common form of inflammatory arthritis that is painful but non-life-threatening). The scientists believe that one of the several components present in the plant may have medicinal properties that relieve gout; however, the concentration of this component is low.
   a) What are the different possible ways that this observation can be used to develop a new drug?
   b) What are the challenges associated with each of these possibilities?
2. Reserpine is a drug that can be used for the treatment of high blood pressure. Though it was originally found in the roots of Indian snakeroot, currently, reserpine is produced entirely through synthetic routes.
   Discuss any four of the possible reasons for the preference of synthetic routes for reserpine production over extracting it from Indian snakeroot.

3. It has been reported that the bark of a rare tree in South America has been used by the locals to cure certain deadly diseases. The scientists believe that one of the several components present in the tree bark may have medicinal properties; however, the concentration of this component is low.
   a) What are the different possible ways that this observation can be used to develop a new life-saving drug?
   b) What are the challenges associated with each of the possibilities?
4. Sunscreen protects the skin from harmful UV rays, reducing the risk of sunburn, premature ageing and skin cancer. It also helps maintain an even skin tone by preventing sun-induced dark spots and hyperpigmentation. It is essential to develop a water-resistant sunscreen for prolonged protection during swimming or sweating. Generate as many ideas as possible to safeguard the skin during water activities.
5. A list of ideas generated for the design of a new liquid soap is given in Table 4.13. Identify the redundancies and obviously infeasible ideas from the list, and categorize the rest of the ideas under appropriate titles and subtitles:

TABLE 4.13
List of Ideas for Liquid Soap

1. Just use water
2. Live in a cold country to wash less
3. Use hand sanitizers
4. Wear gloves
5. Use lotion to clean hands
6. Protection up to 24 hours
7. Smell lasting up to 24 hours
8. Small liquid pallet in dissolvable packaging
9. Small liquid pallet in squeezable packaging
10. Small solid tablet
11. Use hand wipes
12. Dispensing pump containers
13. Automated sensor dispenser with user-specified amount
14. User-specified amount using phones
15. Use UV lights
16. Isolate yourself
17. Do not touch dirty objects
18. Activated charcoal as hand wash
19. Glitter in soap
20. Edible soap
21. On-the-go cleaning
22. One-L hand wash
23. Voice-activated soap dispenser
24. Completely natural ingredients
25. Illuminating hand wash
26. Playdough sticky consistency
27. All-in-one mixed flavours
28. Use tea tree oil for natural fragrance

6. There has been an increase in demand for cold coffee among coffee enthusiasts because it offers a refreshing and smooth taste, especially on hot days, without the bitterness that sometimes comes with hot brews. Its versatility allows for creative customizations, with milk, syrups and ice. A product development team has come out with ideas to preserve the coldness, along with the freshness and taste of cold coffee if it is not consumed immediately. Table 4.14 presents a part of the generated ideas:
   a) Remove redundancies and unrealistic ideas from the list of ideas.
   b) Sort the ideas into appropriate categories and apply product screening to identify the most promising ideas for further analysis.

TABLE 4.14
List of Ideas to Preserve Freshness of Cold Coffee

1. Put it back in the refrigerator.
2. Add ice to the beverage.
3. Just let the beverage be.
4. Finish the beverage as soon as possible.
5. Do not drink a cold beverage.
6. Add a coolant packet beside the can.
7. Change the shape of the can to a spiral shape.
8. Change the shape of the can to a rectangular shape.
9. Add another layer of coating to the can.
10. Add an insulating jacket around the can.
11. Use a thicker label that will be stuck around the can.
12. Use coextruding polystyrene foam shrink-label stock that insulates the can.
13. Put the can in an outdoor cooler.
14. Put the can and some salt in a cooler.
15. Put the can in an insulated cooler backpack.
16. Put the can in an ice box.
17. Put the can in an ice box along with ice packs.
18. Fill the cooler with ice and water, along with a good amount of salt that lowers the freezing point of water.
19. Wrap a blanket around the can.
20. Put the can or bottle in the water of a lake or stream nearby.
21. Bury the can or bottle in the sand if you are at the beach.
22. Add edible coolants to the beverage.
23. Invent a self-cooling beverage.
24. Drink the beverage in rooms with air-conditioning.
25. Use a tumbler with vacuum insulation as packaging material.
26. Use aluminium alloy as packaging material.
27. Use tin-plated steel as packaging material.
28. Use glass as packaging material.
29. Use paper as packaging material.
30. Use plastic as packaging material.
31. Increase the wall thickness of the packaging material.
32. Minimize the surface area of the can.

TABLE 4.15
List of Ideas for Better Dishwashing

1. Recyclable detergent
2. Self-cleaning utensils
3. Colouring to differentiate detergent
4. Baking soda
5. Edible cutleries
6. Microwave/ultrasonic dishwasher
7. Fast-drying sponge
8. Multiple detergents of different strengths in one bottle
9. Aloe vera (moisturizing effect)
10. Lightweight
11. Non-stick utensils
12. Gaseous detergent
13. Odour-free dishcloth/sponge
14. Edible
15. Dish wipes
16. Sanitizing effect
17. Compostable packaging
18. Scrubbing sponge at mouth of bottle
19. Flower smell
20. Reusable dish wipes
21. Spray and wipe
22. Use vinegar
23. Automatic detergent dispenser
24. Disassemble utensils for washing
25. Vacuum cleaning
26. Detergent with scrubbing beads
27. Dry detergent
28. Nano sponge
29. Detergent dispensing gloves
30. Non-grease plate
31. Dishwasher + dryer
32. Use enzyme

7. A product design team has developed a list of ideas for a better dishwashing experience. Table 4.15 contains some of the ideas from the generated list.
   a) Remove redundancies and unrealistic ideas from the list of ideas.
   b) Sort the ideas into appropriate categories and apply product screening to identify the most promising ideas for further analysis.

5 Tools for Selection

The step begins with the identification of property targets for the product. These targeted properties must be satisfied in order for a product to meet the customer's needs. Once the property targets are identified, different property prediction models can be applied to estimate them. The choice of property prediction methods depends on the types of product properties and how accurate the estimation needs to be. In cases where traditional property prediction models are not available, machine learning tools can be used to link the target attributes or properties to the molecular structure. We then explore various molecular design tools that can be used to generate molecules and estimate their properties. In addition to technical performance, safety, health and environmental aspects of a product should also be considered during the selection stage. Selection matrix is another useful tool for comparing the shortlisted products and determining the most suitable option. To ensure a robust decision, sensitivity analysis may also be applied to identify the most influential criteria and assess how they affect the final decision. It is also important to conduct a risk assessment as the product may not perform as expected. In this chapter, we will provide an overview of these quantitative tools that can be used to support the final selection of ideas in chemical product design.

5.1 PROPERTY TARGETS AND DESIGN CONSIDERATIONS

As discussed in Chapter 2, chemical products can be categorized into commodity chemicals, molecular products and microstructured products. Regardless of its type, all chemical products perform the desired functions through their specific properties. To ensure these products achieve their functionality, property targets must be identified and considered during the product design stage. These property targets represent the range of properties that the product must have to meet its performance. At this stage, relevant chemical engineering principles such as thermodynamics, reaction kinetics and transport properties that govern the behaviour of chemical products should be identified.

For example, one may be interested in identifying an alternative refrigerant to replace dichlorodifluoromethane $\left(CCl_2F_2\right)$ refrigerant, commonly known as R12 in industry, with the design objective of minimizing the quantity of refrigerant needed. This performance can be translated to heat of vaporization, which is a measurable property. To ensure that less amount of refrigerant is required compared to R12, the alternative refrigerant must have a higher heat of vaporization than that of R12. In this case, property targets can be defined using R12 as a benchmark refrigerant. Therefore, if the refrigerant is to be used in a refrigeration cycle operating at an evaporation temperature of 272.04 K, the heat of vaporization can be defined as a property target, with a required value of greater than or equal to 18.4 kJ/mol (Sahinidis et al., 2003). The numerical value of this property can be either obtained through

DOI: 10.1201/9781032708911-5

experiments or estimated through property estimation methods. Property estimation methods are useful when experimental values are not available or when it is not feasible to conduct experiments to determine them. Before we explore property estimation methods, it is important to first understand the different types of properties.

5.1.1 Type of Properties

In most cases, a product must satisfy multiple property targets to perform a specific function. Properties are classified into three tiers: primary, secondary and functional (Kontogeorgis and Gani, 2004), as shown in Figure 5.1. These three categories are used to describe properties of pure components. Primary properties are properties that can be determined directly based on their molecular structural information, with each chemical having a single specific value. Using the previous example, the heat of vaporization of an alternative refrigerant at 298 K can be predicted using its molecular structure. Unlike primary properties, secondary properties are derived from other properties and cannot be calculated solely based on molecular structure. For instance, the density of a product at a certain temperature and pressure depends on its mass and volume. Critical compressibility factor is another example of secondary properties where its value depends on critical volume, critical temperature and critical pressure. Functional properties are used to describe those pure component properties which may change with pressure and/or temperature. Some examples include vapour pressure, liquid density and thermal conductivity.

Apart from pure component properties, chemical product design often also requires consideration of properties related to mixtures. This is especially important when a single component molecule cannot meet all the desired product performance requirements. Mixture properties refer to the properties of a mixture that depend on the composition of its components. Mann et al. (2023) further categorized mixture properties into type-A and type-B, representing the properties of homogeneous mixtures and phase equilibrium properties, respectively. Some examples of type-A mixture properties include liquid density, kinematic viscosity and thermal conductivity. The ideal mixing rule can be applied to estimate these types of properties based on

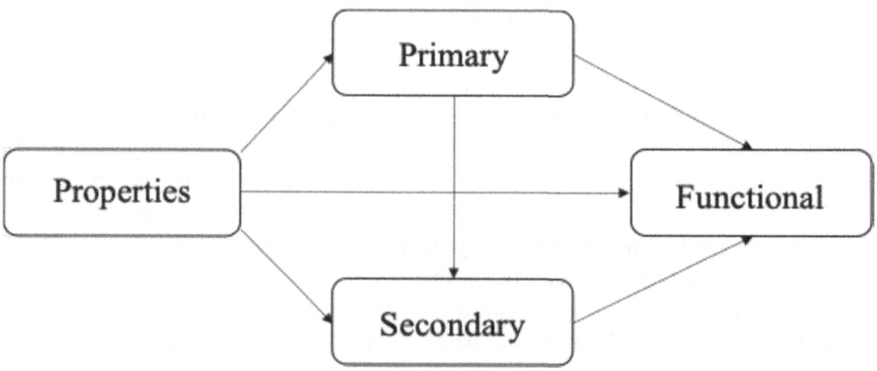

FIGURE 5.1 Classification of properties.

TABLE 5.1
Examples of Pure Component and Mixture Properties

Property Type	Category	Property Examples
Pure component	Primary	Normal boiling point, normal melting point, heat of vaporization at 298K, heat of fusion at 298K, Gibbs free energy of formation at 298K, critical temperature, critical volume, critical pressure
	Secondary	Density, viscosity, surface tension, heat capacity, critical compressibility factor
	Functional	Liquid density, vapour pressure, thermal conductivity, diffusion coefficient
Mixture	Homogenous mixture	Kinematic viscosity, liquid density, thermal conductivity, solid solubility, liquid solubility, gas solubility, diffusivity
	Phase equilibrium	Activity coefficient, fugacity coefficient, saturation temperature, saturation pressure

the known properties of pure components if the mixture is assumed to behave ideally. In cases where the ideal mixing rule cannot be assumed, then the properties have to be estimated from specific mixture property models. On the other hand, phase equilibrium properties usually involve solid-liquid, liquid-liquid and vapour-liquid equilibria. Saturation pressure and/or saturation temperature play an important role in the estimation of phase equilibrium properties. Additional examples of each property type are presented in Table 5.1.

5.1.2 Property Prediction Tools

One of the key challenges in chemical product design is the availability of the numerical values of the required set of properties. Since there are different types of properties, various types of property prediction methods are developed to predict these values when experimental data is not available. It is important to note that property estimation methods cannot entirely substitute experimental work, but they allow rapid estimation of properties and enable users to make comparisons among the alternatives. The promising selected candidates can then be validated through experiments. Hence, property prediction methods play an important role during the selection process.

The selection of a suitable property estimation method depends on the types of product properties and the desired level of estimation accuracy. Property estimation methods are grouped into two main categories: approximate methods and reference methods (Gani and Constantinou, 1996). Approximate methods involve fitting theoretical model to a limited range of experimental data to estimate properties. These methods require less computational effort but offer lower accuracy. On the other hand, reference methods are used to verify a theory or to validate approximate methods. Although they provide high accuracy, reference methods are more

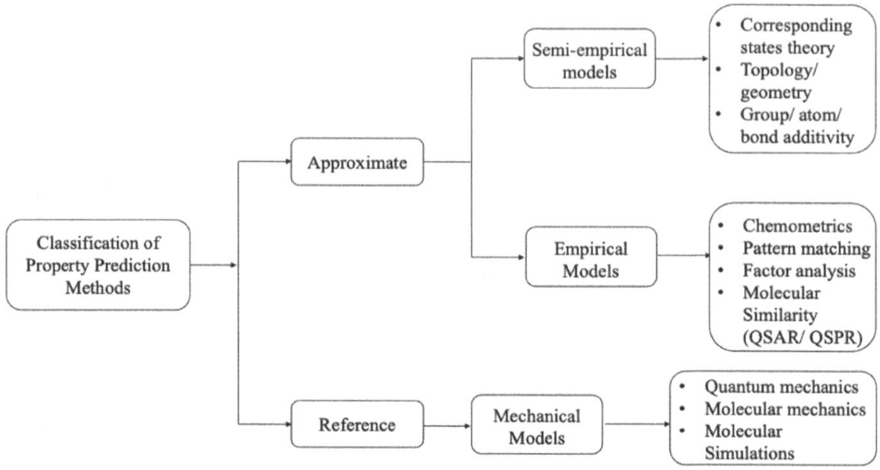

FIGURE 5.2 Classification of property prediction methods.

computationally intensive. Under these two categories, property prediction models can be further classified into mechanical models, semi-empirical models and empirical models. Examples of each model type (Gani and Constantinou, 1996) are illustrated in Figure 5.2.

Majority of the property estimation methods used in product design are either semi-empirical or empirical (Gani and Constantinou, 1996). Some of the traditional property prediction methods include group contribution methods, quantitative structure-activity relationship (QSAR) and quantitative structure-property relationship (QSPR) models. Group contribution methods estimate the property of a molecule by expressing it as a function of the contributions from each molecular group that makes up the molecule (Constantinou and Gani, 1994). On the other hand, QSAR/QSPR are prediction models that mathematically correlate biological activity or molecular property of interest (physical or chemical properties) with molecular structural properties, commonly known as descriptors (Dearden, 2017). The following sections focus on discussing the use of molecular groups in the development of group contribution models as well as the role of descriptors in QSAR/QSPR models.

5.1.3 Group Contribution Methods

Group contribution (GC) methods are a type of semi-empirical property model developed based on the principle of group, atom or bond additivity. They are widely employed in the development of property prediction models. GC methods are particularly useful for predicting properties of molecules from their structure when experimental data are not available as the methods are simple and can give reasonably accurate results for many properties. In GC methods, a molecule is treated as a combination of simple building blocks called molecular groups (Joback and Reid, 1987). Each molecular group has a known contribution to a property such as normal boiling point and melting point. These methods estimate the properties of a compound by breaking it down into simpler molecular groups and adding up their contributions, with each one multiplied by the number of times the group appears in the molecule. This can be illustrated through the general equation shown in Equation 5.1.

$$f(P) = \sum_i N_i C_i \tag{5.1}$$

where $f(P)$ represents a function of the actual property P, C_i denotes the contribution of molecular group i, which appears N times in the molecule. This contribution, C_i, is determined using regression analysis applied to large experimental datasets. GC methods are suitable for preliminary screening as they enable fast property estimation without the need for intensive computational resources (Constantinou et al., 1993).

GC methods have been widely applied to estimate the physicochemical and thermodynamic properties of organic compounds. Most of the primary properties listed in Table 5.1 can be estimated using GC methods. For example, Joback and Reid (1987) developed a GC method to predict liquid viscosity, heat capacity and several pure component properties, including the critical volume, critical pressure, critical temperature, normal boiling point and normal melting point. Their approach assumes that there are no interactions between the molecular groups. As a result, their GC approach cannot differentiate between isomers and may produce less accurate results for such compounds. To address these limitations, the general GC methods are extended by introducing two levels of molecular groups, including first-order and second-order groups (Constantinou and Gani, 1994). The inclusion of second-order groups enables differentiation among isomers and contributes to improved accuracy in estimating properties of complex compounds. Nevertheless, these two levels of molecular groups remain insufficient to accurately describe poly-ring and open-chain polyfunctional compounds with main chains containing more than four carbon atoms (Marrero and Gani, 2001). To overcome this shortcoming, Marrero and Gani (2001) introduced third-order groups into GC methods, which allows a better representation of compounds with multiple functional groups, multi-ring compounds and fused ring compounds. Their GC model was expressed by Equation 5.2.

$$f(P) = \sum_i N_i C_i + y\sum_j M_j D_j + z\sum_k Q_k E_k \tag{5.2}$$

In Equation 5.2, $f(P)$ represents a function of the actual property P. The term C_i represents the contribution from the first-order group i that appears N times in the molecule. First-order groups comprise a large range of basic molecular groups, such as CH_3, CH_2, CH, C, OH, CHO and $CH_2 = CH$. D_j refers to the contribution of the second-order group j that occurs M times. Some examples of second-order groups include $(CH_3)_2CH$, $(CH_3)_3C$, CH_3COCH_2 and CHOH. Similarly, E_k corresponds to the contribution of the third-order group k that is present Q times in a compound. The constants y and z are set to either zero or one, depending on whether second and third-order groups are considered in the property estimation. For instance, when only first-order molecular groups are considered, both y and z are set to zero. If the estimation includes second-order groups, y is set to one and z remains zero. For the

TABLE 5.2
Examples of Pure Component Properties and Their GC Models

Property, P	Property Function, $f(P)$	GC Models
Critical volume, V_c	$V_c - V_{c0}$	$\sum_i N_i V_{c1i} + y\sum_j M_j V_{c2j} + z\sum_k Q_k V_{c3k}$
Critical pressure, P_c	$(P_c - P_{c1})^{-0.5} - P_{c2}$	$\sum_i N_i P_{c1i} + y\sum_j M_j P_{c2j} + z\sum_k Q_k P_{c3k}$
Critical temperature, T_c	$\exp(T_c / T_{c0})$	$\sum_i N_i T_{c1i} + y\sum_j M_j T_{c2j} + z\sum_k Q_k T_{c3k}$
Normal boiling point, T_b	$\exp(T_b / T_{b0})$	$\sum_i N_i T_{b1i} + y\sum_j M_j T_{b2j} + z\sum_k Q_k T_{b3k}$
Normal melting point, T_m	$\exp(T_m / T_{m0})$	$\sum_i N_i T_{m1i} + y\sum_j M_j T_{m2j} + z\sum_k Q_k T_{m3k}$
Standard enthalpy of formation at 298K, H_f	$H_f - H_{f0}$	$\sum_i N_i H_{f1i} + y\sum_j M_j H_{f2j} + z\sum_k Q_k H_{f3k}$
Standard enthalpy of fusion, H_{fus}	$H_{fus} - H_{fus0}$	$\sum_i N_i H_{fus1i} + y\sum_j M_j H_{fus2j} + z\sum_k Q_k H_{fus3k}$
Standard enthalpy of vaporization at 298K, H_v	$H_v - H_{v0}$	$\sum_i N_i H_{v1i} + y\sum_j M_j H_{v2j} + z\sum_k Q_k H_{v3k}$
Standard Gibbs energy at 298K, G_f	$G_f - G_{f0}$	$\sum_i N_i G_{f1i} + y\sum_j M_j G_{f2j} + z\sum_k Q_k G_{f3k}$

TABLE 5.3
Values of Adjustable Parameter

Adjustable Parameter	Value
V_{c0}	7.95 cm³/mol
P_{c1}	5.9827 bar
P_{c2}	0.108998 bar$^{-0.5}$
T_{c0}	231.239 K
T_{b0}	222.543 K
T_{m0}	147.45 K
H_{f0}	5.549 kJ/mol
H_{fus0}	−2.806 kJ/mol
H_{v0}	11.733 kJ/mol
G_{f0}	−34.967 kJ/mol

third level of estimation, where all three levels of molecular groups are considered, both y and z are set to one. Some examples of pure component properties that can be estimated using the GC methods developed by Marrero and Gani (2001), along with their corresponding property functions, are listed in Table 5.2. The adjustable parameter for each property function is provided in Table 5.3.

Example 5.1. Prediction of Properties using Group Contribution Models

Estimate the normal boiling point, T_b, and normal melting point, T_m, of isopropanol (molecular structure shown in Figure 5.3) using the following group contributions (Marrero and Gani, 2001) presented in Table 5.4. Assume first-order groups for the calculation.

Solution

There are two (2)-CH_3, one (1)-CH- and one (1)-OH group in isopropanol. Since only the first-order group is considered, y and z are set to zero. Thus, normal boiling point can be estimated using group contribution equation as follows:

$$T_b = T_{b0}\ln\left(\sum N_i T_{b1i}\right)$$

where T_{b0} represents the adjustable parameter for T_b and has a value of 222.543 K (Marrero and Gani, 2001).

$$T_b = 222.543\ln\left(2\times 0.8491 + 1\times 0.2925 + 1\times 2.5670\right) = 337.557\text{K}$$

Since only first-order group is considered, y and z are set to zero. Thus, normal melting point can be estimated using group contribution equation as follows:

$$T_m = T_{m0}\ln\left(\sum N_i T_{m1i}\right)$$

TABLE 5.4
First-Order Molecular Groups and Their T_b and T_m Contributions

Molecular Group	T_b Contribution, T_{b1i}	T_m Contribution, T_{m1i}
-CH_3	0.8491	0.6953
-CH-	0.2925	−0.3730
-OH	2.5670	2.7888

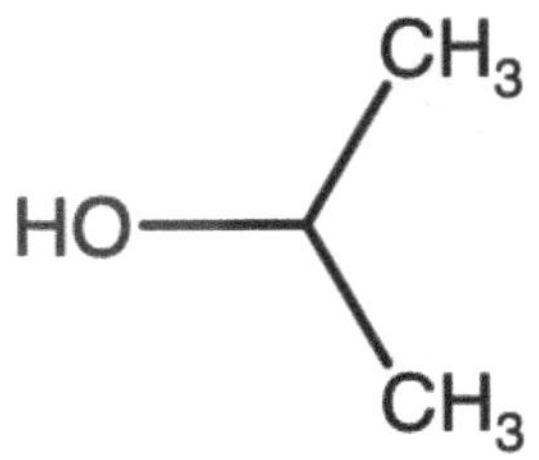

FIGURE 5.3 Molecular structure of isopropanol.

where T_{m0} represents the adjustable parameter for T_m and has a value of 147.450 K (Marrero and Gani, 2001).

$$T_m = 147.450\ln(2 \times 0.6953 + 1 \times -0.3730 + 1 \times 2.7888) = 197.094\text{K}$$

5.1.4 Topological Indices

In addition to representing a molecule as a combination of molecular groups, it can also be described using descriptors such as topological indices (TIs). TIs are commonly applied in QSAR/QSPR property prediction models. In QSAR/QSPR models, these descriptors translate molecular structures into numerical values that are associated with a certain biological activity or other physicochemical properties of molecules. Both QSAR and QSPR models are developed based on the molecular similarity principle, which assumes that molecules with similar structures exhibit similar properties. TIs have been successfully applied in QSAR/QSPR models to predict heat capacity, boiling point, flash point, refractive index, biodegradability, aqueous solubility and enthalpy of formation.

TIs models consider how atoms are connected and how they interact within a molecule. TIs are numerical values derived from chemical graph theory, where atoms of a molecule are represented as vertices (or nodes) and the chemical bonds as edges (or lines) in a molecular graph (Wilson, 1986). A molecular graph is generally defined as $G = (V, E)$, where V is the set of vertices and E is the set of edges. These molecular graphs are usually represented as hydrogen-suppressed graphs where hydrogen atoms are omitted from the molecular skeletons (except those bonded to heteroatoms). Besides, double and triple bonds are also not explicitly shown in a hydrogen-suppressed graph. The difference between a standard molecular structure and its hydrogen-suppressed graph is illustrated using 2-Pentanone as an example in Figure 5.4.

In Figure 5.4, we can see that the numbers 1 to 5 correspond to the carbon atoms without hydrogen atoms, while number 6 represents the oxygen atom. The bonds between these atoms are depicted as edges labelled 'a' through 'e'. Since we cannot distinguish between single and double bonds in Figure 5.4 (b), the bond indices introduced later are defined to account for such differences. These indices are then incorporated into the calculation of TI of a molecule, which eventually yields a

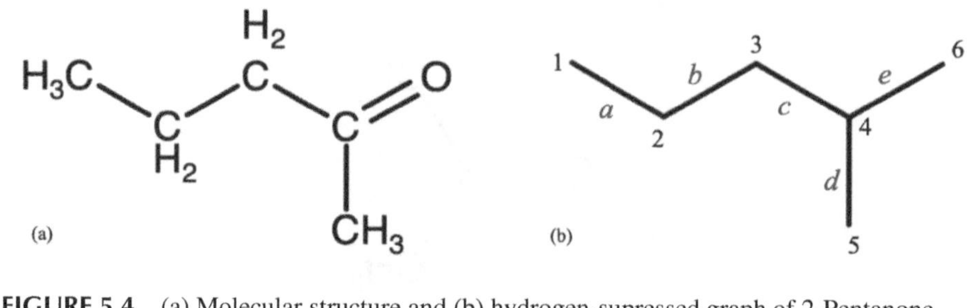

FIGURE 5.4 (a) Molecular structure and (b) hydrogen-supressed graph of 2-Pentanone.

single numerical value that characterizes the molecular graph. Considering standard molecular graph properties such as connectivity between atoms, path length between atoms, atomic types and degree of a vertex, TIs can be expressed in various forms. Thus, there are various types of TIs, each capturing different structural features or properties of molecules. In the following sections, we will discuss several commonly used TIs and how it was applied in QSAR/QSPR models to predict some properties.

5.1.4.1 Connectivity Indices

Connectivity index (CI) is a form of TI that considers the structural formula of a molecule and relates it to its properties. The first CI, known as the branching index, was introduced by Randic (1975), who used it to quantify the degree of branching in alkane molecules. The branching index, also known as the Randić index, of a molecular graph G, $R(G)$ can be defined mathematically as shown in Equation 5.3.

$$R(G) = \sum_{i,j \in E(G)} \left[(v_i)(v_j) \right]^{-0.5} \tag{5.3}$$

where $E(G)$ is the set of edges in graph G, and v_i and v_j represent the valency of the connected vertices i and j. Because of their inverse relationship, an increase in branching within a molecular structure leads to a lower branching index. This index was found to be strongly correlated with some physical properties of alkanes, such as boiling points. An example calculation of the branching index is provided in Example 5.2.

Example 5.2. Estimation of Branching Index

Calculate the branching index value for the following molecules:

a) 2,2 Dimethylhexane
b) 2,4 Dimethylhexane

Solution

Figures 5.5 and 5.6 illustrate the molecular skeleton of 2,2-Dimethylhexane and 2,4-Dimethylhexane, respectively. In both figures, the labels *a, b, c, d, e, f* and *g* represent different bonds within the molecules, whereas numbers next to each carbon atom represent their valency.

a) Using Equation 5.3, the branching index for the 2,2-Dimethylhexane molecule can be determined.
Bond $a: (1 \times 2)^{-0.5} = 0.7071$
Bond $b: (2 \times 2)^{-0.5} = 0.5000$
Bond $c: (2 \times 2)^{-0.5} = 0.5000$
Bond $d: (2 \times 4)^{-0.5} = 0.3536$

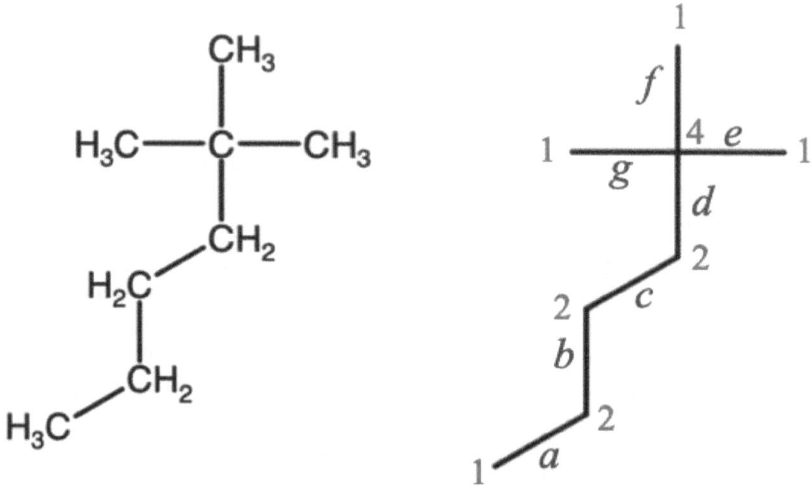

FIGURE 5.5 Molecular skeleton of 2,2-Dimethylhexane molecule.

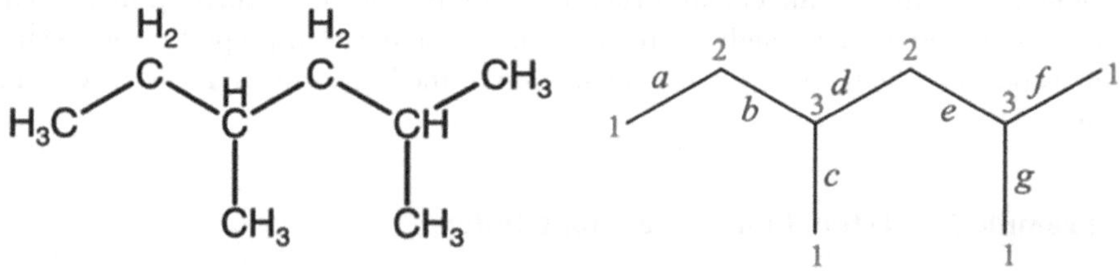

FIGURE 5.6 Molecular skeleton of 2,4-Dimethylhexane molecule.

Bond $e: (4\times 1)^{-0.5} = 0.5000$

Bond $f: (4\times 1)^{-0.5} = 0.5000$

Bond $g: (4\times 1)^{-0.5} = 0.5000$

Branching index of 2,2-Dimethylhexane molecule
= 0.7071 + 0.5000 + 0.5000 + 0.3536 + 0.5000 + 0.5000 + 0.5000 = 3.5607

b) Using Equation 5.3, the branching index for the 2,4-Dimethylhexane molecule can be determined:

Bond $a: (1\times 2)^{-0.5} = 0.7071$

Bond $b: (2\times 3)^{-0.5} = 0.4082$

Bond $c: (3\times 1)^{-0.5} = 0.5774$

Bond $d: (3 \times 2)^{-0.5} = 0.4082$
Bond $e: (2 \times 3)^{-0.5} = 0.4082$
Bond $f: (3 \times 1)^{-0.5} = 0.5774$
Bond $g: (3 \times 1)^{-0.5} = 0.5774$

Branching index of 2,4 Dimethylhexane molecule
= 0.7071 + 0.4082 + 0.5774+ 0. 4082 + 0.4082 + 0.5774 + 0.5774 = 3.6639
Nonetheless, the branching index is only suitable for alkanes as it cannot distinguish between saturated and unsaturated bonds, cyclic structures or the presence of heteroatoms. To address this gap, Kier and Hall (1976) extended the CI to account for the presence of heteroatoms in molecular structures. They introduced a new structural descriptor known as the valence delta, δ^v, which allows atoms to be differentiated according to their valence electrons. Kier and Hall (1976) defined δ^v as

$$\delta_i^v = Z_i^v - h_i \tag{5.4}$$

where Z_i^v denotes the number of valence electrons for atom i, and h_i represents the number of hydrogen atoms attached to atom i. However, Equation 5.4 is less suitable for heavier atoms as it does not capture the number of core electrons. It is important to consider the number of non-valence core electrons when dealing with high atomic weight atoms as these electrons have a substantial impact on properties. Thus, δ^v was redefined by considering the atomic number of an atom i, Z_i, as expressed in Equation 5.5.

$$\delta_i^v = \frac{Z_i^v - h_i}{Z_i - Z_i^v - 1} \tag{5.5}$$

Equation 5.5 now includes both valence and core electron count in an atom. These modified δ^v values were then used to define n^{th} order connectivity indices, ${}^n\chi^v$. The zeroth-order CI of an atom, ${}^0\chi^v$, does not account for bonding and can be estimated using the individual δ^v values (Kier and Hall, 1986), as shown in Equation 5.6.

$${}^0\chi^v = (\delta_i^v)^{-0.5} \tag{5.6}$$

The zeroth-order CI of a molecule can then be calculated by adding up all the CI values of each atom:

$$\left({}^0\chi^v\right)_{molecule} = \sum_{i=1}^{m} ({}^0\chi^v)_m \tag{5.7}$$

Unlike zeroth-order CI, the first-order CI considers the bonding (edge) between two atoms. For a bond connecting atoms i and j, the first-order CI, ${}^1\chi^v$ (Kier and Hall, 1986) is defined as:

$${}^1\chi^v = \left(\delta_i^v \delta_j^v\right)^{-0.5} \tag{5.8}$$

The first-order CI of a molecule is then calculated by the summation of CI values of all edges within the molecule:

$$\left({}^{1}\chi^{v}\right)_{molecule} = \sum_{i=1}^{m}\left({}^{1}\chi^{v}\right)_{m} \quad (5.9)$$

For higher-order CI of a molecule, the calculation can be performed using the same principle, using Equations 5.10 and 5.11.

$${}^{n}\chi^{v} = \left(\delta_i^v \delta_j^v \ldots \delta_n^v\right)^{-0.5} \quad (5.10)$$

$$\left({}^{n}\chi^{v}\right)_{molecule} = \sum_{i=1}^{m}\left({}^{n}\chi^{v}\right)_{m} \quad (5.11)$$

An example of a CI application in QSAR study is its use in predicting the odour similarity of odorant molecules (Kier et al., 1977), as illustrated in Equation 5.12.

$$Odour\,Similarity = 7.47 - 1.84\,{}^{2}\chi + 1.34\,{}^{3}\chi_{c};$$

$$N = 15; r^2 = 0.848; s = 0.395 \quad (5.12)$$

where N is the number of observations, r is the correlation coefficient and s is the standard deviation.

5.1.4.2 Wiener Index

Wiener index, also known as the Wiener number, is the oldest and one of the most extensively studied TIs (Wiener, 1947). It is originally described as the summation of the bonds between all pairs of non-hydrogen atoms in a saturated acyclic hydrocarbon. To calculate this index, Wiener introduced the concept of a 'path number', W, which involves evaluating each bond by calculating the number of atoms n_p and n_q on both sides of each bond (Wiener, 1947). The mathematical expression is given by Equation 5.13.

$$W = \sum_{all\ bonds} n_p n_q \quad (5.13)$$

A sample calculation of the Wiener index using 'path number' concept is demonstrated in Example 5.3. A lower path number indicates a more compact molecular structure. Wiener used this index together with Wiener polarity index, P_w to establish a correlation with the boiling points of acyclic hydrocarbons. P_w refers to the number of pairs of carbon atoms that are exactly three bonds apart. He observed that a more compact acyclic hydrocarbon has a lower boiling point. Using these TIs, boiling point of acyclic hydrocarbons can be estimated through the following equations (Wiener, 1947):

$$\Delta T_b = \frac{98}{n^2}\Delta W + 5.5\Delta P_w \quad (5.14)$$

$$T_b = T_0 - \Delta T_b \quad (5.15)$$

$$\Delta W = W_0 - W \tag{5.16}$$

$$\Delta P_w = P_0 - P_w \tag{5.17}$$

$$T_0 = 745.42 \log(n + 4.4) - 689.4 \tag{5.18}$$

$$W_0 = \frac{1}{6}(n-1)(n)(n+1) \tag{5.19}$$

$$P_0 = n - 3 \tag{5.20}$$

where n represents the number of carbon atom, ΔT_b is the difference in boiling point between a straight and a branched chain isomer. T_0 denotes the boiling point of the straight chain isomers, which have structural variables W_0 and P_0. These structural variables quantify the effect on all atoms resulting from a change in position of an atom or of a group within the molecule (Wiener, 1947).

Example 5.3. Estimation of Wiener Index

Calculate the Wiener index of 2-Methylpentane using 'path number' concept. Following that, estimate its boiling point. The molecular structure of 2-Methylpentane is shown in Figure 5.7.

Solution

Before calculating the Wiener index, we should first identify all the bonds in the molecule. To do so, we begin by labelling the carbon atoms in the molecule, as depicted in Figure 5.8.

From Figure 5.8, it can be clearly seen that the carbon-carbon bonds involved are C1-C2, C2-C6, C2-C3, C3-C4 and C4-C5. Using path number approach, we calculate the number of carbon atoms on both sides of the bond. Basically, for each bond, we 'cut' the bond and examine how many carbon atoms lie on each side. For example, if we look into C3-C4 bond, there are four carbon atoms on the left side and 2 carbon atoms on the right. Table 5.5 illustrates the calculation of Wiener index for 2-Methylpentane molecule.

Based on the calculation, Wiener index, W, of 2-Methylpentane is 32.

To estimate its boiling point, Equations 5.14–5.20 are employed. Since there are 6 carbon atoms in 2-Methylpentane, hence $n=6$. Thus, $P_0=3$. $P_w=3$ as there are 3 pairs of carbon atoms separated by three bonds (C1-C4, C2-C5 and C4-C6). By substituting $n=6$ into Equation 5.19, W_0 can be calculated. Hence,

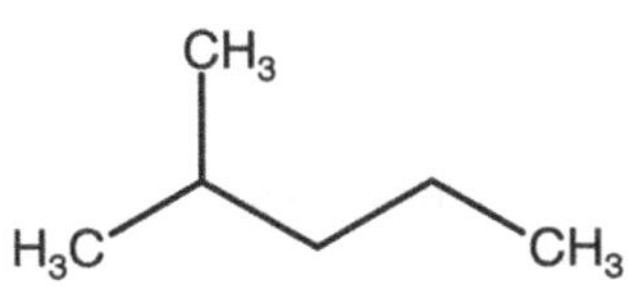

FIGURE 5.7 Molecular structure of 2-Methylpentane.

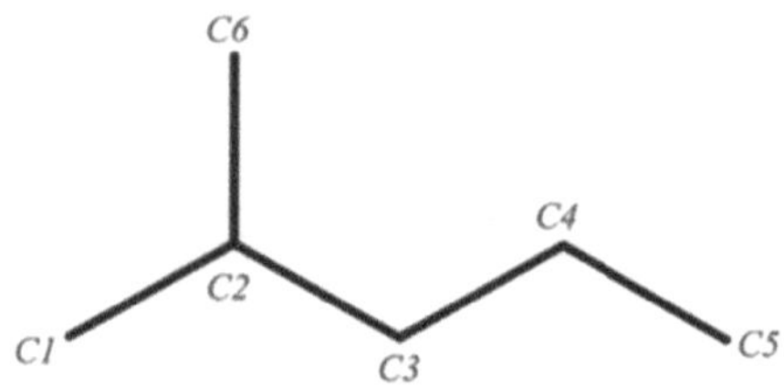

FIGURE 5.8 Carbon atom numbering of 2-Methylpentane.

TABLE 5.5
Calculation of Wiener Index of 2-Methylpentane Using 'Path Number' Concept

Bond	Number of Carbon Atom on Left Side of the Bond, n_p	Number of Carbon Atom on Right Side of the Bond, n_q	$n_p \times n_q$
C1-C2	1	5	5
C2-C6	5	1	5
C2-C3	3	3	9
C3-C4	4	2	8
C4-C5	5	1	6
Total			32

$\Delta W = 35 - 32 = 3$. $\Delta T_b = 8.17°C$. $T_b = 68.72 - 8.17 = 60.5°C$ Boiling point of 2-Methylpentane estimated using this correlation is 60.5°C.

Later in 1971, Hosoya discovered that the Wiener index can be interpreted from the molecular graph, where the Wiener index, $W(G)$, of a molecular graph G is defined as half the sum of all the off-diagonal elements of the distance matrix, D_{ij} (Hosoya, 1971), as shown in Equation 5.21.

$$W(G) = \frac{1}{2}\sum_{i=1}^{N}\sum_{j=1}^{N} D_{ij} \tag{5.21}$$

Hosoya's work further extended Wiener's definition, which originally applied only to acyclic hydrocarbons, to include cyclic molecules. When used together with other molecular descriptors, the Wiener index can be employed to predict properties such as heat of vaporization, heat of formation, boiling point, molar volume and molar refractions (Nikolić et al., 1995).

5.1.4.3 Shape Indices

The TIs discussed in earlier sections quantify the features of molecular structure based on distances or the valence degree of atoms. However, these TIs do not account for the shape of the molecule. Molecular shape plays an important role in determining the chemical reactivity and biological activity. To address such limitations, Kier (1985) introduced shape indices, known as the Kappa shape indices, κ based on hydrogen-suppressed molecular graph. In general, a shape index is a type of molecular descriptor developed to represent molecular shape using numerical values. To cover the different aspects of the shape, different orders of Kappa indices were developed. The first-order (${}^{1}\kappa$), second-order (${}^{2}\kappa$) and third-order (${}^{3}\kappa$) Kappa shape index (Kier, 1986b) are defined as shown in Equations 5.22–5.25, respectively.

$$ {}^{1}\kappa = \frac{n(n-1)^2}{({}^{1}P)^2} \tag{5.22} $$

$$ {}^{2}\kappa = \frac{(n-1)(n-2)^2}{({}^{2}P)^2} \tag{5.23} $$

$$ {}^{3}\kappa = \frac{(n-1)(n-3)^2}{({}^{3}P)^2} \quad \text{if n is odd number} \tag{5.24} $$

$$ {}^{3}\kappa = \frac{(n-3)(n-2)^2}{({}^{3}P)^2} \quad \textit{if}\text{ n is even number} \tag{5.25} $$

where n is the number of non-hydrogen atoms, and ${}^{i}P$ is the number of paths of both length i. Kier later also introduced a modification term α to account for heteroatoms (Kier, 1986a). Example 5.4 illustrates the calculation of the three orders of Kappa shape index for the pentane molecule.

Example 5.4. Estimation of Kappa Indexes of Different Order

Calculate ${}^{1}\kappa$, ${}^{2}\kappa$ and ${}^{3}\kappa$ of pentane molecule.

Solution

For better visualization, a hydrogen-suppressed molecular graph is used to represent the pentane molecule, with its carbon atoms labelled to facilitate the calculation of bond-length paths in subsequent steps. Figure 5.9 shows the carbon atom numbering of pentane molecule.

There are five carbon atoms. Thus, n=5. Since n is an odd number, Equation 5.24 will be used for the third-order calculation later.

FIGURE 5.9 Carbon atom numbering of pentane molecule.

Calculation of $^1\kappa$

The number of paths of bond length 1, $^1P = 4$ as there are four possibilities: C1-C2, C2-C3, C3-C4 and C4-C5.

$$\therefore\ ^1\kappa = \frac{5(5-1)^2}{(4)^2} = 5$$

Calculation of $^2\kappa$

The number of paths of bond length 2, $^2P = 3$ as there are three possibilities: C1-C2-C3, C2-C3-C4 and C3-C4-C5.

$$\therefore\ ^2\kappa = \frac{(5-1)(5-2)^2}{(3)^2} = 4$$

Calculation of $^3\kappa$

The number of paths of bond length 3, $^3P = 2$ as there are two possibilities: C1-C2-C3-C4 and C2-C3-C4-C5.

$$\therefore\ ^3\kappa = \frac{(5-1)(5-3)^2}{(2)^2} = 4$$

For example, the Kappa shape index has been used to correlate the molecular structure of polyhydric alcohol with flash point. The flash point of polyhydric alcohols, $T_{f,polyhydric\ alcohols}$, (Patel et al., 2009) can be estimated using Equation 5.26.

$$T_{f,polyhydric\ alcohols} = 13.064(^2\kappa) - 8.665(^3\kappa) + 385.328 \quad (5.26)$$

5.2 MACHINE LEARNING TOOLS

In the previous sections, we explored traditional property prediction methods such as GC methods and QSPR/QSAR models, which are widely used to estimate physicochemical and thermodynamic properties of chemical compounds. We also discussed how molecular descriptors such as TIs are incorporated into QSPR/QSAR models to establish correlations and predict target properties. Most of the physicochemical, thermodynamic and transport properties that determine the performance of chemical products can usually be estimated using these property prediction methods. However, for some product attributes and more complex chemical formulations, property prediction models are not well developed because the theory is not fully understood (Zhang et al., 2020). Examples include the odour characteristics of a molecule for its use in perfumes, the taste or flavour of molecules in food products, the colour of a product or the mosquito-repelling attributes of a molecule. To overcome these limitations, machine learning (ML) tools can be used to develop prediction models for these attributes based on collected data. ML-based models help to establish the correlation between molecular structures and their properties, including those subjective descriptors such as odour. Before we explore the application of ML in modelling product attributes, let us first introduce the concept of ML and its main types.

ML is a field of artificial intelligence (AI) and computer science that uses statistical models and algorithms to enable machines to learn patterns from data and improve their performance without being explicitly programmed. It can analyse large and fuzzy datasets to reveal patterns, determine underlying structures and relationships, and use this information to develop robust predictive models. There are four main types of ML: supervised, unsupervised, semi-supervised and reinforcement learning. Supervised learning learns the underlying relationship between input and output values using labelled training dataset. In the context of ML, labelled data refers to a dataset in which each input is paired with its corresponding known output. Both training and validation dataset is required for supervised learning process. This trained model can then be used to predict the outputs of new, unseen data points based on their input values. Some examples of supervised learning algorithms include Artificial Neural Network (ANN), Convolutional Neural Network (CNN), Extreme Gradient Boosting (XGBoost), Gaussian Process Regression (GPR), Random Forest (RF) and Support Vector Machines (SVM).

In contrast, unsupervised learning learns from raw and unlabelled data to identify patterns and group similar data points into clusters without any explicit guidance. The identified hidden patterns can then be analysed to make manual predictions. Examples of unsupervised learning algorithms include Principal Component Analysis (PCA) and clustering techniques. Likewise, semi-supervised learning combines features of both supervised and unsupervised learning algorithms, in which it learns from datasets comprising both labelled and unlabelled data. Reinforcement learning is a goal-oriented learning process in which an agent interacts with the environment through trial-and-error, adapting its actions and learning based on the reward-punishment signals to improve its performance.

Among the various types of ML, supervised learning techniques such as ANN, CNN and SVM are widely used to establish the relationship between molecular descriptors and their target properties. A notable example is the use of CNN to develop an odour prediction model that relates molecular structures to perceived odour, which was then used to design fragrance molecules for shampoo additives (Zhang et al., 2018). The model achieved 92.9% accuracy for predicting odour descriptors within the database and 75% for those outside the database. There are also an increasing number of studies that employ ML algorithms to develop QSPR/QSAR models. QSPR modelling, as discussed earlier, often relies on linear correlations to describe the relationship between molecular descriptors and their target properties (Zhou et al., 2020). Linear GC models, which are a special case of QSPRs, are commonly used to predict the physical properties of solvents (Gani, 2019). Although first-order GC models offer quick estimation of molecular properties, their accuracy is generally limited to simple molecules. As noted earlier, when dealing with more complex molecular structures such as multi-ring compounds, higher-order GC models should be used to perform prediction. Nonetheless, when the relationship between molecular descriptors and target properties becomes complex and non-linear, conventional GC or linear QSPR methods may no longer be sufficient. In such cases, more advanced correlation methods, such as ML approaches, can be employed to capture the non-linearity and improve prediction accuracy.

For instance, ANN has been integrated into GC models to predict acid dissociation constants of organic compounds (Zhou et al., 2018). Similarly, SVM was applied in the development of QSAR models to investigate the relationships between molecular structures of hydroxyl benzoic esters and their antibacterial activity (Wen et al., 2017). The model can then be used to predict antibacterial activity of structurally similar compounds. Their study demonstrated that SVM effectively addresses key challenges in QSAR modelling, such as the complex non-linear relationship between molecular descriptors and the issue of multicollinearity. In another contribution, two ML-based GC models, namely ANN-GC and SVM-GC models, were developed to predict carbon dioxide (CO_2) solubility in ionic liquids (Song et al., 2020). These models correlated CO_2 solubility with molecular structure of ionic liquid as well as temperature and pressure. Their result showed that both models provided reliable predictions on CO_2 solubility, with the ANN-GC model slightly outperforming SVM-GC model.

Supervised ML often relies on the availability of large and well-labelled training datasets. In chemical product design, such data are not always readily available. Moreover, most commonly used ML tools, such as ANN and SVM, are considered as black-box models as they generate outputs without providing explanations on their decisions. Hence, their results are harder to interpret and translate into insights for decision-making. This lack of interpretability may pose a limitation in chemical product design as it is important to understand the relationship between molecular structure and product properties to perform design and selection. In situations where only a small dataset is available, other ML approaches such as Rough Set Machine Learning (RSML) and Hyperbox ML can be employed. Unlike black-box models, these methods are able to produce transparent and interpretable results to predict product attributes. For example, Hyperbox ML has been applied to predict mosquito-repelling attribute of molecules (Hatamleh et al., 2022). Using TIs and functional groups as molecular descriptors, RSML has also been applied to predict the odour attributes of molecules (Tiew et al., 2023).

Although ML-based models have shown great potential in property estimation, their success is highly dependent on the quality of the training data. Poor data quality may result in inaccurate predictions and poor generalization to new data. When dealing with ML, users must be aware of the risk of overfitting and underfitting. Overfitting means that the trained model fits the training data well but fails to generalize when applied to new and unseen data. Underfitting happens when the model does not have enough parameters to describe the input-output relationship accurately. We can mitigate the risk of overfitting by applying cross-validation during model training. A commonly used approach is k-fold cross-validation, where the dataset is divided into *k* parts. It is a technique used to evaluate the performance of an ML model by performing multiple training and validation cycles with different portions of the original dataset. It ensures every part of the dataset is used to minimize the risk of data variability and helps detect overfitting. Besides, it is also recommended to keep a separate testing dataset to obtain a more realistic measurement of the model's performance.

5.3 MOLECULAR DESIGN TOOLS

Traditionally, the selection of chemical products is determined using a bottom-up approach, which relies heavily on the trial-and-error experimental studies, design heuristics and expert judgments. These approaches can be costly, time consuming and inefficient. For example, many molecular compounds have to be synthesized in the laboratory, after which their properties are evaluated based on analytical results to determine whether they meet the desired needs. If the requirements are not met, then the process has to be repeated. Given the vast number of possible molecular compounds, this trial-and-error process becomes a big challenge. Although experimental approaches often yield safe and reliable products, it is impractical to evaluate all possible options because it is still expensive and time consuming, even with the Design of Experiment (DOE) to speed up the process (Zhang et al., 2020).

Design heuristics are normally developed based on available information and knowledge, experience and insights. They help identify potential solutions, but the rules can sometimes be contradictory, difficult to apply and only valid for certain products (Zhang et al., 2020). Another method for solvent selection is database screening, which provides a cheaper and faster option to identify potential candidates in chemical product design. It should be noted that this method does not generate the final design. Instead, a set of molecular candidates has to be predefined before screening through the database. Most existing databases only cover a limited range of common molecules and physical properties. Thus, the options for product design are limited by the availability and size of database. In addition, a lack of experience or bias from the user can lead to choosing solvent that is not the best option.

These limitations highlight the need for a more systematic way to guide chemical product design and selection. An alternative is to approach chemical product design problems using top-down method based on Computer Aided Molecular Design (CAMD) techniques if property prediction models are available. In the earlier section where we introduced property prediction models, the examples illustrate forward engineering problems as we estimate the properties based on the given molecular structure. This may yield suboptimal solutions as the search space is limited. In contrast, CAMD techniques solve molecular design problems from the opposite direction. In CAMD, property prediction models such as GC and QSPR models are used to link the properties of molecules to their structure and composition. It is a form of reverse engineering since a CAMD problem is the inverse of a property prediction problem, where its main goal is to determine the molecular structure that meets the target properties. CAMD has emerged as a powerful method in the field of chemical product design since it is able to design novel molecules that satisfy a specified set of target performance from a pool of molecular building blocks.

Various CAMD techniques have been developed and applied to solve chemical product design problems. CAMD techniques can be mainly grouped into three categories such as generate-and-test method, deterministic optimization method and stochastic optimization method. In generate-and-test method, CAMD problem is solved using a combinatorial approach; that is, every possible molecular structure is generated, followed by estimating its properties using property prediction model and testing each of them against the specified properties (Gani and Brignole, 1983). This sequential generate-and-test method is also known as enumeration techniques in literature. It works particularly well in the selection of an optimal molecule from a small design space. However, combinatorial explosion problem may occur when this method is applied to molecular design problems with large search space since it considers all possible combinations of the selected molecular groups. For instance, a total of 135,710 molecules can be generated even if we limit the molecules to contain two to four groups from the available 40 molecular groups (Joback and Stephanopoulos, 1995). Thus, different combination rules based on heuristic or knowledge have been incorporated when generating the molecules to manage the size of the search space.

Besides the generation-and-test method, deterministic optimization or mathematical programming approach is also one of the widely used techniques for solving CAMD problems. In this approach, molecular design problem is formulated as a mathematical optimization model to optimize a performance objective subjected to the structural constraints, mass balances, as well as process and design limitation (Odele and Macchietto, 1993). It is necessary to have structural constraints to ensure that only structurally feasible molecules are generated from the selected molecular building blocks. For example, the octet rule is often included to ensure that the generated molecules have no free attachment, that is, zero valency. The CAMD problem is often solved as a mixed-integer non-linear programming (MINLP) problem since the number of structural groups assumes integer values. Unlike the sequential generate-and-test method, deterministic optimization approach generates potential molecules with their properties estimated at the same time. However, it faces the same combinational explosion challenge as the generation-and-test method.

An alternative strategy to solve CAMD problems is through stochastic optimization method. This method is particularly useful when the design space becomes too large and results in combinational complexity. Some examples of the stochastic optimization methods include genetic algorithm (GA), simulated annealing (SA) and Tabu search (TS). Similar to deterministic optimization method, stochastic approach also aims to determine the optimal solution for the specified performance objective, but it does not require any gradient information. Because of this, it allows discontinuous properties to be specified as design targets in CAMD model. In stochastic optimization, solutions are generated through successive pseudo-random processes. One notable contribution in this area is by Venkatasubramanian et al. (1994), who applied GA for large CAMD problems to overcome issues of combinatorial complexity and local optima traps.

It should be noted that there are other CAMD techniques available beyond those discussed here. Examples include decomposition methods, property clustering techniques and algebraic techniques. With the advances in chemical modelling, computational power and artificial intelligence, CAMD techniques incorporating more accurate methods continue to emerge. There are also some software tools available to design the solvents effectively. These include Integrated Computer-Aided System (ICAS) (Gani et al., 1997), Automated MOlecular DEsign using Optimization (AMODEO)(Samudra and Sahinidis, 2013), ProCAPD (Kalakul et al., 2018), etc. The following section illustrates how a CAMD problem can be formulated mathematically before solving it using the available numerical techniques.

5.3.1 General Mathematical Expression of Molecular Design Problem

As discussed earlier, the ultimate goal of a CAMD problem is to determine a molecule or a set of molecules that exhibit the desired properties. To achieve this goal, the molecular design problem can be translated into a mathematical optimization model, allowing the identification of optimal candidates. There are three main stages in solving CAMD problem using mathematical programming approach: problem formulation, model development and molecular design.

5.3.1.1 Problem Formulation

The process starts with identification of design objective for molecular design problem by identifying the product needs, which may arise from customer requirements or from the operating condition of industrial processes. These product needs are usually expressed by thermophysical properties, which are responsible for the functionality of the product. Once the product needs are determined, the desired target properties can then be chosen as the design objectives to ensure that the designed molecules fulfil the intended function. The identified qualitative product needs are then translated into measurable quantitative target properties. These identified target properties can either serve as an objective function for optimization or be imposed as constraints during the design stage.

Before solving any CAMD problem, it is necessary to formulate property constraints (Achenie et al., 2002). Identifying the target properties and their ranges is important to ensure the success of a CAMD model as the model will identify the wrong products if the key properties are not correctly identified. Property constraints can be defined by considering factors such as, but not limited to, operational limits for the product, product performance criteria, safety margins and environmental regulations. Upper and lower bounds are normally defined for these properties to ensure that the designed molecules achieve similar characteristics. It can be represented mathematically using Equation 5.27.

$$V_p^L \leq p \leq V_p^U \; \forall p \in P \tag{5.27}$$

where p is target property, V_p^L and V_p^U represents lower and upper limits of target property.

5.3.1.2 Model Development

The next stage involves the estimation of the target properties identified in stage 1 through appropriate property prediction model. As discussed in Section 5.1.2, one can select a property prediction model based on the types of identified target properties and the level of estimation accuracy required. Among various models, GCM is widely considered in most CAMD techniques. For example, if boiling point and heat of vaporization are identified as design objectives, then we can apply the GC models presented in Table 5.2 to estimate the properties.

5.3.1.3 Molecular Design

In the final stage, we have to select potential molecular building blocks based on the nature of the design problem. Consider the case where we are tasked to design alcohol-based molecules, then hydroxyl group (-OH) must be chosen as one of the molecular building blocks. This step is important to ensure that the generated molecules share similar molecular structures and properties with conventional product.

After identifying the potential molecular building blocks, structural constraints must be applied to ensure the formation of a feasible molecular structure. To generate a molecule, the sum of occurrences of all selected molecular groups must be greater than zero. This rule is expressed mathematically by Equation 5.28.

$$\sum_{i=1} N_i > 0 \tag{5.28}$$

where N_i represents the number of occurrences of molecular group i. To avoid unbounded solutions, an upper limit constraint may be imposed on the N_i. Moreover, a structurally feasible molecule should not have any free bonds. To ensure that the final molecule does not contain any free attachment, the octet rule of structural feasibility (Odele and Macchietto, 1993), as shown in Equation 5.29, can be applied.

$$\sum_{i=1} N_i (2 - v_i) = 2g \tag{5.29}$$

where v_i is the valence of group i and g is 1, 0 or –1 for acyclic, monocyclic and bicyclic compounds, respectively. In the cases where acyclic compound is considered, then Equation 5.29 is reduced to Equation 5.30.

$$\sum_{i=1} N_i (2 - v_i) - 2 = 0 \tag{5.30}$$

To satisfy the structural constraints, each selected molecular building block should have a valence of at least 1 to be able to connect with other molecular groups and form a feasible structure. For instance, the $-CH_3$ group has a valence of 1, meaning that it can form one bond with another molecular group. The $-CH_2$ group has a valence of 2, meaning that it can connect to another two groups. By considering a simple example such as ethanol and CH_3CH_2OH, we can make use of Equation 5.30 to verify its structural feasibility since it is a simple acyclic group. Ethanol comprises three molecular groups, including $-CH_3$, $-CH_2$ and -OH, and their valences are 1, 2 and 1, respectively. We can compute the left-hand side of Equation 5.30 to check whether it yields the value of 0. The step-by-step calculation is illustrated as follows:

Expand Equation 5.30, and the equation becomes

$$\sum_{i=1} (2 \times N_i) - \sum_{i=1} (N_i \times v_i) - 2;$$

$$2\times\left[N_{CH3}+N_{CH2}+N_{OH}\right]-\left(N_{CH3}\times v_{CH3}+N_{CH2}\times v_{CH2}+N_{OH}\times v_{OH}\right)-2;$$

$$2\times\left[1+1+1\right]-\left(1\times1+1\times2+1\times1\right)-2=0$$

We prove that the structural feasibility is satisfied for ethanol molecule.

The CAMD problem can be modelled using a set of generalized mathematical expressions. A basic CAMD formulation is expressed as:

$$\min\, or \max F\left(P\right) \tag{5.31}$$

subject to:

$$P=f\left(N\right) \tag{5.32}$$

$$V_p^L \le p \le V_p^U \forall p \in P \tag{5.33}$$

$$N_i^L \le N_i \le N_i^U \ \forall i \tag{5.34}$$

$$\sum\nolimits_{i=1} N_i\left(2-v_i\right)=2g \tag{5.35}$$

In this formulation, P represents the vector of target properties. F(P) is the objective function of the CAMD problem, which quantifies the performance of the designed molecule based on its predicted properties. F(P) can be either maximized or minimized depending on the molecular design problem. Equation 5.32 estimates P from number of molecular groups i. Equation 5.33 represents the property constraints, whereas Equations 5.34 and 5.35 represent structural constraints, respectively. It is important to note that N is a vector of integer variables as the number of molecular groups can only take integer values. Hence, CAMD problem is often solved as MILP or MINLP problem. By solving the optimization model, optimum molecular candidates can then be determined.

5.4 SAFETY, HEALTH AND ENVIRONMENTAL IMPACTS

We have seen that a chemical product must possess specific physicochemical and thermodynamic properties in order to perform its function in the desired application. In most traditional design problems, the objective is to minimize the cost while meeting the target performance. In other words, it is important to select a chemical product that achieves the desired technical aspects for a certain application at its lowest cost. Nonetheless, it is also important to ensure that the chemical products produced are safe to both human and environment. Let us take chemical solvents as examples. The selection of chemical solvents has significant impact on chemical processes. Its physical and thermodynamic properties determine the degree of difficulty of a chemical process. Besides, the use of solvents in industry has been found to be closely related to various indirect environmental impacts (Capello et al., 2007). For example, solvents with a low boiling point or high volatility can often be recovered more easily through evaporation. But their high volatility poses health risks to workers as the solvents can easily escape into the work environment. The rapid development of industrialization and urbanization has also led to detrimental impacts on the environment. As the number of industries increases, it is inevitable that solvents become a major contributor to waste since they play an important role in the chemical industries.

There is also an increased emphasis on health and safety aspects in chemical industries as mishandling hazardous chemicals often leads to industrial catastrophe. For example, one of the worst chemical disasters in history – the Bhopal Disaster, which happened in the year of 1984 – was caused by the accidental leakage of more than 40 tonnes of methyl isocyanate (MIC) gases from a pesticide plant (Broughton, 2005). MIC gas is extremely toxic to humans. As a result, this toxic release killed at least 3800 people immediately after the accident. Moreover, due to stringent environmental regulations, the concepts of sustainability and green engineering have become pressing concerns across all chemical industries. To increase public awareness on environmental and health issues, federal agencies such as the United States Environmental Protection Agency (EPA) recognize and reward industries that comply with safety, health and environmental (SHE) standards. Hence, industries are making more efforts to seek safer and sustainable alternatives to replace the hazardous chemicals.

Since there is an ongoing transition towards greener chemical products, it is important to ensure that the chemical product design strategies are able to identify chemical products with desirable target properties and adhere to SHE regulations. To incorporate SHE aspects into chemical product design, some property prediction tools have been developed over the years. One significant contribution in this field is by Hukkerikar et al. (2012), who developed GC+ (combined GC method and atom CI method) models to predict 22 environmental-related properties of organic chemicals, such as not limited to global warming potential (GWP), acidification potential (AP), permissible explosion limit (PEL), bioconcentration factor (BCF), photochemical oxidation potential (PCO) and oral rat LD_{50}. Table 5.6 shows the GC models of some environmental properties together with their universal constants (Hukkerikar et al., 2012). Another work developed GC models to correlate molecular functional groups with life cycle assessment (LCA) indices (Baxevanidis et al., 2022). The LCA indices considered in their work include GWP, cumulative energy demand and Eco-Indicator 99.

Similarly, GC models are also available to predict some safety-related properties, such as flash point (Albahri, 2014) and LC_{50} (Hukkerikar et al., 2012). In addition, the inherent safety and health performance of chemical product can be estimated using the framework developed by Ten et al. (2016). In their work, inherent safety and occupational health aspects were integrated into chemical product design framework. The safety and health performance of a molecule can be quantified using safety and health indices, which were developed to account for all factors that contribute to occupational health and safety hazards. The safety impact of a chemical product can be evaluated using safety indices, namely explosiveness-based index, I_{EX}, and flammability-based index, I_{FL}. The index scores for I_{EX} and I_{FL} were taken from Inherent Safety Index (Heikkilä, 1999) and National Fire Protection Association (National Fire Protection Association, 2007), respectively, as presented in Tables 5.7 and 5.8.

TABLE 5.6
Examples of Environmental Properties and Their GC Models

Property, P	Property Function, $f(P)$	GC Models	Universal Constant
BCF	$\log BCF$	$\sum_i N_i BCF_{1i} + y\sum_j M_j BCF_{2j} + z\sum_k Q_k BCF_{3k}$	-
GWP	$\log GWP$	$\sum_i N_i GWP_{1i} + y\sum_j M_j GWP_{2j} + z\sum_k Q_k GWP_{3k}$	-
PCO	$-\log PCO$	$\sum_i N_i PCO_{1i} + y\sum_j M_j PCO_{2j} + z\sum_k Q_k PCO_{3k}$	-
PEL	$-\log PEL$	$\sum_i N_i PEL_{1i} + y\sum_j M_j PEL_{2j} + z\sum_k Q_k PEL_{3k}$	-
LD_{50} (mol/kg)	$-\log LD_{50} - A_{LD50} - B_{LD50} MW$	$\sum_i N_i LD_{501i} + y\sum_j M_j LD_{502j} + z\sum_k Q_k LD_{503k}$	A_{LD50} = 1.9372 B_{LD50} = 1.9372

TABLE 5.7
Index Score for I_{EX}

Explosiveness, I_{EX}	Penalty Score
Non-explosive	0
$0 \le S < 20$	1
$20 \le S < 45$	2
$45 \le S < 70$	3
$70 \le S < 100$	4

where S (vol%) = upper explosion limit (UEL) – lower explosion limit (LEL)

TABLE 5.8
Index Score for I_{FL}

Flammability, I_{FL}	Penalty Score
Non-flammable	0
Flash point ≥ 93.4°C	1
Flash point < 93.4°C	2
Flash point < 37.8°C	3
Flash point < 22.8°C & boiling point < 37.8°C	4

Two indices, namely Process Route Healthiness Index (PRHI) (Hassim and Edwards, 2006) and Inherent Occupational Health Index (IOHI) (Hassim and Hurme, 2010), can be considered to evaluate the health performance of a chemical product (Ten et al., 2016). Based on these two health indices, some relevant chemical properties are selected to assess the health hazards of molecules. As suggested by Ten et al. (2016), exposure limit, I_{EL}; material phase, I_{MS}; and volatility, I_V, are selected from IOHI, whereas viscosity, I_η, is taken from PRHI. Besides, Ten et al. (2016) also suggested including acute health hazard, I_{AH}, parameter to assess the health performance of a molecule. The index score for acute health hazard is taken from the National Fire Protection Association health hazard rating (National Fire Protection Association, 2007). The index scores for these five health sub-indices are presented in Tables 5.9–5.13. Chemical products with higher safety and health index values are generally less favourable as these indicate a greater potential for hazards.

Similar to other property prediction models, these models related to SHE aspect can be integrated into CAMD tools. By doing so, it allows SHE aspects of a chemical product to be considered during the early stages of molecular design, along with other physicochemical properties.

TABLE 5.9
Index Score for I_η

Viscosity, $I\eta$	Penalty Score
$0.1\ cp \le \eta < 1\ cp$	1
$1\ cp \le \eta < 10\ cp$	2
$10\ cp \le \eta < 100\ cp$	3

TABLE 5.10
Index Score for I_{MS}

Material Phase, I_{MS}	Penalty Scores
Gas	1
Liquid	2
Solid	3

TABLE 5.11
Index Score for I_{EL}

Exposure Limit, I_{EL}	Penalty Score
PEL > 1000 ppm	0
PEL ≤ 1000 ppm	1
PEL ≤ 100 ppm	2
PEL ≤ 10 ppm	3
PEL ≤ 1 ppm	4

TABLE 5.12
Index Score for I_V

Volatility, I_V	Penalty Score
Boiling point > 150°C	0
150°C ≥ boiling point > 50°C	1
50°C ≥ boiling point > 0°C	2
Boiling point ≤ 0°C	3

TABLE 5.13
Index Score for I_{AH}

Acute Health Hazard, I_{AH}	Penalty Score
LD_{50} (mg/kg) > 2000	0
500 < LD_{50} (mg/kg) ≤ 2000	1
50 < LD_{50} (mg/kg) ≤ 500	2
5 < LD_{50} (mg/kg) ≤ 50	3
LD_{50} (mg/kg) ≤ 5	4

5.5 SELECTION MATRIX

From the previous sections, we can see that the decisions on product design depend on several factors. In general, these factors are assessed using different yardsticks, which makes the direct comparison of performance in different factors challenging. Therefore, selection matrices are employed at this stage to make a final comparison between the shortlisted products (Cussler and Moggridge, 2011). The selection matrices are prepared in the same way as the screening matrices. However, since selection matrices are applied for making the final selection, they need to apply quantitative tools in estimating the scores and weightages. The intuitive approach used for screening is not sufficient at this stage because the cost of manufacturing is extremely high compared to the previous stages, and a wrong step from here onward can be costly for the companies.

5.5.1 Estimation of Scores

One of the differences between screening and selection matrices is the level of accuracy in the estimation of the scores. At this stage, we need to apply quantitative tools, experimental results and sophisticated survey results in assigning the scores. Depending on the specific problem, the quantification methods may differ.

Example 5.5. Liquid Detergents

Environmentally conscious consumers look for alternatives to liquid detergents to wash their clothes. Each time we wash clothes using liquid detergents, we release chemicals, microfibres and large amounts of water into the drainage system. This harms the environment by contaminating both our waterways and landfills. However, liquid detergents are extremely effective in their primary function – which is cleaning clothes – and reasonably inexpensive. Therefore, how well different needs are accomplished by promising alternatives needs to be studied before a decision is made.

Solution

In the initial stages of the product design, the following needs were identified:

Washing ability, fragrance diffusivity, amount of plastic waste per wash, antimicrobial properties and portability.

Laundry capsules, laundry sheets and laundry powder are identified as the potential replacements for liquid laundry detergents. Therefore, the target properties for the short-listed ideas need to be estimated using reliable methods.

It has been found that liquid detergents perform much better than the alternatives in their ability to wash. It is to be noted that the laundry sheets are known to be least effective cleaning method compared to the rest. This is particularly true in cleaning with cold water and for removing different types of stains. Therefore, the alternatives need to be better in the rest of the categories to be considered as a potential replacement. For fragrance, a higher diffusivity suggests that fragrance is more easily detectable by the human olfactory system. The diffusivity can be estimated by the Wilke and Lee equation (Wilke and Lee, 1955):

$$D_{AB} = \frac{\left[3.03 - \left(\frac{0.98}{M_{AB}^{0.5}}\right)\right]\left(10^{-3}\right)T^{\frac{3}{2}}}{P\left(M_{AB}\right)^{0.5}\left(\sigma_{AB}^{2}\right)W_{D}}$$

where D_{AB} = diffusion coefficient, cm^2/s

M_{AB} = a function of molecular weights of A and B

T = temperature, K

P = pressure, bar

σ_{AB} = collision diameter, a Lennard–Jones parameter, Å

Ω_D = diffusion collision integral, dimensionless

The antimicrobial performance of the three options is shown in Table 5.14. The antimicrobial agents are chosen based on their ability to reduce colony-forming units for *E. coli*. For instance, MI is found to have an ability to accomplish a reduction of 100 times in colony-forming units for *E. coli* (Bingham et al., 2019), whereas this is 6000 times for TAED (Betz and Cerny, 2000).

TABLE 5.14
Antimicrobial Performance of Different Products

Ideas	Common Antimicrobial Agent	Adverse Effects
Liquid	Methylisothiazolinone (MI) (Castanedo-Tardana and Zug, 2013)	Dermal irritation Acute toxicity when ingested (Castanedo-Tardana and Zug, 2013)
Capsule	No specific agent – same as liquid	Varies
Powder	Tetraacetylethylenediamine (TAED)	N/A
Sheet	Micro-Fresh 1911	N/A

TABLE 5.15
Selection Matrix

Selection Criteria	Weighting Factor	Benchmark: Liquid Detergent	Laundry Powder	Laundry Capsule	Detergent Sheets
Effective washing	0.4	5	4	3	2
Portability	0.2	5	5	7	8
Fragrance diffusivity	0.1	5	3	4	5
Environmental	0.3	5	6	6	7
Total score		5.0	4.7	4.8	5.0

The plastic containers used in the liquid detergent and powder are not biodegradable, whereas the capsule and detergent sheet produce minimal plastic waste. However, the detergent sheets may require hot water for effective cleaning, which indirectly contributes towards environmental impact. However, in terms of portability, they are best as it is easier to make sheets in a more portable form. Laundry capsule is the next best option, followed by laundry powder (Table 5.15).

From this analysis, it is clear that the alternatives to laundry liquid may perform better only in certain situations, such as customers needing more portability. For a normal consumer, it may still be advantageous to use laundry detergents. However, to improve the sustainability side of the product, the efforts should be focused on developing ecofriendly containers that reduce the plastic usage for liquid laundry products.

5.6 SYSTEMATIC ESTIMATION OF WEIGHTAGE

The final selection of the shortlisted idea depends heavily on the weights allocated to the selection criteria. We often require evaluating alternatives based on various factors, including safety, health, environmental impact, cost and technical performance. Some of these criteria, such as safety and costing, may conflict with one another, which further complicates the decision-making process. Unlike costing, some criteria are intangible, which are difficult to measure and compare with each other.

Analytic Hierarchy Process (AHP) is a widely used Multi-Criteria Decision Making (MCDM) tool in such problems (Saaty, 1977). It is developed based on relative measurement theory, where the emphasis is not on obtaining the absolute measurements of certain quantities, but on understanding the proportional relationships between them. It is important to note that pairwise comparisons are the core of AHP. Pairwise comparison, a technique employed by psychologists like Thurstone (1927) and Yokoyama (1921), has been in use for many years. Psychologists often argue that it is easier and more precise for individuals to compare two elements at a time rather than assessing more than two elements simultaneously (Ishizaka and Labib, 2011). For example, when comparing the length of two rods based on their exact measurements, the pair of measurements (3, 1) is incorrect unless the lengths of the first and second rods are 3 m and 1 m, respectively. However, in the context of relative measurement, if we were comparing how much longer a rod is compared to another, the pairs of measurements (3, 1), (6, 2), (3/5, 1/5) are all valid provided that the first rod is three times longer as compared to the second rod. In most cases, one can determine the best alternative by having the knowledge of relative measurements between alternatives. Hence, this theory of relative measurement is mostly well-suited for situations involving the selection of best alternative.

The attractive part of AHP is that it allows both tangible and intangible criteria to be combined under the same analysis. AHP provides a systematic mechanism to quantify the subjective judgements of decision makers when making comparisons. By having pairwise comparison, ratio-scale priority weights are derived to provide a rating for each alternative. Besides, the consistency of derived priority weights is also assessed to verify that the solution accurately represents the decision maker's preferences. Thus, the AHP approach is capable of treating the inherent subjectivity in the selection process, especially when there is no obvious superior choice when making a decision.

5.6.1 Develop Decision Hierarchy

The initial step in AHP is to define the decision problem and identify the criteria and/or sub-criteria, as well as the alternatives. To structure a decision problem in a hierarchical manner, one should first imagine it like a pyramid. At the top is the primary goal that one is trying to achieve. The middle layer contains criteria and/or sub-criteria that help to evaluate the options. The bottom layer includes the set of alternatives that can be used to achieve the primary goal. Organizing a decision problem in such a hierarchical manner allows one to visualize how each level connects to the others, making it easier to evaluate and select the most suitable alternative. Figure 5.10 illustrates the general decision structure in AHP.

5.6.2 Construct Pairwise Comparison Matrix

Next, the preference, priorities or relative importance of criteria/elements located at the same level are compared in a pairwise manner with respect to a specific element

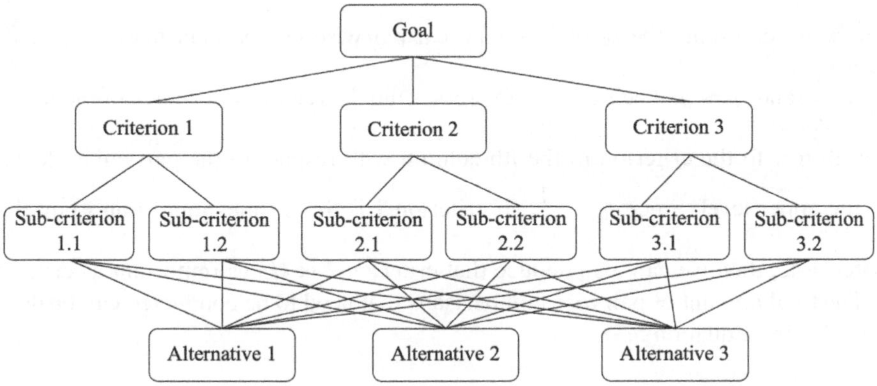

FIGURE 5.10 General decision structure in AHP.

TABLE 5.16
Saaty's Nine-Point Scale and Linguistic Equivalent

Numerical Score	Linguistic Equivalent for Comparison of Criteria
1	Equally important
3	Moderately more important
5	Strongly more important
7	Very strongly more important
9	Extremely more important
Note that 2, 4, 6, 8 are intermediate values	

in the immediate upper level. The relative importance/preference between two criteria can be deduced based on expert's judgement or by referring to literature review. By referring to Figure 5.10, instead of comparing three criteria simultaneously, one can initially compare criterion 1 with criterion 2 by asking the question: 'How much more important/preferred is criterion 1 compared to criterion 2 in achieving the goal of the decision problem?' The step is then followed by comparing criteria 1 and 3, as well as criteria 2 and 3 by asking the same question. The verbal description can be transformed into numerical quantities representing the values of a_{ij} using Saaty's fundamental nine-point scale (Saaty, 1980) presented in Table 5.16.

All the pairwise comparisons can then be gathered into a pairwise comparison matrix, $A = a_{ij}$ of size $n \times n$ as shown in Equation 5.36.

$$A = \begin{bmatrix} a_{11} & a_{12} & \cdots & a_{1n} \\ a_{21} & a_{22} & \cdots & a_{2n} \\ \vdots & \vdots & \ddots & \vdots \\ a_{n1} & a_{n2} & \cdots & a_{nn} \end{bmatrix}, \text{ for } i, j = 1, 2, \ldots, n \tag{5.36}$$

where A represents the positive reciprocal pairwise comparison matrix, $a_{ij} = \frac{w_i}{w_j}$, $a_{ji} = \frac{1}{a_{ij}}$ and $a_{ij} > 0$. The ratio $\frac{w_i}{w_j}$ expresses the degree of importance of element in the ith row to the criterion in the jth column with respect to the particular element in the upper level. Since $a_{ji} = \frac{1}{a_{ij}}$, if criterion 1 is three times more important than criterion 2, then we can also deduce that criterion 2 is 1/3 as important as criterion 1. The total amount of pairwise comparisons that need to be conducted can be determined using Equation 5.37.

$$\text{Number of pairwise comparisons} = \frac{n(n-1)}{2} \tag{5.37}$$

where n indicates the order of matrix.

5.6.3 Synthesize Pairwise Comparison

Once pairwise comparison matrix is constructed, the priority vector (also known as weighting factor) of each criterion/alternative can be derived by the principal Eigenvalue Method (Saaty, 1980). The priority vector is implied by the principal eigenvector, w, of the matrix A in Equation 5.38.

$$Aw = \lambda_{max} w \tag{5.38}$$

where λ_{max} represents the principal eigenvalue, a constant that indicates the consistency of the matrix. The matrix is considered inconsistent if the difference between λ_{max} and n is large. Ideally, $\lambda_{max} = n$ for perfect consistency.

w can be calculated using power method through a spreadsheet by following these steps:

a) Raise the pairwise comparison matrix to the power of two
b) Calculate and normalize the sums of rows to compute the eigenvector
c) Repeat the process until there are no significant changes in the value of eigenvector as compared to the previous calculation, that is, up to four decimal places of accuracy

5.6.4 Verify Consistency of Pairwise Comparison

After compute λ_{max}, one can determine the Consistency Index, CI, using Equation 5.39.

$$CI = \frac{\lambda_{max} - n}{n-1} \tag{5.39}$$

TABLE 5.17
Random Index, *RI*

n	1	2	3	4	5	6	7	8	9	10
RI	0	0	0.52	0.89	1.11	1.25	1.35	1.4	1.45	1.49

The step is then followed by comparing CI with a random index, RI, to derive Consistency Ratio, CR. CR is used to measure the degree of consistency of pairwise comparison and can be calculated using Equation 5.40.

$$CR = \frac{CI}{RI} \tag{5.40}$$

RI represents the consistency index derived from a randomly generated reciprocal matrix with sizes ranging from $n = 1$ to $n = 10$, using a nine-point scale. Table 5.17 presents the RI table, which was developed by Saaty (1980) using a sample size of 500.

Inconsistency is deemed acceptable when CR is less than 10%. Nonetheless, one has to investigate and re-elicit the pairwise comparison if the calculated CR is higher than 10%.

5.6.5 Develop Overall Priority Ranking of Alternatives

Once all pairwise comparisons are conducted at each level and their consistency is within the acceptable level, the overall priority ranking of each alternative can be calculated by summing the product of the criterion priority vector with the priority vector of the decision alternative with respect to that criterion. The alternative with the highest overall priority vector value will be the most preferred choice.

The overall procedure involved in the AHP framework is summarized in Figure 5.11.

Example 5.6. Illustration of AHP Framework

A product design team is considering a new shampoo product. The team decided to choose Pantene Shampoo as the benchmark to compare with the alternatives, including scalp balancing shampoo, dry shampoo, exfoliating scalp scrub, no foam shampoo and shampoo with natural ingredients. They had identified the key selection criteria to be cost, environmental friendliness, user-friendly ingredients, cleaning effectiveness and comfort. However, the primary challenge faced by the team was to assign weighting factor to each selection criterion as they are a mixture of tangible and intangible factors. It is difficult to decide whether a criterion should have a weighting factor of 0.1, 0.25, 0.30 or other values. In such a case, AHP can be used to derive the weighting factor of each criterion. This will then be followed by calculating the total score for each alternative, allowing the team to identify the best option.

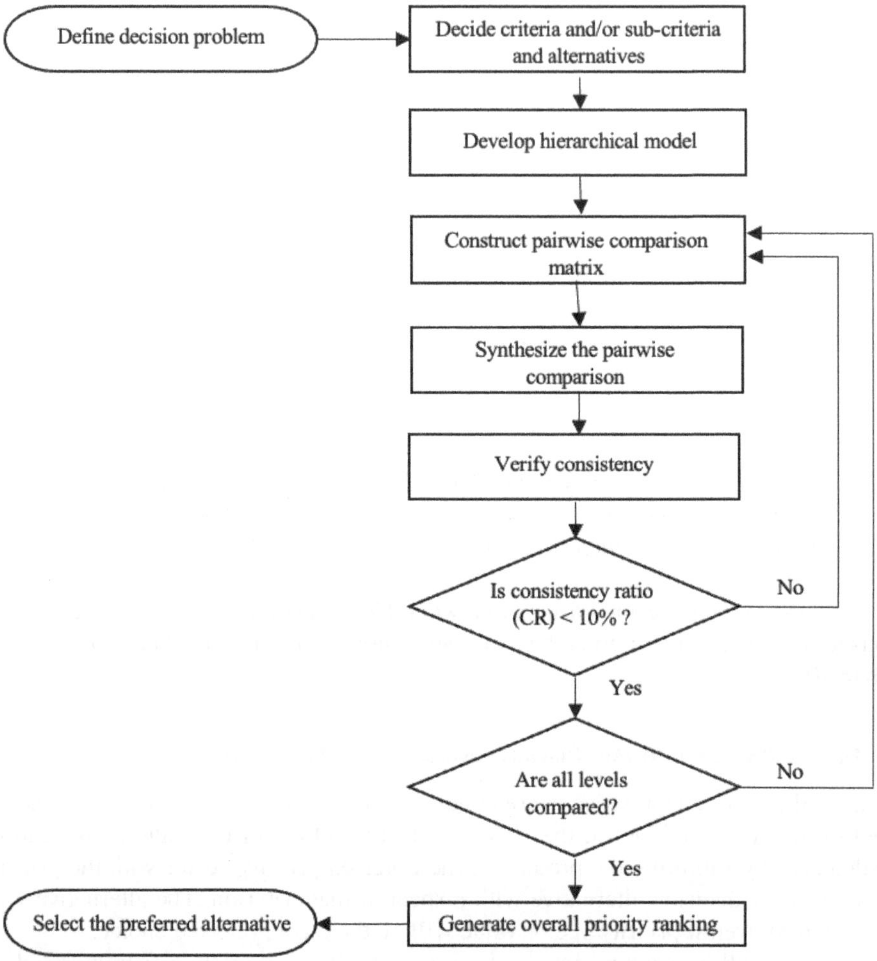

FIGURE 5.11 General procedure in AHP framework.

Solution

To begin with, a three-level hierarchy model, with the aim of choosing a new shampoo product, is constructed and shown in Figure 5.12. The criteria selected for choosing a new shampoo product include cost, environmental friendliness, user-friendly ingredients, cleaning effectiveness and comfort, which are positioned in the middle. The bottom layer consists of all the alternatives.

The step is then followed by assessing the relative importance of each selection criterion with respect to the goal. In this example, the judgements are performed based on literature sources. To compare the selection criteria, the process involves asking the question: 'which criterion is more important and how much more important is it

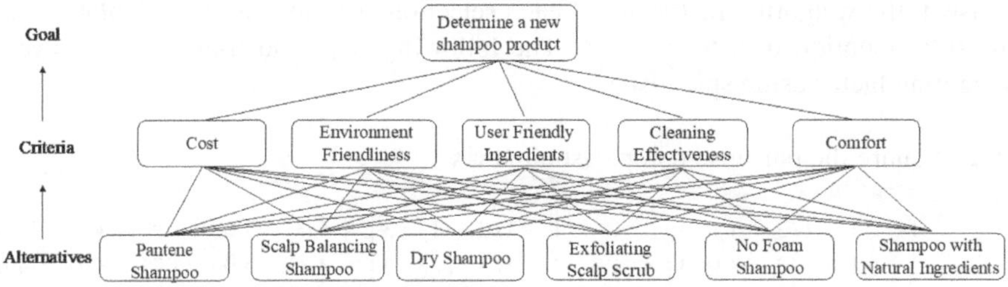

FIGURE 5.12 Hierarchy model for choosing a new shampoo product.

to fulfil a requirement in considering a new shampoo product?' User-friendly ingredients are assumed to be moderately more important than cost as safety should always be prioritized, especially when choosing personal care products. Some chemicals, such as parabens, sulphates and synthetic fragrances, have a tendency to cause skin irritations and other health risks. Since people have become more aware of the side effects of chemical-based shampoo formulation, there has been an increasing market demand for shampoos that are derived from natural sources (Sang et al., 2023).

It is assumed that cost and cleaning effectiveness of a shampoo product are equally important as most consumers prefer shampoo with good cleaning ability at the cheapest price (Khanam and Rahman, 2022). Cost and environmental friendliness are assumed to be equally important, driven by a growing number of environmentally conscious consumers. Consumers are willing to purchase eco-friendly products if the prices of eco-friendly and non-eco-friendly options are similar (Khuzaimah et al., 2020). When comparing comfort with cost, it is assumed that cost is moderately more important than comfort. Comfort refers to the overall sensory experience of using the shampoo, including, but not limited to, fragrance, foaming ability and texture. A study reported that aroma is relatively less significant compared to cost when it comes to selecting a shampoo (Khanam and Rahman, 2022).

Table 5.18 shows the pairwise comparison matrix of key selection criteria.

TABLE 5.18
Pairwise Comparison Matrix of Key Selection Criteria

	Cost	Environmental Friendliness	User-Friendly Ingredients	Cleaning Effectiveness	Comfort
Cost	1	1	1/3	1	3
Environmental friendliness	1	1	1/3	1/3	3
User-friendly ingredients	3	3	1	3	5
Cleaning effectiveness	1	3	1/3	1	3
Comfort	1/3	1/3	1/5	1/3	1

Next, the weighting factor, w, of each selection criterion can be calculated using the steps mentioned in Section 5.6. The following steps illustrate how to solve for weighting factor using spreadsheet:

a) Square the pairwise comparison matrix

$$\begin{bmatrix} 1 & 1 & 1/3 & 1 & 3 \\ 1 & 1 & 1/3 & 1/3 & 3 \\ 3 & 3 & 1 & 3 & 5 \\ 1 & 3 & 1/3 & 1 & 3 \\ 1/3 & 1/3 & 1/5 & 1/3 & 1 \end{bmatrix} \times \begin{bmatrix} 1 & 1 & 1/3 & 1 & 3 \\ 1 & 1 & 1/3 & 1/3 & 3 \\ 3 & 3 & 1 & 3 & 5 \\ 1 & 3 & 1/3 & 1 & 3 \\ 1/3 & 1/3 & 1/5 & 1/3 & 1 \end{bmatrix} = \begin{bmatrix} 5.00 & 7.00 & 1.93 & 4.33 & 13.67 \\ 4.33 & 5.00 & 1.71 & 3.67 & 11.67 \\ 13.67 & 19.67 & 5.00 & 11.67 & 37.00 \\ 7.00 & 9.00 & 2.60 & 5.00 & 19.67 \\ 1.93 & 2.60 & 0.73 & 1.71 & 5.00 \end{bmatrix}$$

b) Calculate and normalize the sums of rows to compute the eigenvector

$$\begin{bmatrix} 5.00 + 7.00 + 1.93 + 4.33 + 13.67 \\ 4.33 + 5.00 + 1.71 + 3.67 + 11.67 \\ 13.67 + 19.67 + 5.00 + 11.67 + 37.00 \\ 7.00 + 9.00 + 2.60 + 5.00 + 19.67 \\ 1.93 + 2.60 + 0.73 + 1.71 + 5.00 \end{bmatrix} = \begin{matrix} \textit{Sums of row} \\ 31.9333 \\ 26.3778 \\ 87.0000 \\ 43.2667 \\ 11.9778 \\ \text{Total} = \underline{200.5556} \end{matrix} \xrightarrow{\textit{normalise}} \begin{matrix} \textit{Eigenvector} \\ 0.1592 \\ 0.1315 \\ 0.4338 \\ 0.2157 \\ 0.0597 \\ \underline{1.0000} \end{matrix}$$

c) Repeat the above process by squaring matrix $\begin{bmatrix} 5.00 & 7.00 & 1.93 & 4.33 & 13.67 \\ 4.33 & 5.00 & 1.71 & 3.67 & 11.67 \\ 13.67 & 19.67 & 5.00 & 11.67 & 37.00 \\ 7.00 & 9.00 & 2.60 & 5.00 & 19.67 \\ 1.93 & 2.60 & 0.73 & 1.71 & 5.00 \end{bmatrix}$ and normalizing the sums of rows to obtain the eigenvector. Then, compute the differences between the two sets of eigenvectors, as shown in Table 5.19.

TABLE 5.19
Differences in Values between Two Sets of Eigenvectors

Eigenvector (1st Set)	Eigenvector (2nd Set)	Differences between Two Sets of Eigenvectors
0.1592	0.1603	0.0011
0.1315	0.1332	0.0017
0.4338	0.4341	0.0003
0.2157	0.2115	−0.0043
0.0597	0.0609	0.0012

Since there are differences between the two sets of eigenvectors, the process needs to be repeated until there are no significant changes in the value of the eigenvector, as compared to the previous calculation, that is, up to four decimal places of accuracy. After two more iterations, the values of eigenvector reach a steady state. λ_{max}, *CI* and *CR* can then be calculated. Table 5.20 shows the weighting factor, w, of each selection criterion together with values of λ_{max}, *CI* and *CR*. The inconsistency of the pairwise comparison is considered acceptable since the calculated *CR* value is 4%, which is below the limit of 10%.

Since a benchmark was chosen, the comparison of alternatives based on the selection criteria was done by evaluating how well each one performs compared to the benchmark using a score of 1 to 10 (Cussler and Moggridge, 2011). A score of 10 means that the option is much better than the benchmark, whereas a score of zero being much worse than the benchmark. The benchmark product was assigned an average score of five. The total score of each alternative can then be calculated using Equation 5.41.

$$Score = \sum_{i=1}^{n} w_i c_{ij} \tag{5.41}$$

where w_i represents the weighting factor of each selection criterion, i, and c_{ij} represents the score of each alternative, j, with respective to the selection criterion, i. The alternative with the highest total score will rank first. The score of each shampoo alternative is presented in Table 5.21.

Note that if a benchmark is not available, one can utilize AHP method, that is, to conduct pairwise comparison at each level and obtain the overall priority ranking of alternatives by following the steps mentioned in Section 5.6.

From the result, scalp balancing shampoo is the first choice, followed by shampoo with natural ingredients. Nonetheless, their scores are close, and if shampoo with natural ingredients is slightly cost-effective, then it will be ranked first. On the other hand, dry shampoo will be the last option when it comes to choosing a new shampoo product.

If the product design team has identified their target market, it is recommended to create a survey form that allows its target consumers to assess the relative importance of the criteria in choosing a new shampoo product. Similarly, a consumer can provide their judgements on comparing criteria in pairs using Saaty's nine-point scale on the

TABLE 5.20
Weighting Factors of Selection Criteria

Selection Criteria	Weighting Factor
Cost	0.16
Environmental friendliness	0.13
User-friendly ingredients	0.44
Cleaning effectiveness	0.21
Comfort	0.06

λ_{max} = 5.19; CI = 0.05; CR= 0.04

TABLE 5.21
Score of Shampoo Alternatives

Selection Criteria	Weighting Factor	Benchmark: Pantene Shampoo	Scalp Balancing Shampoo	Dry Shampoo	Exfoliating Scalp Scrub	No Foam Shampoo	Shampoo with Natural Ingredients
Cost	0.16	5	4	2	1	2	2
Environment friendliness	0.13	5	5	3	3	4	5
User-Friendly ingredients	0.44	5	6	2	3	7	8
Cleaning effectiveness	0.21	5	7	2	8	5	4
Comfort	0.06	5	7	2	5	3	7
Total score		**5**	**5.8**	**2.1**	**3.9**	**5.1**	**5.7**

survey form. An additional step to take is to aggregate their individual judgements, a_{ij}, using the geometric mean method before synthesizing the weighting factor.

5.7 SENSITIVITY ANALYSIS

The development of the selection matrix is based on fixed weighting factors for the selection criteria and the scores assigned to different ideas. However, there are subjective elements involved in the scoring process. Additionally, it is possible that one of the selection criteria may become irrelevant, or a new criterion may emerge as an important factor. The subjectivity in the weighting factors can be addressed using the MCDM methods described in the previous section. To address the subjectivity involved in the scoring process and the effect of new criteria, sensitivity analysis may be used. A commonly used method is the One-at-a-Time (OAT) analysis, in which only one of the factors affecting the decision is changed at a time, while the others are kept constant. This process is repeated for all the factors to analyse their impact on the outcome. For more complex scenarios, it is necessary to study the effects of simultaneous changes in multiple factors using global sensitivity analysis, based on techniques such as the Monte Carlo method.

It is important to note that the primary purpose of sensitivity analysis is to address the subjectivity inherent in scoring and decision-making by systematically evaluating the impact of individual or combined variables. Parameters set by the designers are not treated as sources of uncertainty and are therefore excluded from the sensitivity analysis. Furthermore, the scores are not varied randomly, as it is assumed that designers possess prior knowledge about the relative ranking of the alternatives at this stage. Instead, the focus of sensitivity analysis is on examining how variations in scoring – while maintaining the relative ranking of options within each criterion – affect the overall outcome.

Example 5.7. Sensitivity Analysis on the Design of a New Shampoo

Apply sensitivity analysis to the previous example on the design of a new shampoo to evaluate the impact of subjectivity in scoring different parameters.

Solution

During sensitivity analysis, the scores of different ideas under each selection criterion are modified without altering their initial ranking. For example, under the 'Cost' category, all ideas performed worser than the benchmark. Among the five ideas, the scalp-balancing shampoo received the highest score, the exfoliating scalp scrub the lowest, while the other three ideas shared the same score. Therefore, during sensitivity analysis, this ranking should be preserved, and no score should be raised to match or exceed the benchmark, as all ideas were found to underperform in this category. The same approach is applied across the remaining categories, and the results are summarized in Table 5.22.

The results show that the ranking remains unchanged even after the scores were revised. Therefore, it can be concluded that the outcome is robust and not significantly influenced by the subjectivity involved in assigning the scores.

Limitations

The selection matrix is a simple yet powerful tool for comparing ideas and guiding the final selection. Carefully chosen selection criteria, combined with appropriate scoring and weighting, can effectively differentiate between ideas. However, an

TABLE 5.22
Sensitivity Analysis on Scores

Selection Criteria	Weighting Factor	Benchmark	Scalp Balancing Shampoo	Dry Shampoo	Exfoliating Scalp Scrub	No Foam Shampoo	Shampoo with Natural Ingredients
Cost	0.16	5	4	3	2	3	3
Environment friendliness	0.13	5	5	3	3	4	5
User-friendly Ingredients	0.44	5	7	3	4	8	9
Cleaning effectiveness	0.21	5	7	2	8	5	3
Comfort	0.06	5	7	2	6	3	7
Total score		5	6.3	2.7	4.5	5.8	6.1

implicit assumption in this method is the linearity of the selection criteria (Cussler and Moggridge, 2011), which may be unrealistic in certain situations.

1. The criteria are either binary or vary in discrete intervals. For example, the toxicity of chemicals is assessed within specific limit ranges, where any value within a given range is considered to have the same significance.
2. One of the ideas is a game changer. This product may have extremely unique attributes in one category, which makes the performance in the other categories irrelevant.

The product development team must be aware of these scenarios and need to overwrite the decisions from the selection matrix if necessary.

5.8 RISK ASSESSMENT AND MANAGEMENT

During the final stage of a chemical product's development, it is essential to assess the level of risk associated with the product. Product risks refer to potential problems or failures that may arise during the development, launch or operation of the product. These risks are factors or events that can negatively impact the achievement of the product's goals, its success in the market or user satisfaction. Designers often need to compare the performance of various products along with their associated risk levels. Uncertainties at various stages of the design process contribute significantly to the overall risk. These uncertainties may arise from a lack of data or information on potential side reactions, contamination, effects of contaminants, raw material availability, political instability, insufficient scientific knowledge, change in customer attitude, government policies and regulations and more. In some cases, these risks may not be a major concern, whereas in others, risk may be the most critical attribute to consider. Because of this, the risk level of each idea needs to be evaluated before finalizing the selection of a product idea, and further actions need to be taken.

Risk assessment is carried out in the following stages. First, the potential risks are identified, followed by a risk analysis to evaluate the impact of these risks on the selected product. After this step, risk management is considered to determine how to address the risks-whether they are worth the effort to mitigate or if the project should be cancelled altogether.

5.8.1 Risk Assessment

At this stage, potential risks from a project are identified. The risks could be from a variety of courses, and the focus at this stage is to ensure that none of the possible risks are overlooked. The product design team may consult experts, practitioners, customers, marketing team and consultants to cover all sources of risks. Generally, the type of risk depends on the category of product. For example, in pharmaceutical products, the biggest risk is the side effects of the new products and the effect of contaminants on some patients. In the case of consumer products, the biggest risk is rejection from customers. Though there are potential risks from other categories are well, the biggest risks are often the primary focus. We may also anticipate the risks of a competitor developing a better product before the organization, rejection by the regulatory bodies and unexpected issues during process scale-up.

Even though companies explore different sources of risk, there have been many cases where unexpected incidents after product launches have led to serious disasters. Perhaps the most well-known of these was the thalidomide tragedy that occurred in Europe in the late 1950s. Thalidomide was introduced to the market as a treatment for morning sickness in pregnant women. The drug became widely popular and was commonly prescribed by doctors at the time. However, it was soon observed that many babies were being born with severe phocomelia (malformed or missing limbs), as well as other defects such as deformed organs and sensory impairments. In the following years, these deformities were linked to the use of thalidomide by their mothers during pregnancy. By that time, thousands of babies had been born with severe deformities, nearly half of whom died shortly after birth. This incident severely affected public trust in the pharmaceutical industry, as there were obvious omissions during clinical trials which, if conducted properly, could have prevented the entire tragedy. This also led to the development of much more stringent controls and regulations for drug development across the world.

While there are several potential and unanticipated risks associated with all products, it may not be practical to eliminate these risks entirely. A more realistic approach involves the quantification and management of risk. Risk quantification is typically based on two factors: the probability of occurrence and the severity of consequences. These are usually represented on a scale ranging from zero to one. If the probability of an undesirable outcome is low, a score of 0.3 or lower is assigned, whereas higher probabilities of undesirable outcomes receive higher scores. Similarly, ideas that lead to relatively minor consequences are given scores below 0.3 in the consequence category. For example, the consequence of an unexpected development only leads to a decrease in sales, which is considered a minor consequence. At the same time, failure to identify a risk may lead to project cancellation, product withdrawal from the market, loss of public image or serious damage to public property – all of which are considered major consequences and are assigned higher scores in the consequence category.

Once the risk probability and consequence scores are agreed upon by the product development team, the overall risk level of the project is estimated as the product of these two scores. In general, a risk level of 0.3 or above is considered significant and must be addressed during the risk management stage.

5.8.2 Risk Management

In chemical product design, systematic risk management plays a critical role in evaluating new ideas. The process begins with identifying and quantifying potential risks. Even though some factors (such as consumer acceptance or government policies) may lie beyond direct control, understanding their impact allows for informed decision-making. Once risks are characterized, designers can adopt appropriate strategies guided largely by economic and competitive considerations.

The available courses of action span from abandoning the idea entirely to accepting the risks outright or implementing measures to reduce them. Practical methods include *eliminating* the risk (e.g. through a design change), *reducing* the risk (e.g. lowering the likelihood or consequence), *transferring* the risk (e.g. purchasing insurance) or *accepting* the risk (e.g. by allocating contingency funds). In real-world product development, the choice often hinges on cost-benefit trade-offs: how expensive is it to reduce the risk, and would mitigation delay time-to-market enough to allow competitors to gain an advantage? Integrating such risk-based thinking early in chemical product design improves decision-making and supports the delivery of safer, more robust products.

Economic risks in product development are treated differently by different companies. Larger organizations tend to take more risks in this area, as losses on some products can be offset by major breakthroughs made possible only through calculated risks taken at certain stages. Smaller companies, however, are often more conservative in this regard. At the same time, ideas involving inherent risks related to safety, health, the environment, societal impacts and international image are taken more seriously by major corporations, as mistakes in these areas can lead to long-lasting consequences and damage the company's reputation. In addition, government regulations often limit excessive risk-taking by companies in these areas, in order to protect citizens from the consequences of unexpected failures (Cussler and Moggridge, 2011).

Example 5.8. Risk Identification

The company's R&D department has proposed to develop a novel product to use as a face cleaning agent. This new product will form a smooth suspension with water and remove the dirt and dead skin. This face wash powder has no known side effects.

What are the potential risks in developing this face wash powder as a commercial product? What are the possible measures the company can take to reduce/manage the risks? Why may it be beneficial for the company to accept the risks and proceed with manufacturing without eliminating the risks completely?

Solution

The potential risks are as follows:

- Side effects after prolonged use, sensitivity to certain skin types and concerns from some consumers about potential effects on people with specific health conditions, along with immature technology and customers potentially not perceiving the value of the new product, are all possible risks.
- To minimize these risks, the company could conduct extensive tests on people from different age groups, skin types and geographical locations. While the possibility of unexpected side effects remains, the risk of waiting too long is even greater. For example, a competitor may develop a similar product.

TABLE 5.23
Risk Description in the Design of Scalp-Balancing Shampoo

Risk	Description
Consumer acceptability	Willingness of consumers to try out new hair shampoo
Environmental impact	Effect of products on the environment (focusing on the ingredients, packaging and manufacturing process)
Technology reliability	How reliable/difficult is the manufacturing process?
Competition	Presence of similar products in the market, which may affect the product sales

TABLE 5.24
Risk Analysis

Risk	Probability	Consequences	Risk Level
Consumer acceptability	0.40	0.40	0.16
Environmental impact	0.20	0.40	0.08
Technology reliability	0.10	0.50	0.05
Competition	0.50	0.40	0.20

Example 5.9. Risk Analysis on New Shampoo Product

Conduct a risk analysis on the new shampoo design covered in Example 5.6.

Solution

In this case, the main risks are consumer acceptability, environmental impact, technology reliability and competition (Table 5.23).

Risk analysis is then performed as illustrated in Table 5.24.

The biggest risk is competition, as the existing shampoos on the market may affect the economic feasibility of the product. Consumer acceptability of the new product also poses significant risks. However, the overall risk level remains within acceptable limits, and it is recommended to explore ways to reduce these risks before proceeding to the manufacturing stage.

5.9 CONCLUSIONS

In this chapter, the various steps involved in the final stage of a project idea are discussed. While there are several specific tools for solving different types of problems, four representative approaches that address a large portion of chemical product design problems have been introduced. The concept of the screening matrix, introduced in the previous chapter, has been extended into the development of a selection matrix, which is used to finalize the selection. The key difference between the screening and selection matrices lies in the level of rigour and quantification involved in scoring and estimating weightages. Therefore, more quantitative tools than those used in the screening stage are generally applied during the selection stage. Currently, several multi-criteria decision-making tools are available to make this analysis more robust. One of the most commonly used tools was covered in this chapter. Finally, the risks involved in the product design process have been defined, and methods to quantify and manage these risks have been discussed.

This chapter concludes the steps involved in the design of new chemical products prior to manufacturing. At this stage, the most promising ideas for advancement to manufacturing have been identified. Companies may now perform detailed database searches, laboratory-scale experiments, patent searches, and rigorous modelling and simulation. However, before proceeding to manufacturing, it is important to explore the intellectual property value of the new idea and the ways to protect it. A detailed economic analysis that takes into account potential risks is also crucial for the success of the project. These topics will be discussed in the next chapter.

REFERENCES

Achenie, L., Venkatasubramanian, V., & Gani, R. (2002). *Computer Aided Molecular Design: Theory and Practice*. Elsevier Science. https://books.google.com.my/books?id=djgsyXdAO8EC

Albahri, T. A. (2014, January). MNLR and ANN Structural Group Contribution Methods for Predicting the Flash Point Temperature of Pure Compounds in the Transportation Fuels Range. *Process Safety and Environmental Protection*, *93*, 182–191. https://doi.org/10.1016/j.psep.2014.03.005

Baxevanidis, P., Papadokonstantakis, S., Kokossis, A., & Marcoulaki, E. (2022). Group Contribution-Based LCA Models to Enable Screening for Environmentally Benign Novel Chemicals in CAMD Applications. *AIChE Journal*, *68*(3), e17544. https://doi.org/10.1002/aic.17544

Betz, M., & Cerny, G. (2000). Antimicrobial Effects of Bleaching Agents. *Tenside Surfactants Detergents*, *37*(4), 230–235. https://doi.org/10.1515/tsd-2000-370404

Bingham, J., Cartner, T. J., Mays Suko, P. A., & Leslie, R. A. (2019). Multifactor Assessment of Non-Antimicrobial Soap Performance. *Open Forum Infectious Diseases*, *6*(5), ofz151. https://doi.org/10.1093/ofid/ofz151

Broughton, E. (2005). The Bhopal Disaster and Its Aftermath: A Review. *Environmental Health*, *4*(1), 6. https://doi.org/10.1186/1476-069X-4–6

Capello, C., Fischer, U., & Hungerbuhler, K. (2007). What is a Green Solvent? A Comprehensive Framework for the Environmental Assessment of Solvents. *Green Chemistry*, *9*(9), 927–934. https://doi.org/10.1039/B617536H

Castanedo-Tardana, M. P., & Zug, K. A. (2013, January–February). Methylisothiazolinone. *Dermatitis*, *24*(1), 2–6. https://doi.org/10.1097/DER.0b013e31827edc73. PMID: 23340392

Constantinou, L., & Gani, R. (1994). New Group Contribution Method for Estimating Properties of Pure Compounds. *AIChE Journal*, *40*(10), 1697–1710. https://doi.org/10.1002/aic.690401011

Constantinou, L., Prickett, S. E., & Mavrovouniotis, M. L. (1993). Estimation of Thermodynamic and Physical Properties of Acyclic Hydrocarbons Using the ABC Approach and Conjugation Operators. *Industrial & Engineering Chemistry Research, 32*(8), 1734–1746. https://doi.org/10.1021/ie00020a030

Cussler, E. L., & Moggridge, G. D. (2011). Chemical Product Design. In *Cambridge Series in Chemical Engineering* (2nd ed.). Cambridge University Press. https://doi.org/10.1017/CBO9781139035132

Dearden, J. C. (2017). *The Use of Topological Indices in QSAR and QSPR Modeling BT – Advances in QSAR Modeling: Applications in Pharmaceutical, Chemical, Food, Agricultural and Environmental Sciences* (K. Roy (ed.), pp. 57–88). Springer International Publishing. https://doi.org/10.1007/978-3-319-56850-8_2

Gani, R. (2019). Group Contribution-Based Property Estimation Methods: Advances and Perspectives. *Current Opinion in Chemical Engineering, 23*, 184–196. https://doi.org/10.1016/j.coche.2019.04.007

Gani, R., & Brignole, E. A. (1983). Molecular Design of Solvents for Liquid Extraction Based on UNIFAC. *Fluid Phase Equilibria, 13*, 331–340. https://doi.org/10.1016/0378-3812(83)80104-6

Gani, R., & Constantinou, L. (1996). Molecular Structure Based Estimation of Properties for Process Design. *Fluid Phase Equilibria, 116*(1), 75–86. https://doi.org/10.1016/0378-3812(95)02874-9

Gani, R., Hytoft, G., Jaksland, C., & Jensen, A. K. (1997). An Integrated Computer Aided System for Integrated Design of Chemical Processes. *Computers & Chemical Engineering, 21*(10), 1135–1146. https://doi.org/10.1016/S0098-1354(96)00324-9

Hassim, M. H., & Edwards, D. W. (2006). Development of a Methodology for Assessing Inherent Occupational Health Hazards. *Process Safety and Environmental Protection, 84*(5), 378–390. https://doi.org/10.1205/psep.04412

Hassim, M. H., & Hurme, M. (2010). Inherent Occupational Health Assessment During Process Research and Development Stage. *Journal of Loss Prevention in the Process Industries, 23*(1), 127–138. https://doi.org/10.1016/j.jlp.2009.06.009

Hatamleh, M., Chong, J. W., Tan, R. R., Aviso, K. B., Janairo, J. I. B., & Chemmangattuvalappil, N. G. (2022). Design of Mosquito Repellent Molecules via the Integration of Hyperbox Machine Learning and Computer Aided Molecular Design. *Digital Chemical Engineering, 3*, 100018. https://doi.org/10.1016/j.dche.2022.100018

Heikkilä, A. M. (1999). *Inherent Safety in Process Plant Design. An Index-Based Approach* (p. 384). VTT Publications.

Hosoya, H. (1971). Topological Index. A Newly Proposed Quantity Characterizing the Topological Nature of Structural Isomers of Saturated Hydrocarbons. *Bulletin of the Chemical Society of Japan, 44*(9), 2332–2339. https://doi.org/10.1246/bcsj.44.2332

Hukkerikar, A. S., Kalakul, S., Sarup, B., Young, D. M., Sin, G., & Gani, R. (2012). Estimation of Environment-Related Properties of Chemicals for Design of Sustainable Processes: Development of Group-Contribution+ (GC+) Property Models and Uncertainty Analysis. *Journal of Chemical Information and Modeling, 52*(11), 2823–2839. https://doi.org/10.1021/ci300350r

Ishizaka, A., & Labib, A. (2011). Review of the Main Developments in the Analytic Hierarchy Process. *Expert Systems with Applications, 38*(11), 14336–14345. https://doi.org/10.1016/j.eswa.2011.04.143

Joback, K., & Reid, R. C. (1987). Estimation of Pure-Component Properties from Group-Contributions. *Chemical Engineering Communications, 57*(1–6), 233–243. https://doi.org/10.1080/00986448708960487

Joback, K., & Stephanopoulos, G. (1995). Searching Spaces of Discrete Solutions: The Design of Molecules Possessing Desired Physical Properties. *Advances in Chemical Engineering, 21*(C), 257–311. https://doi.org/10.1016/S0065-2377(08)60075-7

Kalakul, S., Zhang, L., Fang, Z., Choudhury, H. A., Intikhab, S., Elbashir, N., Eden, M. R., & Gani, R. (2018). Computer Aided Chemical Product Design – ProCAPD and Tailor-Made Blended Products. *Computers & Chemical Engineering*, *116*, 37–55. https://doi.org/10.1016/j.compchemeng.2018.03.029

Khanam, U. S., & Rahman, M. B. (2022). Preference of Global Shampoo Brands: A Study on Women Consumers. *Open Journal of Social Sciences*, *10*(12), 241–254. https://doi.org/10.4236/jss.2022.1012016

Khuzaimah, A. B., Taslim, F. A., Zabri, I. F., Jia Jia, K., Ganamoorthi, K. R., En Ling, L., Abu Samah, M. A., Isai, K. I. A., Muthutamilselvan, K., Naidu Kanan, V., Ramalingam, S., & Maruthai, E. (2020). Consumer Behaviour in Green Purchasing: A Case Study in Petaling Jaya, Malaysia. *Malaysian Journal of Social Sciences and Humanities (MJSSH)*, *5*(5), 44–59. https://doi.org/10.47405/mjssh.v5i5.403

Kier, L. B. (1985). A Shape Index from Molecular Graphs. *Quantitative Structure-Activity Relationships*, *4*(3), 109–116. https://doi.org/10.1002/qsar.19850040303

Kier, L. B. (1986a). Distinguishing Atom Differences in a Molecular Graph Shape Index. *Quantitative Structure-Activity Relationships*, *5*(1), 7–12. https://doi.org/10.1002/qsar.19860050103

Kier, L. B. (1986b). Shape Indexes of Orders One and Three from Molecular Graphs. *Quantitative Structure-Activity Relationships*, *5*(1), 1–7. https://doi.org/10.1002/qsar.19860050102

Kier, L. B., Di Paolo, T., & Hall, L. H. (1977). Structure-Activity Studies on Odor Molecules Using Molecular Connectivity. *Journal of Theoretical Biology*, *67*(3), 585–595. https://doi.org/10.1016/0022-5193(77)90059–5

Kier, L. B., & Hall, L. H. (1976). Molecular Connectivity VII: Specific Treatment of Heteroatoms. *Journal of Pharmaceutical Sciences*, *65*(12), 1806–1809. https://doi.org/10.1002/jps.2600651228

Kier, L. B., & Hall, L. H. (1986). *Molecular Connectivity in Structure-Activity Analysis*. Wiley. https://books.google.com.my/books?id=qpA0AAAAMAAJ

Kontogeorgis, G. M., & Gani, R. (2004). *Computer Aided Property Estimation for Process and Product Design: Computers Aided Chemical Engineering*. Elsevier Science. https://books.google.com.my/books?id=mE1R5Q-s-48C

Mann, V., Gani, R., & Venkatasubramanian, V. (2023). Group Contribution-Based Property Modeling for Chemical Product Design: A Perspective in the AI Era. *Fluid Phase Equilibria*, *568*, 113734. https://doi.org/10.1016/j.fluid.2023.113734

Marrero, J., & Gani, R. (2001). Group-Contribution Based Estimation of Pure Component Properties. *Fluid Phase Equilibria*, *183–184*, 183–208. http://dx.doi.org/10.1016/S0378-3812(01)00431–9

National Fire Protection Association. NFPA 704: (2007). Standard System for the Identification of the Hazards of Materials for Emergency Response. In *National Fire Protection Association*. Quincy, MA: National Fire Protection Association.

Nikolić, S., Trinajstić, N., & Mihalić, Z. (1995). The Wiener Index: Development and Applications *. *Croatica Chemica Acta CCACAA*, *68*(1), 105–129.

Odele, O., & Macchietto, S. (1993). Computer Aided Molecular Design: A Novel Method for Optimal Solvent Selection. *Fluid Phase Equilibria*, *82*, 47–54. http://dx.doi.org/10.1016/0378-3812(93)87127-M

Patel, S. J., Ng, D., & Mannan, M. S. (2009). QSPR Flash Point Prediction of Solvents Using Topological Indices for Application in Computer Aided Molecular Design. *Industrial & Engineering Chemistry Research*, *48*(15), 7378–7387. https://doi.org/10.1021/ie9000794

Randic, M. (1975). Characterization of Molecular Branching. *Journal of the American Chemical Society*, *97*(23), 6609–6615. https://doi.org/10.1021/ja00856a001

Saaty, T. L. (1977). A Scaling Method for Priorities in Hierarchical Structures. *Journal of Mathematical Psychology*, *15*(3), 234–281. https://doi.org/10.1016/0022-2496(77)90033–5

Saaty, T. L. (1980). *The Analytic Hierarchy Process: Planning, Priority Setting, Resource Allocation*. McGraw-Hill International Book Company. https://books.google.com.my/books?id=Xxi7AAAAIAAJ

Sahinidis, N. V., Tawarmalani, M., & Yu, M. (2003). Design of Alternative Refrigerants via Global Optimization. *AIChE Journal*, *49*(7), 1761–1775. https://doi.org/10.1002/aic.690490714

Samudra, A. P., & Sahinidis, N. V. (2013). Optimization-Based Framework for Computer-Aided Molecular Design. *AIChE Journal*, *59*(10), 3686–3701. https://doi.org/10.1002/aic.14112

Sang, S.-H., Akowuah, G. A., Liew, K. Bin, Lee, S.-K., Keng, J.-W., Lee, S.-K., Yon, J.-A.-L., Tan, C. S., & Chew, Y.-L. (2023). Natural Alternatives from Your Garden for Hair Care: Revisiting the Benefits of Tropical Herbs. *Heliyon*, *9*(11), e21876. https://doi.org/10.1016/j.heliyon.2023.e21876

Song, Z., Shi, H., Zhang, X., & Zhou, T. (2020). Prediction of CO2 Solubility in Ionic Liquids Using Machine Learning Methods. *Chemical Engineering Science*, *223*, 115752. https://doi.org/10.1016/j.ces.2020.115752

Ten, J. Y., Hassim, M. H., Ng, D. K. S., & Chemmangattuvalappil, N. G. (2016). A Molecular Design Methodology by the Simultaneous Optimisation of Performance, Safety and Health Aspects. *Chemical Engineering Science*, *159*, 140–153. https://doi.org/10.1016/j.ces.2016.03.026

Thurstone, L. L. (1927). A Law of Comparative Judgment. *Psychological Review*, *34*(4), 273–286. https://doi.org/10.1037/h0070288

Tiew, S. T., Chew, Y. E., Lee, H. Y., Chong, J. W., Tan, R. R., Aviso, K. B., & Chemmangattuvalappil, N. G. (2023). A Fragrance Prediction Model for Molecules Using Rough Set-based Machine Learning. *Chemie-Ingenieur-Technik*, *95*(3), 438–446. https://doi.org/10.1002/cite.202200093

Venkatasubramanian, V., Chan, K., & Caruthers, J. M. (1994). Computer-Aided Molecular Design Using Genetic Algorithms. *Computers & Chemical Engineering*, *18*(9), 833–844. https://doi.org/10.1016/0098-1354(93)E0023-3

Wen, L., Li, Q., Li, W., Cai, Q., & Cai, Y.-M. (2017). A QSAR Study Based on SVM for the Compound of Hydroxyl Benzoic Esters. *Bioinorganic Chemistry and Applications*, *2017*(1), 4914272. https://doi.org/10.1155/2017/4914272

Wiener, H. (1947). Structural Determination of Paraffin Boiling Points. *Journal of the American Chemical Society*, *69*(1), 17–20. https://doi.org/10.1021/ja01193a005

Wilke, C. R., & Lee, C. Y. (1955). Estimation of Diffusion Coefficients for Gases and Vapors. *Industrial & Engineering Chemistry*, *47*(6), 1253–1257.

Wilson, R. J. (1986). *Introduction to Graph Theory*. John Wiley & Sons, Inc.

Yokoyama, M. (1921). The Nature of the Affective Judgment in the Method of Paired Comparisons. *The American Journal of Psychology*, *32*(3), 357–369. https://doi.org/10.2307/1414000

Zhang, L., Mao, H., Liu, L., Du, J., & Gani, R. (2018). A Machine Learning Based Computer-Aided Molecular Design/Screening Methodology for Fragrance Molecules. *Computers & Chemical Engineering*, *115*, 295–308. https://doi.org/10.1016/j.compchemeng.2018.04.018

Zhang, L., Mao, H., Liu, Q., & Gani, R. (2020). Chemical Product Design – Recent Advances and Perspectives. *Current Opinion in Chemical Engineering*, *27*, 22–34. https://doi.org/10.1016/j.coche.2019.10.005

Zhou, T., Jhamb, S., Liang, X., Sundmacher, K., & Gani, R. (2018). Prediction of Acid Dissociation Constants of Organic Compounds Using Group Contribution Methods. *Chemical Engineering Science*, *183*, 95–105. https://doi.org/10.1016/j.ces.2018.03.005

Zhou, T., McBride, K., Linke, S., Song, Z., & Sundmacher, K. (2020). Computer-Aided Solvent Selection and Design for Efficient Chemical Processes. *Current Opinion in Chemical Engineering*, *27*, 35–44. https://doi.org/10.1016/j.coche.2019.10.007

5.10 PROBLEMS

1. Estimate the normal boiling point, T_b, normal melting point, T_m, and standard heat of vaporization at 298K, H_v of 2-Methylpentane using the following group contributions (Marrero and Gani, 2001):

TABLE Q5.1
First-Order Molecular Groups and Their T_b, T_m and H_v Contributions

Molecular Group	T_b Contribution, T_{b1i}	T_m Contribution, T_{m1i}	H_v Contribution, $H_{v1,i}$
$-CH_3$	0.8491	0.6953	0.2170
$-CH_2-$	0.7141	0.2515	0.4910
-CH-	0.2925	−0.3730	7.962

2. Calculate the branching index value of 3-ethylheptane.
3. Calculate the Wiener index of 2,2-Dimethylhexane using 'path number' concept.
4. Calculate $^1\kappa$, $^2\kappa$ and $^3\kappa$ of 2,2-Dimethylbutane molecule.
5. In a lithographic printing process, ink is conveyed to an impression plate by means of a train of rubber rollers known as 'blankets'. These blankets must be cleaned regularly since their cleanliness is crucial for the production of high-quality images. It is desired to replace the current solvent, 1,1,1-trichloroethane (CH_3-C-Cl_3), with an environmentally friendly solvent. The new solvent should also have the ability to rapidly dissolve dried ink and have short drying time. Finally, the new solvent's ability to clean the rollers should be comparable to the current solvent. A new solvent, methyl propyl ketone ($CH_3(CH_2)_2COCH_3$), is evaluated for the aforementioned properties and found to be a promising candidate in terms of the ability to clean the rollers. Group contribution technique may be used to estimate the desired properties, with the contribution of each molecular group given as in Tables Q5.1 and Q5.2. Please note that some data is not needed, whereas some missing group contributions can be obtained from the smaller groups in the table.
   a) To ensure the solvent is in liquid phase at ambient pressure, it should have a boiling temperature (T_b) higher than 323 K. To save utility cost, a latent heat of vaporization (ΔH^v) of less than 50 kJ/mol is desirable. Check whether the new solvent satisfies these requirements.
   b) A useful measure of solvents' health impact is bioconcentration factor, which relates the likelihood of a solvent accumulating in and harming living tissue. Octanol-water partition coefficient (K_{ow}) is used to reflect the bioconcentration factor. The factor is sufficiently low when $\log_{10} K_{ow}$ is less than 3. Determine if the new solvent satisfies this criterion.

TABLE Q5.2
First-Order Group Contributions

Groups	t_b (K)	t_m (K)	h_v (kJ/mole)	v_m(m³/kmol)	X_i^0	X_i^1
CH_3-	0.849	0.695	0.217	0.026	0.89	1
-CH_2-	0.714	0.252	4.91	0.016	0.66	1
-CH-	0.293	−0.373	7.962	0.007	0.43	1
-C-	−0.067	0.026	10.73	−0.004	–	–
-Cl-	1.515	1.545	–	0.021	–	–
-CO-	0.498	−0.445	13.853	0.012	−1.33	1
CH_3-CO-	3.118	2.959	18.99	0.037	−0.44	2
-C-Cl-	1.705	1.842	–	0.020	–	–

c) To avoid ignition in air at typical operating conditions, candidate solvents are selected such that the flash point >323 K.
Useful equations

$T_b(K) = 222.543 \times \ln\left(\sum_i n_i t_{bi}\right)$; T_b is boiling point, n_i = number of group i in the solvent

$$T_m(K) = 147.45\ln\left(\sum_i n_i t_{mi}\right)$$

$$\Delta H^v(kJ/mol) = 0.0121 + \sum_i n_i \cdot h_{vi} + \sum_{i2} n_{i2} \cdot h_{i2}$$

$$\log_{10} K_{ow} = \sum_i n_i \chi_i^0 + 0.12\sum_i n_i \chi_i^1$$

where χ_i^0 and χ_i^1 are fragment and factor of group i.

$$T_f = 1.477 T_b^{0.8} \times (\Delta H^v)^{0.17} \times n_c^{-00}$$

n_c is the total number of carbon atoms in the molecule.

6. Example 4.4 in Chapter 4 presented an idea screening matrix to select a new insect repellent product. Apply idea selection methods to choose the best product idea, and evaluate the robustness of the product by performing a sensitivity analysis. Also, perform a risk analysis on the selected product.
7. A company is exploring new ideas to develop a new face cleaning product specifically targeting oily skin. Identify some of the key needs, propose ideas to meet those needs and select the most promising idea.

6 Economics of Chemical Products

6.1 INTELLECTUAL PROPERTY

Businesses are highly competitive, and to be successful, organizations need to have distinct selling points over their competitors. For many businesses, the intellectual property they possess provides the edge over other organizations. Intellectual property (IP) can be broadly defined as a category of property created by the human minds using their intellectual capabilities. It is a form of intangible property. The ability to acquire, manage and protect IP can determine the success or failure of a business today. In addition, it is necessary to have knowledge of IP to avoid accidental or unintentional infringement of IP owned by third parties. Litigation and subsequent penalties for violations can be costly for an organization, both financially and in terms of reputation. Intellectual property rights can be kept, sold or licensed. Needed registration of IP in the appropriate forms to own legal rights over the IP.

6.1.1 Trade-Related Aspects of IP Rights (TRIP)

TRIPS is an international agreement administered by the World Trade Organization (WTO), which sets minimum standards for different forms of IP regulations. These were agreed upon by the member states at the General Agreement on Tariffs and Trade (GATT) in 1994. TRIPS stipulated that the member states must develop laws that meet various forms of IP, such as copyrights, patents, trademarks, trade dress, confidential information, industrial design, integrated circuit layout designs, traditional knowledge and geographical indication. TRIPS also proposed strategies to enforce the developed laws,

6.1.2 Why IP Must Be Protected

Organizations and individuals spend a significant amount of time, money and resources to develop new knowledge and create new ideas, which need to be rewarded to encourage the development of new knowledge. IP can prevent 'me too' approach used by organizations that wish to copy the innovations done by other companies. Because of this, the companies that developed a culture to be innovative and regularly produce IP can improve the economic sustainability of their businesses and become the world leaders in their sector.

The IP can be protected through various approaches based on the nature of development. The most common IP protection approaches are as follows:

- Patents
- Trademarks

DOI: 10.1201/9781032708911-6

- Copyrights
- Trade secrets
- Industrial design
- Traditional knowledge
- Geographical indication

Protection of IP is important because it rewards the original creator and thus promotes innovation. For that reason, it is important to prevent the infringement of others' IP rights. To do these, it is important to know the underlying principles and rules behind the different forms of IP. IP can provide the right direction to research, a revenue generation tool, a branding of an organization. IP rights are legal rights given to the creator or owner of IP for a limited period of time and are subjected to certain conditions. The spirit of IPR is that the creator of knowledge gets the ownership of their work. In general, inventions are protected under patents. For instance, books, stories and music are protected under copyrights, while the name of a business is secured under trademark. In general, IP can be classified as industrial property and copyright. Patents, trademarks, industrial designs and geographical indicators are covered under industrial property while the work of authorship falls under copyright.

6.1.3 Patents

The word 'patent' originates from the Latin word 'patere', which means 'to lay open'. This indicates that the complete disclosure of the details of the invention is an integral part of patenting. To get a patent approved, detailed descriptions of the invention need to be submitted to the patent office in a specific format. Once the patent office is convinced that the invention is new, the inventor will be granted a set of exclusive rights by the state for a period of 20 years.

It is important to know that patent rights are territorial. To obtain patent coverage in different countries, the inventor has to apply for patents in different countries. The purpose of a patent is to encourage industrial development of theoretical curiosities, growth and sharing of knowledge by making the details of the invention publicly available, preventing the duplication of work, preventing exploitation and generating revenue for the original inventors upon commercialization. The patents are given to an invention which should be a *novel* product or process, which is a *non-obvious* development from an existing process/product with a capability for *industrial applications*. The key emphasis is on the three terms in this description, which are novelty, non-obviousness and industrial applications.

While submitting a patent application, the inventor must convince the patent office that the new development is novel by comparing the new development with the existing state of the art. The patent office will conduct a patentability search in which an extensive search on the patent databases and literature to confirm that the new development is indeed novel. The search includes the developments made anywhere in the world and the inventors' own previous publications and patents.

Once the patent office is convinced that the development is novel, the non-obviousness of it is checked. Non-obviousness means that innovation is not a simple combination of two or more of the already established ideas for a person skilled in the art. For example, the combination of two drugs to treat two diseases that typically occur together (such as common flu and headache) may not be patented even if the combination drug is developed for the first time. However, if the combination effects are different due to a different mechanism from the individual drugs, the new product may be patented. Proving the non-obviousness can be tricky at times, and it can be achieved through different ways. The inventors may elaborate on the demand for the developed invention for a long period of time and present the failure of others to achieve the desired results. The inventor may also highlight the complexity in the new innovation to strengthen their case for non-obviousness. It is to be noted that proving the non-obviousness of the invention is an essential condition for a patent to be granted in all major jurisdictions. However, the interpretation of the non-obviousness may vary across jurisdictions.

A patent can only be given to those inventions with commercial applications. So, the new theorems, mathematical models, methods of doing business, etc., cannot be protected using a patent; however, novel and non-obvious those developments may be, though the boundaries are often vague. In addition, patents are given only for inventions that are useful for humankind and improve the quality of life.

6.1.3.1 Inventions Not Patentable

In general, discoveries are not patentable – only inventions are. In addition, there are certain restrictions placed on certain types of developments to prevent monopoly, which are against national defence and security, and anything considered against public policy or morality.

Some of the developments that are not usually patentable include the following:

- A scientific or mathematical discovery, theory or method.
- New forms of existing products or adaptations from the existing products.
- Mixtures that do not produce a significant change in the performance of the individual constituents.
- A way of doing business.
- The presentation of information (this includes some computer programs).
- An animal or plant variety that exists in nature.
- Completely biological process.
- A method of medical treatment or diagnosis. This is different from the medical products or devices, which are patentable.
- Anything which is considered against public policy or morality.

It is to be noted that there are occasional differences between the interpretation of these general guidelines in several countries. At the same time, many countries have additional restrictions on what can/cannot be patented.

The procedure for the application of patents is given in Section 6.1.3.2.

6.1.3.2 Patent Filing

To file a patent, a patent specification is required. A patent specification is a document describing the technical details of an invention, for which legal protection is sought. While the layout of patent specifications varies between different countries, the essential content is the same everywhere in the world. A patent application is submitted with either a provisional or complete patent specification. A provisional specification is typically filed to establish the invention's priority when the invention is still in the conceptual stage, and a delay is anticipated in providing a full and detailed description of the invention. This is also a legal document on which no amendments are allowed in the later stages of the filing process. In order to complete the patent filing, a complete specification must be submitted within 12 months of the initial filing. The complete specification must be a logical expansion of the provisional specification (do not use provisional specification as a 'place holder') and is a more detailed description of the claims made in the preliminary specification.

Before writing the specification, it is important to develop a theme, and the specification should be directed to that theme. The complete specification must have the following content:

1. Title of the invention
2. Preamble
3. Technical field of the invention
4. Background of the invention
5. Summary of the invention
6. Brief description of drawings
7. Detailed description of the invention and claims
8. Drawings

6.1.3.3 Patent Prosecution

Patent prosecution is the process of drafting, filing and obtaining the patent office in a country. This is not to be confused with patent litigation, which is the suing of another person or organization for patent violation. A patent application can be filed by the true and first inventor or their legal representative. The prime stages in the process of patenting, from filing application till grant, are:

- Filing application
- Publication
- Examination
- Grant of patent

Once a patent application is filed, it will be published by the patent office 18 months after the filing date or the priority date, whichever comes first, allowing the public to view the application. At this stage, anyone can oppose the application by clearly demonstrating the reasons for objection to the patent office. In the next stage, the patent office examines whether the patent application is prepared according to the requirements. The patent may be awarded if the patent application fulfils the requirements. However, it is also possible to file a post-grant opposition to an awarded patent by showing the evidence for opposition in a timeframe set by the patent office.

6.1.4 Copyrights

Copyrights are another form of IP given to the works of authorship, such as literature, music, artistic contributions and motion pictures.

Copyright law does not protect ideas but only the expression of the idea in various ways. The idea is considered to be a mental image as a conception of theory, whereas the expression is the implementation of that idea. For example, the suffering of people at the end of a war could be an idea which cannot be protected. However, the presentation of suffering differs in pictures, literature and movies. Therefore, protection is only applied how the way it is presented in these mediums and not to the theme of suffering.

To obtain copyright protection, the work must meet the requirement of originality and fixation. For the originality, as long as the work is independently created with substantial amount of skill, the criterion has been satisfied. To meet the fixation requirement, the work must be fixed in a tangible medium of expression. The medium could be anything that a human or machine can perceive. So, a song written on paper or recorded in a device is considered to be fixed on a medium. It could also be an artwork in various forms.

Copyright protection is granted for different durations based on when the work was created and varies across different countries. For example, in the United States (for the work created after 1 January 1978) and the European Union, protection is granted for a period of 70 years, which is counted from the year following the death of the creator of that work. In the case of joint authorship, this is based on the death of the last surviving creator. However, several countries grant the protection only up to 50 years after the death of the original author. Moreover, several countries have different durations of protection for different categories of work.

In general, the person who created the work automatically owns the copyright to that creation. However, if work is 'made for hire', the person/organization that hired the creator is considered the owner of copyright. In addition, if the work is created as part of their employment, the employer is considered the owner of copyright. At the same time, the copyright of the work developed by the creator is not related to employment even while employed is generally given to the creator. On many occasions, the decisions on the recognition of copyright by the employees depend on whether the work was developed when the employee was under a contract *of* service or a contract *for* service. There have been several notable and often ridiculous lawsuits on the interpretation of the copyright rules. Some of these include the following:

1) 'Naruto vs Slator' in which PETA filed a lawsuit to claim the copyright for a monkey (in the United States)
2) 'Nichols vs Universal Pictures Corporation', in which non-literal copying of a dramatic work was challenged (in the United States)
3) 'V.T. Thomas vs Malayala Manorama', in which the ownership of fictional characters was challenged (in India)

6.1.5 Trademark

A trademark is a distinctive sign or symbol or collection of words which are used to represent certain products or businesses and to differentiate them from similar organizations or services. Trademarks can be a single letter or digit, a combination of letters and digits, drawings or symbols. For example, the way the letter 'H' is written in Toyota and Hyundai helps the customers differentiate between those two car brands. Therefore, a trademark helps companies to identify and differentiate themselves from the rest, guarantees consistency in services and creates an image for the company. Thus, preserving their trademarks is critical for companies to safeguard their brand identities, enhance consumer trust, facilitate market expansion and increase their overall value.

A trademark is considered to be infringed if another company or person uses a registered trademark or uses a registered trademark with minor alterations. In addition, usage of another company's trademark in advertisements is also considered a form of infringement.

6.1.6 Trade Secrets

Trade secrets are information or data related to businesses that the owners try to keep secret within the organization or management groups. Trade secrets typically involve information that gives a company a competitive edge over its competitors. The term 'confidential information' is occasionally used in the context of trade secrets, though there are subtle differences between these two terms. Confidential information refers to any type of information that an organization wants to keep secret, and it may or may not have commercial value. For example, financial records, employee's personnel data and results of internal surveys are considered to be confidential for an organization. However, a trade secret is a type of information that has economic value, and a competitor who possesses that information can use it to produce comparable products. Well-known examples of trade secrets include the Coca-Cola formula, L'Oreal skincare formula and Google's algorithm.

In general, the legal protection available for trade secrets is limited, unlike other forms of IP. Companies must make all possible efforts to prevent the information from being public. Unlike other forms of IP, many trade secrets are not even disclosed, and once the secret is revealed to the public (means to competitors), it is impossible to retrieve, and company may lose their edge in the market. Therefore, there is no fixed duration for the trade secret. As long as the company can keep the secret, it is protected.

It should be noted that trade secrets are often used to create significant marketing hype by corporate companies.

6.1.7 Industrial Design

An industrial design right protects the visual design of the products without considering the application of that product.

6.1.8 Geographical Indicators

A geographical indication is a sign used to highlight the origin of a product from a specific geographical region that possesses a reputation for superior quality for that kind of service or product. Some famous examples include Scotch whisky, Colombian coffee and Sabah tea.

6.1.9 Formulation of IP Strategy

The IP strategy has to be in line with the life cycle of a product. The speed of development in digital technologies allowed the information exchanges at a fast pace to a large audience. This makes the protection of IP more challenging than before.

Example 6.1. Protection of Intellectual Property

1. Classify the possible scientific developments in the following table into the following categories with clear justification. For the developments that are not patentable, suggest a way to protect the intellectual property. Also, propose whether any alternate plan exists for the patentable development (s), which may be more advantageous. If such a plan exists, explain clearly where the advantage lies.
   i) May not be patentable based on ethical issues
   ii) May be patentable but can upset the public perception of morality
   iii) Cannot be patentable because of other reasons
   iv) Can be patentable with no ethical issues

Scientific Development
1. New drug to treat malaria.
2. An extracted protein from the human body for medical applications.
3. A new kind of explosive suitable for military applications.
4. A new drug that can remove a wart (a small benign growth on skin) in a few hours.
5. A unique logo designed for a new product that can increase public attention to the product.
6. Exact solution of Navier-Stokes equation.
7. A combination of bacteria with beneficial properties.
8. A new set of operating conditions for a chemical reaction that increase the conversion to the desired product.
9. An essential oil extracted from coriander that can be used in perfumes with a unique fragrance.
10. A novel insecticide that is effective against termites.
11. A new molecule synthesized in the laboratory. It needs to be tested for antifungal properties.
12. A novel alloy with superior mechanical properties.

Solution

Scientific Development

1. New drug to treat malaria – II (may patent and then choose to license to the companies in poorer countries at a significantly reduced rate).
2. An extracted protein from the human body for medical applications-III (protein already exists – either keep it as trade secret or, if the structure is unique, can be synthesized and then patented).
3. A new kind of explosive suitable for military applications-I (cannot patent an application that's not beneficial to the world. Can negotiate with government on how to commercialize it).
4. A new drug that can remove a wart (a small benign growth on skin) in a few hours (IV- no issue).
5. A unique logo designed for a new product that can increase public attention to the product (III – logo cannot be patented, register the trademark).
6. Exact solution of Navier-Stokes equation (III – models can't be patented – may be published or develop an actual application with the model and then patent it).
7. A combination of bacteria with beneficial properties (III – bacteria already exist – keep it as a trade secret).
8. A new set of operating conditions for a chemical reaction that increase the conversion to the desired product (III – conditions cannot be patented – trade secret).
9. An essential oil extracted from coriander that can be used in perfumes with a unique fragrance (III – oil already exists – trade secret).
10. A novel insecticide that is effective against termite (IV).
11. A new molecule synthesized in the laboratory. It needs to be tested for antifungal properties (III – need to have a commercial application. Study the properties, and if it is found to have commercial values, it can be patented).
12. A novel alloy with superior mechanical properties (IV – no issue).

6.2 PROJECT ECONOMIC EVALUATION

Since the main aim of investing in a project is to earn profit, it is vital and essential that an economic analysis of the project is conducted to anticipate the total investment needed and assess the economic viability of the project (Towler and Sinnott, 2013). In the first step, the expected market size is estimated through a bottom-up approach. The conceptual estimation of the capital cost, total production cost and revenues of the project is carried out. Various economic analysis tools, such as cash flow diagram, return on investment (ROI), net present value (NPV), payback period and discounted cash-flow rate of return (DCFROR), are utilized to ensure that this project will be able to generate substantial profit for the company and shareholders. A cash flow diagram is used to present the forecast cumulative net cash flow over the lifespan of a project based on the best estimates of capital investment, operating costs and the sales volume and price. Next, the payback period is useful as it gives potential stakeholders a basic idea of the duration required for them to wait until they begin receiving a net positive return on their investments. If the payback period is shorter, then the investment is more attractive. The ROI of the project shows potential stakeholders the annual returns on their investment. In addition, the net present value of a project is the total amount of the present value of the future cash flows. Lastly, Discounted Cash Flow Rate of Return (DCFROR) is a financial tool used to determine the maximum discount rate at which a project's net present value (NPV) becomes zero at the end of its lifetime (Towler and Sinnott, 2013).

The specific steps followed in the economic analysis of chemical products are illustrated in Example 6.2.

6.2.1 Economic Analysis of Chemical Products

While the economic analysis tools used in chemical product design are similar to that of bulk chemicals, there are some important differences. Firstly, unlike bulk chemicals, the chemical products are not capital-intensive. The major expenses are R&D costs, marketing costs, cost of separation and purification and in some cases, raw material costs. It is to be noted that the cost of equipment, energy and utility costs may also be significant, but they are seldom the major cost component. In addition, many chemical products have significantly shorter life cycles than commodity chemicals (Cussler and Moggridge, 2011).

Another important aspect of chemical product economics is the final product cost. Unlike commodity chemicals, these prices are not fixed by the market. This is much more challenging for new products for which there are no current benchmarks available. There are several cost models developed by various researchers to guide the companies to set the most appropriate price (Bagajewicz, 2007). One of the most comprehensive models is shown in Equations 6.1 and 6.2.

$$p_1 d_1 = \left(\frac{\alpha}{\beta}\right)^{\rho} p_2 \left(\frac{Y - p_1 d_1}{p_2}\right)^{1-\rho} d_1^{\rho} \tag{6.1}$$

$$d_2 = \frac{Y - p_1 d_1}{p_2} \tag{6.2}$$

where p_1 is the product price, p_2 is the price of the current product in the market, d_1 and d_2 are the demands of the new and existing products, α is the level of awareness of the new product among the potential consumers, β is the previously defined preference score, and Y is the total budget of the consumers. ρ is known as elasticity, which is a parameter used to describe the shift from one product to another when price shifts.

Therefore, the economic analysis must focus on development costs, expected product costs, time available in the market, product demand over time and marketing costs. All of these factors involve significant risk, making risk quantification crucial for obtaining a realistic picture of profitability. To quantify risk, Monte Carlo simulation is one of the most comprehensive tools used in the industry. It performs risk assessment by running multiple iterations with randomly generated variable inputs, modelling uncertainty, predicting a range of possible outcomes and estimating the probability of different risk scenarios. The different stages involved in the economic analysis are shown in Figure 6.1.

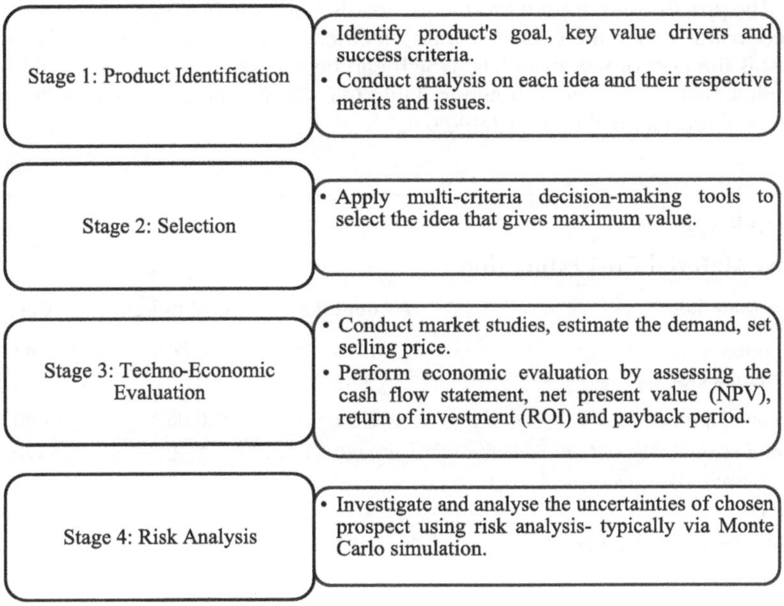

FIGURE 6.1 Stages involved in project costing.

Example 6.2. Economic Analysis of New Shampoo Product

Perform an economic analysis on the development of new shampoo product described in Example 5.6 of Chapter 5.

Solution

The first step is the estimation of market size. A bottom-up approach can be used to estimate an approximate market size (Table 6.1).

TABLE 6.1
Bottom-Up Approach

How many shops do you sell from?	550
How many people will go to 1 store every year?	43,800
How many of those will buy a shampoo?	8760
How many will buy your product?	438
How often will they buy?	0.33
Production rate (bottles per year)	722,700

The process used for the production involves the use of storage tanks and jacketed mixtures, and the utilities involve electricity and steam. Therefore, the major cost is the cost of raw materials, which can be estimated as shown in Table 6.2. Capital cost estimation is shown in Table 6.3. Finally, the economic potential of the new product is shown in Table 6.4.

TABLE 6.2
Raw Material Cost Estimation

Raw Materials	Amount (kg)	Cost ($/kg)	Cost ($)
Phenoxyethanol	4336	10.00	43,362
Potassium Sorbate	2602	5.00	13,009
Cocamidopropyl Betaine	8672	120.00	1,040,688
Sodium Lauryl Glycol Carboxylates with water (Polyglyceryl-4-Lauryl Ether)	69,379	7.50	520,344
Alpha-terpinene (Tea-tree oil extract)	8672	30.00	260,172
Deionized Water	303,534	0.00	0
Willow Bark Extract	13,009	51.90	675,208
Cocamide MEA	8672	2.40	20,772
Citric Acid	867	0.80	692
Polyquaternium-10	5203	15.97	83,082
Panthenol	8672	24.63	213,568
Total			2,871,000

TABLE 6.3
Capital Cost Estimation

Onsite cost	144,570
Offsite cost	65,050
Indirect cost	41,925
Fixed cost	260,225
Startup cost	26,025
Total capital cost	338,295
Working capital	50,745
Equipment life (yrs)	8
Cost of money (inflation & interest)	0
Capital cost per year	21,500

TABLE 6.4
Estimation of Economic Potential Sales Revenue = Production Volume × Price of Product/Bottle ($5) = $3,613,500

Sales revenue	3,613,500
Raw materials cost	2,871,000
Utility cost	2500
Capital cost	21,500
Economic potential	718,500

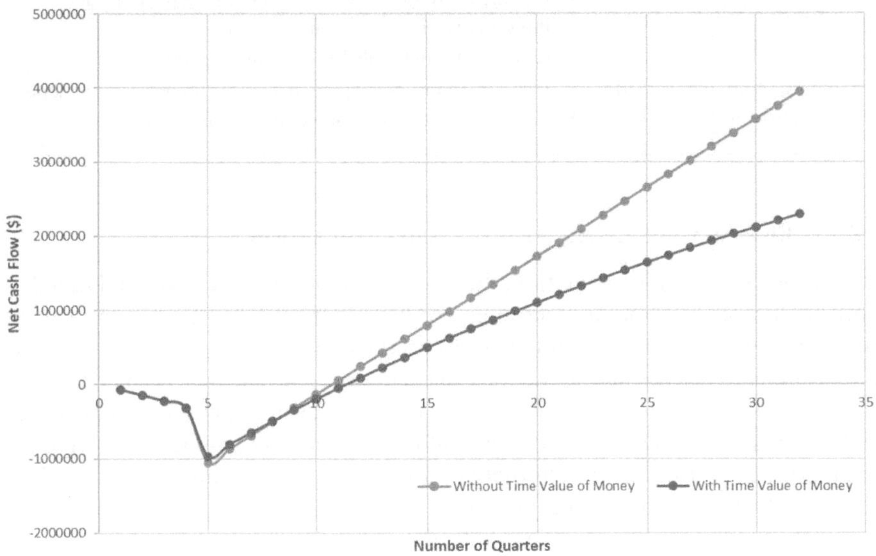

FIGURE 6.2 Cashflow diagram of new shampoo product.

The development cost of the product has been estimated at $72,000. Based on this information, the cash flow diagram is presented in Figure 6.2. The return on investment is only 7%, which is not considered attractive for a chemical product. Therefore, efforts should be made to increase sales beyond the initially anticipated value. Since a conservative approach has been adopted, there remains considerable room for improvement.

A preliminary risk analysis and management plan has been proposed, as shown in Table 6.5.

TABLE 6.5
Economic Risk Management

Risk	Mitigation
Fluctuations in raw materials price	• Set an agreement or contract with the supplier • Purchase raw materials in bulk • Find alternative suppliers
Unstable sales revenue caused by the change in consumer demand	• Find alternative datasets for forecasting market trends of the shampoo industry • Invest in the development of new products to diversify products and consumer base
Foreign currency fluctuations	• Currencies associated with risk should be assessed at the planning phase of the project • Identify both local and international clients to mitigate the severity of currency fluctuations • Obtain raw materials locally wherever possible
Unforeseen incidents leading to the production halt	• Invest in an insurance plan that covers unforeseen incidents • Invest in better safety preventive measurement

6.2.2 Risk Analysis Using Monte Carlo Simulation

In the project economic evaluation, the calculation of NPV is considered as a deterministic approach as the uncertainties are being ignored, indicating an ideal scenario. However, the parameters such as product sales price, production rate and sales volume are subjected to various uncertainties, which will further affect the actual cash flow of a project. For instance, the product sales price might be affected by discounts and allowances, government regulations and competition (Green and Southard, 2019). Hence, the alternative is to treat those parameters as stochastic variables. Instead of assuming just one value for the parameter, these stochastic variables are given in terms of probability distribution, which is a mathematical description of the likelihood for different values that a variable can depict. Monte Carlo simulation, an economic risk analysis tool that employs stochastic variables in its calculation, can be used for risk analysis. As various uncertainties are incurred in a project, a risk analysis is crucial to examine their impacts on the forecasts of the project viability. This allows a decision maker to examine and determine the progression of a project.

6.2.3 Introduction of Monte Carlo Simulation

The definition of Monte Carlo is the process of modelling and simulating a system affected by randomness (Brandimarte, 2014). Towler and Sinnott (2013) mentioned that Monte Carlo simulation is a more sophisticated method to perform the sensitivity analysis for a project compared to the use of simple method of statistical analysis. In the simple method, the mean and standard deviation of each parameter are estimated using Equations 6.3 and 6.4.

$$\textit{Mean Value}, \bar{x} = \frac{(H + 2ML + L)}{4} \tag{6.3}$$

$$\textit{Standard Deviation}, S_x = \frac{(H - L)}{2.65} \tag{6.4}$$

where ML is the most likely value, H is the upper value and L is the lower value. The upper and lower values are estimated using a range of variation. However, there is a likelihood of cost underestimating or overestimating using simple method, as not all parameters are normally distributed. Therefore, Monte Carlo Simulation, which assigns different parameters with different types of probability distribution functions, is more useful.

In a Monte Carlo Simulation, random numbers in the range from 0 to 1 are generated. To explain this, a frequency distribution curve example of the crude oil selling price is shown. There is a certain probability of occurrence for each crude oil price value, and the cumulative probability of that value being realized is a number in the range from 0 to 1. Similarly, the random values of other parameters can be generated in the same manner. All these parameters can be combined to give a random net present value (NPV) for a project. Lastly, a cumulative-probability curve can be plotted by conducting a large number of iterations, such as 10,000 to 50,000 times, for the evaluation of NPV (Green and Southard, 2019; Towler and Sinnott, 2013).

The general steps for conducting a Monte Carlo Simulation are depicted in Figure 6.3.

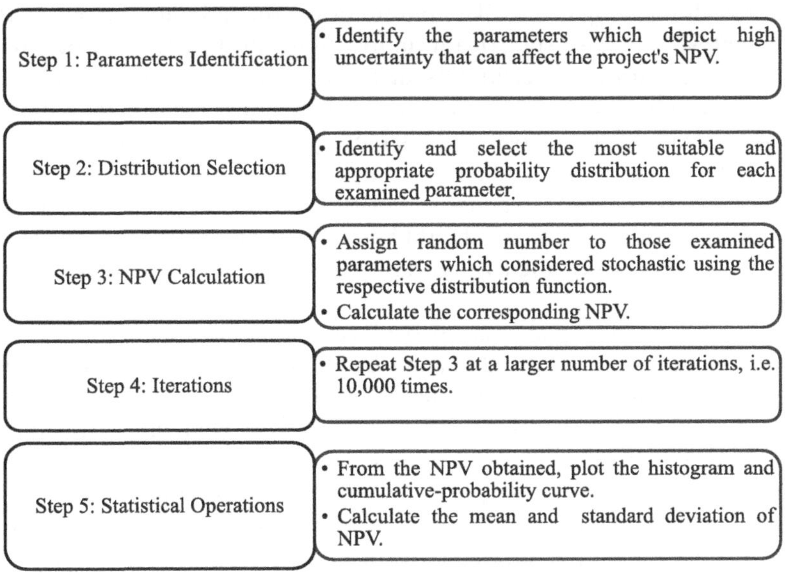

FIGURE 6.3 General steps of Monte Carlo simulation.

The mean of the large sample of NPV is an estimate of the expected value of the project, and this approach has been used by the oil industry since the early 1960s. Besides that, the companies are generally averse to risk. The risk can be characterized by the standard deviation of the NPV. For two projects with similar expected revenue, a risk-averse decision maker would choose the project with smaller standard deviation.

6.3 BUSINESS PLAN

After conducting the economic analysis, risk assessment and the intellectual property protection plan of the project, a business plan needs to be developed to convince the management about the economic feasibility of the project. A business plan is a formal document that outlines the project objectives, business idea, market analysis and organizational structure, including the credentials of the team, economic indicators, risk analysis and operational plans to guide its growth. This document must be written in such a way that it is attractive to an investor from a financial standpoint and also should be convincing and feasible from a scientific viewpoint (Cussler and Moggridge, 2011). Therefore, it should be understandable to non-technical readers but also should have sufficient technical information to convince technical critics.

For commodity chemicals, the key information needed are the return on investment (ROI), net present value (NPV), payback period (PBP) and discounted cash-flow rate of return (DCFROR), as these metrics can give a fairly accurate description of the economic potential of the project. Even for chemical product business, these metrics must be attractive. At the same time, attractive values of these metrics are rarely sufficient to convince the investors to proceed with the project. Unlike commodities, the assumptions made in economic analysis can have significant implications for the accuracy of forecasted profits. For example, uncertainties arise regarding whether consumers will accept the product, how large the market can grow and whether the selling price is acceptable. Therefore, the business plan must provide clear justifications for the assumptions, and a risk analysis considering uncertainties should be presented to estimate a realistic profit range and assess the probability of losses. At the same time, there should be compelling reasons for why investing in this project is worthwhile by highlighting the potential high rewards relative to the risks.

Finally, the business plan must also address potential competition, especially if the product is being launched in a market dominated by similar products. The risk analysis should include an assessment of the competition and mitigation plans outlining how the product will overcome these challenges. It is important to note that the first product in the market usually dominates sales, while new entrants typically capture only a fraction of the available market. However, the uniqueness of the new product can help address some of the shortcomings of existing products, potentially securing a larger market share. All of this information must be presented in the business plan in an assertive yet realistic manner.

6.4 CONCLUSIONS

This chapter introduces some of the preliminary steps in chemical product manufacturing. By this stage, the decision on what product to manufacture has already been made based on the target attributes. However, to succeed in the market, the product must be highly profitable, and the company must have a realistic understanding of the risks involved in its commercial production. Additionally, the company must consider protecting its intellectual property rights and ensuring that the product does not violate any existing intellectual property regulations. Given the short market lifespan and high risks, investors are generally interested in having complete control over the market for the new product. With these preliminary considerations in place, the product design team can develop a business plan to persuade investors to proceed with manufacturing. A well-prepared business plan with realistic market demand estimates and profitability analysis helps investors make informed decisions about moving forward with production.

REFERENCES

Bagajewicz, M. J. (2007). On the Role of Microeconomics, Planning, and Finances in Product Design. *AIChE Journal*, *53*, 3155–3170. https://doi.org/10.1002/aic.11332

Brandimarte, P. (2014). *Handbook in Monte Carlo Simulation: Applications in Financial Engineering, Risk Management, and Economics (Wiley Handbooks in Financial Engineering and Econometrics)*. Wiley.

Cussler, E. L., & Moggridge, G. D. (2011). Chemical Product Design. In *Cambridge Series in Chemical Engineering* (2nd ed.). Cambridge University Press. https://doi.org/10.1017/CBO9781139035132

Green, D. W., & Southard, M. Z. (2019). *Perry's Chemical Engineers' Handbook* (9th ed.). New York: McGraw-Hill Education.

Towler, G., & Sinnott, R. (2013). *Chemical Engineering Design: Principles, Practice and Economics of Plant and Process Design* (2nd ed.). Oxford: Butterworth-Heinemann; Elsevier.

6.5 PROBLEMS

1. A set of scientific developments are shown in Table Q6.1.

TABLE Q6.1
Potential Scientific Developments

Scientific Development
1. A new molecule synthesized in the laboratory. This molecule can be used as a medicine to treat irregular heartbeats with less discomfort to the patients.
2. A unique slogan developed for a product that can increase public attention.
3. A mixture of essential oils having a unique and attractive fragrance that can be recovered from the bark of a tree found in Amazon forests.
4. An artificial intelligence-based model to predict the optimal conditions that increase the yield of a reaction.
5. A new molecule with no known applications but may have biological activities.

Classify the developments in Table Q6.1 as patentable or non-patentable with clear justification. For the developments that are not patentable, suggest a way to protect the intellectual property or a way to help the company make money from the development.

2. A set of possible scientific developments are presented in Table Q6.1.

TABLE Q6.2
Scientific Developments to Be Considered for Patenting

Scientific Development
1. A new group contribution model that can accurately predict ionic liquid properties
2. An extract from a recently found plant that can alleviate the side effects of chemotherapy
3. A new recipe that can produce an ice cream with a unique flavour
4. A lightweight dialysis machine that can be carried easily by patients, which can improve the mobility of patients with poor kidney functions
5. A new molecule synthesized in the laboratory and currently being investigated for its biological activities

Which of the developments in Table Q6.1 are not patentable and why? For the developments that are not patentable, suggest a way to protect the intellectual property or a way to make those developments patentable.

3. Classify the possible scientific developments in Table Q6.3 into the following categories with clear justification. For the developments that are not patentable, suggest a way to protect the intellectual property. Also, propose whether any alternate plan exists for the patentable development(s), which may be more advantageous. If such a plan exists, explain clearly where the advantage lies.
   i) May not be patentable based on ethical issues
   ii) May be patentable but can upset the public perception of morality
   iii) Cannot be patentable because of other reasons
   iv) Can be patentable with no ethical issues

TABLE Q6.3
Scientific Developments to Be Considered for Patenting

Scientific development
1. A vaccine that can prevent common cold
2. A new set of operating conditions for a chemical reaction that increase the conversion to the desired product
3. A new drug that can cure tuberculosis (a lung disease common in developing countries and closely related to poverty) more effectively than current medicines
4. A new equation of state that can predict the behaviour of real gases at different temperature and pressure ranges
5. An essential oil extracted from coriander that can be used in perfumes with a unique fragrance

7 Scale Up

There are many occasions where engineers focus on identifying the size and operating conditions of small-scale vessels as well. This is because large-scale vessels already exist in the plant, and that equipment could be used for certain processes. In order to assess the behaviour of the system at a large scale, it is important to carry out testing at a small scale before large-scale production. So, a scaled-down version of the existing equipment may be used to perform laboratory-level testing to ensure optimum performance at the large scale.

The topic of scale-up is often overlooked in the field of chemical engineering because many design equations and methodologies are size invariants. For example, the number of stages in a distillation column and absorber depends only on the mole fractions and not flow rates. However, several chemical processes that involve the mass, heat or momentum transfer also depend on the scale in which they operate. In such cases, the behaviour of the system differs significantly between small and large scales of operation (Zolkarnik, 1991). Consequently, the major concerns during the scale-up or scale-down of a process are (1) which process parameters change during scale-up/scale-down, (2) how these changes can be quantified, (3) whether it is possible to maintain complete similarity at different scales and (4) if complete similarity cannot be maintained, what are the most important parameters to focus on during scale change (Zolkarnik, 1998).

It is evident that the key considerations for scale-up are rooted in dimensional analysis, through which the relationships between various physical quantities are delineated. Dimensional analysis aids in identifying dimensionless numbers pertinent to a system. These dimensionless numbers represent combinations of physical quantities that offer insights into the behaviour of a system. For example, Reynolds number gives a quantification of pressure forces relative to the shear forces by identifying the trade-offs between the impeller diameter and impeller speed with respect to the fluid properties. Thus, the scale-up process for a small-scale unit commences with the identification of the relevant dimensionless numbers for that particular system.

The fundamentals of scale-up are based on the concept of similarity. A rule has been developed by Zolkarnik (1991) in a famous book on scale-up of chemical processes to establish complete similarity between two systems. The rule has been stated as follows (Zolkarnik, 1991):

> Two processes may be considered completely similar if they take place in similar geometric space and if all the dimensionless numbers necessary to describe them have the same numerical value.

Hence, the aim during scale-up is to uphold similar geometric proportions and keep the pertinent dimensionless numbers constant, thereby ensuring that the performance achieved at the small scale can be replicated in the industrial-scale process.

DOI: 10.1201/9781032708911-7

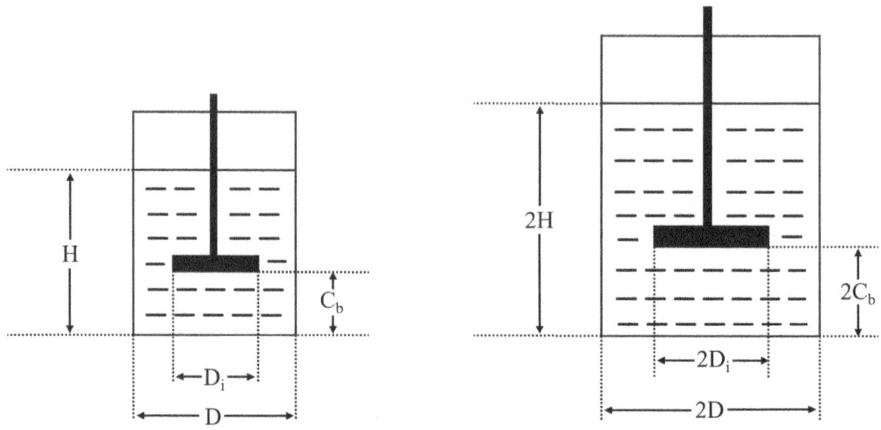

FIGURE 7.1 Geometrically similar vessels.

In the rest of the sections, the strategies to meet these two requirements will be discussed for various types of processes.

7.1 GEOMETRIC SIMILARITY

Typically, the scale-up of chemical processes is conducted in geometrically similar vessels.

Two systems are geometrically similar if all corresponding dimensional ratios are the same in both systems. This can only be applied to similar kinds of systems that differ only in size. Therefore, the scale-up of a laboratory-scale vessel to a pilot plant-scale vessel is only possible when the laboratory scale uses an exact similar system. All linear dimensions of the small-scale equipment must be related to the corresponding dimensions of the production scale equipment by a constant scale-up factor, as shown in Figure 7.1. It can be seen that a constant scale-up factor of two is maintained in all the ratios.

7.2 DIMENSIONAL ANALYSIS

Dimensional analysis can be used to establish a relationship between variables and simplify problems. A number of process variables are represented in terms of dimensionless numbers. Different dimensionless numbers have different physical interpretations. During the scale-up, we try to keep the important parameters constant to get similar outputs.

Dimensional analysis is conducted based on two theorems, which are dimensional homogeneity and Buckingham Pi theorem.

7.2.1 Dimensional Homogeneity Theorem

Any physical quantity is dimensionally a power law monomial – [Any Physical Quantity] = $M^aL^bT^cQ^d$.

7.2.2 Buckingham Pi Theorem

If a system has N physical quantities of relevance that depend on P independent dimensions, then there is a total of N-P independent dimensionless products p_1, $p_2, \ldots, p_{N-P}$. The behaviour of the system is describable by a dimensionless equation.

$$f(p_1, p_2, \ldots, p_{N-P})=0; \; p_1= f(p_2, \ldots, p_{N-P})$$

The following procedure can be used to reduce a system of **N** variables in terms of the dimensionless numbers:

1. List all N variables involved in the problem and express each variable in terms of basic dimensions.
2. Determine the required number of dimensionless parameters (N – P).
3. Select P number of repeating variables. All dimensions must be included in this set, and each repeating variable must be independent of the others.
4. Form a dimensionless parameter π by multiplying one of the non-repeating variables by the product of the repeating variables, each raised to an unknown exponent.
5. Solve for the unknown exponents.
6. Repeat this process for each non-repeating variable.
7. Express result as a relationship among the dimensionless parameters – $\mathbf{F}(\pi_1, \pi_2, \pi_3, \ldots) = \mathbf{0}$.

These steps have been illustrated in Example 7.1.

Example 7.1 Dimensional Analysis

Consider a liquid mixture being produced in an agitated vessel, and we are developing a model to estimate the power of the agitator. It was hypothesized that the power is a function of the diameter of the agitator, speed of agitator and the physical properties of the liquid – mainly the density and viscosity of the liquid. Based on this hypothesis, develop a model to estimate the power of the agitator in a mixing tank.

Solution

Buckingham's theorem states that the total number of these relevant dimensional parameters (N) can be grouped into N-P independent dimensionless groups.

Primary dimensions

Power, P = Work × time^{-1} = (force × displacement) × time^{-1} = $[ML^2T^{-3}]$

Diameter of the agitator, D = [L]

Agitator speed, N = [1/t]

Viscosity. μ = [M/(Lt)]

Density, ρ = $[M/L^3]$

There are five variables (P, D, N, μ and ρ) and three primary dimensions (M, L and t) involved. Therefore, the number of dimensionless groups = N–P = 2. Therefore, the general form of the equation can be written as shown in Equation 7.1.

$$f(\pi_1) = \pi_2 \tag{7.1}$$

Now, we need to select P-dimensional parameters that collectively include all the primary dimensions. We will select three since we have three primary dimensions involved in the problem. We will select ρ, N and D for this example.

The two dimensionless groups can be formed by making each of the non-repeating variables (P and μ) into dimensionless form. We start with P:

$$\pi_1 = \rho^a N^b D^c P \tag{7.2}$$

In order to make π_1 dimensionless, a, b & c exponents are needed to take the values that can non-dimensionalize the group.

$$\left[M/L^3\right]^a [1/t]^b [L]^c \left[ML^2T^{-3}\right] = M^0L^0t^0$$

According to the dimensional homogeneity theorem, the dimensions must be the same on both sides of the equation. Therefore, we need to match the powers of the same dimension.

$$M: a + 1 = 0$$

$$L: -3a + c + 2 = 0$$

$$t: -b - 3 = 0$$

By solving these three equations, we get a = -1, b = -3, c = -5

Therefore,

$$\pi_1 = \rho^{-1}N^{-3}D^{-5}P = \frac{P}{\rho D^5 N^3} \tag{7.3}$$

Similarly, $\pi_2 = \rho^a N^b D^c \mu$

$$\left[M/L^3\right]^a [1/t]^b [L]^c \left[M/(Lt)\right] = M^0 L^0 t^0$$

$$M: a + 1 = 0$$

$$L: -3a + c - 1 = 0$$

$$t: -b - 1 = 0$$

By solving these three equations, we get a = -1, b = -1, c = -2

Therefore,

$$\pi_2 = \rho^{-1}N^{-1}D^{-2}P = \frac{\mu}{\rho D^2 N} \tag{7.4}$$

Now, Equation 7.1 can be used to relate the variables in this problem in the dimensionless form, as shown in Equation 7.5.

$$f\left(\frac{P}{\rho D^5 N^3}\right) = \frac{\mu}{\rho D^2 N} \tag{7.5}$$

The general form of the equation has been shown in Equation 7.6.

$$\frac{P}{\rho D^5 N^3} = k\left(\frac{\mu}{\rho D^2 N}\right)^n \tag{7.6}$$

It can be noticed that the five variables have been represented in terms of two dimensionless numbers. The term on the LHS of Equation 7.6 is the well-known power number, whereas the RHS of Equation 7.6 is the inverse of Reynolds number. The constant k and n must be estimated for the particular type of agitator.

It is possible that the dimensionless forms may be obtained with no obvious physical significance. Moreover, the incorrect selection of variables or the omission of relevant variables may lead to unreliable models. In addition, the estimation of the constants and powers requires experimental analysis. Despite these limitations, dimensional analysis provides a powerful tool to simplify the representation of engineering problems and guide towards the scaling up/down of process equipment.

7.3 SCALE-UP STRATEGY

Usually, the scale-up process begins by estimating the dimensions of the vessel that maintain geometric similarity. There can be different combinations of parameters that make the same dimensionless groups. The physical properties of the fluids usually remain the same during the scale-up. Once the dimensions of the system are fixed to maintain geometric similarity, the rest of the variables need to be estimated to maintain the relevant dimensionless numbers constant. Therefore, one of the key steps at this point is the identification of the most relevant dimensionless numbers or the process. Table 7.1 summarizes some of the most commonly used dimensionless groups in scale-up and their physical significance.

TABLE 7.1
Some Common Dimensionless Groups

Dimensionless Number	Equation	Significance
Reynolds number	$Re = \frac{vD\rho}{\mu}$	Ratio of inertial forces to viscous forces
Euler number	$Eu = \frac{\Delta P}{\rho v^2}$	Ratio of pressure forces to inertial forces
Froude number	$Fr = \frac{v^2}{lg}$	Ratio of inertial forces to gravitational forces
Weber number	$We = \frac{\rho v^2 l}{\sigma}$	Ratio of inertial forces to surface tension forces
Power number	$Np = \frac{P}{\rho N^3 D^5}$	Ratio of resistance force to inertial force for impellers

Example 7.2. Scale-Up of Flow Reactor

A pilot-scale flow reactor that carries an aqueous solution of reagents is a smooth circular pipe with a diameter of 2.5 cm and a length of 30 cm. It is required to scale this reactor up by a factor of 500. If the flow velocity to the pilot scale vessel is 10 m/s, estimate the vessel dimensions, flow velocity and pressure drop for the full-scale reactor. The pressure drop in a frictionless circular pipe can be estimated from Equation 7.7.

$$\frac{\Delta p}{\rho v^2} = 64\left(\frac{\mu}{\rho v D}\right)\left(\frac{L}{D}\right) \tag{7.7}$$

Solution

For scale-up, we need similar geometrical space and same numerical values for relevant dimensionless numbers. So, the first step is to find the dimensions of the pilot scale vessel:

$$\text{Scale-up factor} = 500 \Rightarrow \frac{(Volume)_2}{(Volume)_1} = 500$$

$$\text{Volume, V} = \frac{\pi}{4} D^2 L$$

Since the geometric similarity is maintained, L/D will be a constant for all sizes. If 'm' is the L/D ratio:

$$V = \frac{\pi m}{4} D^3$$

$$\frac{(Volume)_2}{(Volume)_1} = 500 = \left(\frac{D_2}{D_1}\right)^3 = \left(\frac{L_2}{L_1}\right)^3$$

$$L_1 = 30 \text{ cm and } D_1 = 2.5 \text{ cm}$$

Therefore, $D_2 = D_1 \sqrt[3]{500} = 7.94 \times 2.5 = 19.8\ cm$

$$L_2 = L_1 \sqrt[3]{500} = 238\ cm$$

By maintaining geometric similarity, L/D ratio is already maintained constant during the scale-up. Now, $\frac{\Delta P}{\rho v^2}$ and $\frac{\mu}{\rho v D}$ needs to be maintained constant. Since the physical properties would not change during scale-up, these conditions can be shown as:

$$(vD)_1 = (vD)_2 \tag{7.8}$$

$$\left(\frac{\Delta p}{v^2}\right)_1 = \left(\frac{\Delta p}{v^2}\right)_2 \tag{7.9}$$

From (7.8), $V_2 = 1.26$ m/s

$$\frac{\Delta P_2}{\Delta P_1} = \frac{V_2^2}{V_1^2} = \frac{1.26^2}{10^2} = \frac{1}{63}$$

From (7.9),

The lab-scale vessel pressure drop is 63 times higher than the full-scale. This is because the higher flow velocity in the lab-scale tube leads to higher friction, requiring a higher pressure drop for the flow.

7.4 SCALE-UP PRINCIPLES

There are different kinds of similarities that need to be maintained during the scale-up. Those can be classified as follows:

1. Geometric similarity
2. Mechanical similarity
3. Thermal similarity
4. Chemical similarity

However, not all these similarities are relevant for all the systems. Moreover, it is not possible to maintain all these similarities during scale-up. Therefore, one of the key decisions needed for a successful scale-up is to choose the most important similarities to maintain constant. These decisions are usually based on literature and experience, based on the fundamental mechanisms which may be explored through experiments, from mass and energy balances and from constitutive equations that describe the response of system to external forces.

Example 7.3. Stirred Tank Scale-Up

A mixing experiment has been performed in a laboratory-scale stirred tank with a volume of 100 L. The equivalent industrial unit has a volume of 30 m³. Develop a strategy for the scale-up and illustrate the changes to the other variables and dimensionless numbers during scale-up. Point out the major issues we may face during this scale-up.

The typical scale-up targets in a stirred tank and the approach towards scale-up are given in Table 7.2.

In this scale-up, our main concern is to maintain the similar micro-scale turbulence. Therefore, the (P/V) ratio must be maintained constant during scale-up. Equation 7.6 provided an expression to relate power to some of the design variables. According to Equation 7.6, the power number must be constant during the scale-up:

$$\frac{P}{\rho N^3 D^5} = \text{Constant} \Rightarrow P \propto \rho N^3 D^5$$

Since the density is constant and volume, V is proportional to D^3.

TABLE 7.2
Scale-Up Targets and Approach in Stirred Tanks Scale-Up

Target	Approach
Maintain similar micro-scale turbulence	Constant power per unit volume (P/V)
Limit shear damage to sensitive fluids	Constant tip speed (πDN)
Maintain the same flow characteristics	Constant Reynolds number
Constant mixing time or reaction time	Constant average rate of shear (N) (mixing time = N x t) (dimensionless)

$$\frac{P}{V} \propto \frac{N^3 D^5}{D^3} \propto N^3 D^2$$

$$N_1^3 D_1^2 = N_2^3 D_2^2$$

We can estimate the impeller diameter of the large vessel from geometric similarity:

$$Volume \propto D^3 \propto D_i^3$$

$$\frac{30\ m^3}{0.1\,m^3} = \left(\frac{D_{i2}}{0.06\ m}\right)^3$$

$$D_{i2} = 0.4\text{ m} = 40\text{ cm}$$

Now, the impeller speed required to maintain the constant (P/V) ratio has been estimated as:

$$N_2 = \sqrt[3]{\frac{N_1^3 D_1^2}{D_2^2}} = 300\left(\frac{6}{40}\right)^{\frac{2}{3}} = 85\ rpm$$

It is necessary to confirm that the flow is turbulent in both scales. For that, Re>4000.

Reynolds number for the lab-scale vessel:

$$Re = \frac{N\rho D^2}{\mu} = \frac{\left(\frac{300}{60}\right)\times 10^3 \times (0.06)^2}{10^{-3}} = 18000$$

Reynolds number for the large-scale vessel:

$$Re = \frac{N\rho D^2}{\mu} = \frac{\left(\frac{85}{60}\right)\times 10^3 \times (0.4)^2}{10^{-3}} = 226{,}666$$

Both the flows are turbulent. We have also found that the Reynolds number changes significantly during the scale-up.

To see the change of impeller tip speed:

$$\frac{(\pi dN)_1}{(\pi dN)_2} = \frac{6\times 300}{40\times 85} = 0.53$$

The impeller tip speed is almost doubled during scale-up.

To see the change of power during scale-up:

$$\frac{P_1}{P_2} = \frac{(N^3D^5)_1}{(N^3D^5)_2} = 3.34\times10^{-3}$$

The power requirement increased by 300 times during the scale-up.

We have verified that it is not possible to maintain all the dimensionless numbers and scale-up criteria constant during the scale-up process. Therefore, choosing the most appropriate scale-up criteria is the most important decision to be made during the scale-up process. In addition, how the rest of the variables have been affected by the scale-up needs to be evaluated before finalizing the design.

7.5 GAS DISPERSION AND IMPELLER FLOODING

Mechanically agitated vessels are widely used in fermentation and chemical industry. One of the most frequently used impellers in these vessels is the Rushton turbine, well known for its efficient gas dispersion and mixing capabilities. Rushton turbine generates a radial flow pattern in non-aerated broth. In an aerated tank, it can manage relatively high gas flow rates without impeller flooding. Nonetheless, if the gas flow rate were too high, flooding may occur. Flooding is undesirable because it significantly reduces micro-mixing and mass transfer between gas and liquid, primarily due to the extremely uneven distribution of gas in the vessel (Warmoeskerken and Smith, 1985). Thus, extra attention has to be paid to the flow regimes such as flooding, loading and completely dispersion when gas and liquid are mixed together. Please do note that gas can be dispersed by impeller or stirrer, but it does not mean that it is completely dispersed. Figure 7.2 illustrates the flow regimes produced in an aerated tank stirred with a Rushton turbine by either manipulating the gas flow rate, F_g, or impeller speed, N_i. Flooding occurs either at a low impeller speed or when the system is aerated with too high gas flow rate. Likewise, complete gas dispersion can be achieved at a higher impeller speed or low gas flow rate.

To predict whether an aerated tank stirred with Rushton turbine will be flooded or fully dispersed, correlations have been developed for low viscosity fluid. The correlations were determined using ring sparger and are expressed through dimensionless groups, namely the gas flow number FL_g and the Froude number F_r.

$$FL_g = \frac{F_g}{N_i D_i^3} \tag{7.10}$$

$$Fr = \frac{N_i^2 D_i}{g} \tag{7.11}$$

where F_g represents gas flow rate, N_i represents impeller speed, D_i represents impeller diameter and g represents gravitational acceleration.

To prevent flooding when using a Rushton turbine in a vessel, the condition shown in Equation 7.12 must be satisfied (Nienow et al., 1985). Equation 7.12 was developed for vessel with a diameter between 0.29 and 1.2 m and D_i / D_T a ratio between 0.33 and 0.5.

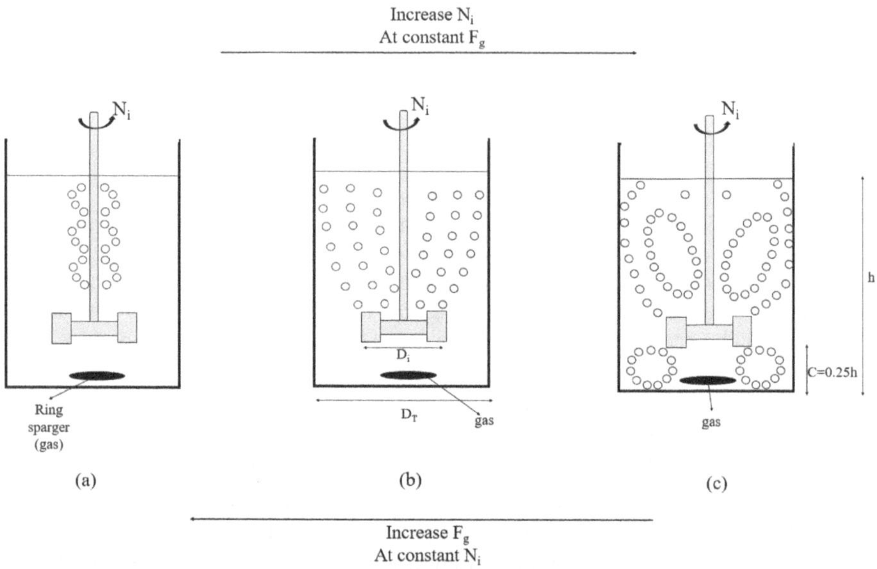

FIGURE 7.2 Flow regimes in an aerated tank stirred with a Rushton turbine: (a) flooding, (b) loading and (c) completely gas dispersion.

$$FL_g \leq 30\left(\frac{D_i}{D_T}\right)^{3.5} Fr \tag{7.12}$$

where D_T represents the vessel diameter. In addition to assessing if flooding will occur, Equation 7.12 also indirectly aids in determining the minimum impeller speed required to prevent flooding.

To ensure that complete gas dispersion occurs in an aerated tank stirred with Rushton turbine, one must satisfy the following condition (Nienow, 1990):

$$FL_g \leq 0.2\left(\frac{D_i}{D_T}\right)^{0.5} Fr^{0.5} \tag{7.13}$$

Any value of FL_g higher than the value obtained from Equation 7.13 results in incomplete gas dispersion. Note that Equation 7.13 applies for an impeller off-bottom clearance to liquid height ratio, *C*/*h*, of 0.25.

Example 7.4. Bioreactor Conditions

A bioreactor of diameter and liquid height of 1.5 m is employed to produce vitamin C using *Acetobacter* sp. The bioreactor is fitted with a Rushton turbine of diameter 0.5 m and off-bottom clearance 0.375m. To ensure aerobic growth of *Acetobacter* sp., air is supplied at a rate of 0.5 m^3/min to provide oxygen to the culture. It was proposed to first operate the Rushton turbine at 200 rpm. Under these operating conditions, evaluate if flooding will occur.

Solution

The first step is to calculate the gas flow number FL_g and the Froude number Fr using Equations 7.10 and 7.11.

$$FL_g = \frac{F_g}{N_i D_i^3} = \frac{\left(0.5\frac{m^3}{min}\right)}{(200\ rpm)(0.5\ m)^3} = 0.02$$

$$Fr = \frac{N_i^2 D_i}{g} \frac{\left(200\ min^{-1} \times \frac{1\ min}{60s}\right)^2 (0.5\ m)}{9.81\frac{m}{s^2}} = 0.5663$$

The step is then followed by checking if condition from Equation 7.12 is satisfied.

$$FL_g = 30\left(\frac{0.5}{1.5}\right)^{3.5}(0.5663) = 0.363$$

Since the value of FL_g calculated using Equation 7.10 is 0.02, which is less than 0.363, flooding will not occur.

Although we know that flooding can be avoided, it is also important to determine if complete gas dispersion can be achieved under this operating condition. Apply Equation 7.13 to evaluate if the condition can be satisfied.

$$FL_g = 0.2\left(\frac{0.5}{1.5}\right)^{0.5} 0.5663^{0.5} = 0.0869$$

Complete gas dispersion can be achieved since the value of FL_g calculated using Equation 7.10 is 0.02, which is less than 0.0869.

Example 7.5. Bioreactor Scale-Up

A pharmaceutical product was produced in a pilot-scale fermenter with liquid height equal to the vessel diameter of 0.5 m. A Rushton turbine with diameter one-third the vessel diameter is fitted 0.125 m from the bottom of the fermenter. The stirring speed is set at 185 rpm. Since the pilot-scale studies were successful, it is proposed to build a larger production-scale fermenter with a capacity of 6 m^3. The production-scale fermenter is assumed to be geometrically similar to the pilot-scale fermenter. Air is supplied at a rate of 2 m^3/min to provide oxygen to the culture. The density and viscosity of the fermentation broth are 1000 kg m^{-3} and 5 x 10^{-4} kg m^{-1} s^{-1}, respectively.

a) Determine the impeller speed (rpm) in the production-scale fermenter if constant power per unit volume (P/V) must be maintained during scale-up.
b) Analyse if air can be fully dispersed in the fermenter.

Solution

a) Assumption

Density of the broth, the h/D_T and D_i/D_T ratios are kept constant for the scaled-up fermenter.

Determine the flow regime in pilot-scale fermenter using Reynolds number:

$$Re = \frac{\rho N_i D_i^2}{\mu} = \frac{1000(185/60)(0.167)^2}{5\times10^{-4}} = 1.72\times10^5$$

Since the flow is turbulent, power number is a constant. Thus, the equation to scale up using constant P/V can be simplified to:

$$N_{i1}^3 D_{i1}^2 = N_{i2}^3 D_{i2}^2$$

We can first estimate the impeller diameter of the production-scale fermenter from geometric similarity:

The pilot-scale fermenter:

$$Volume = \pi\left(\frac{D_T}{2}\right)^2 h = \pi\left(\frac{0.5}{2}\right)^2 (0.5) = 0.0982\ m^3$$

$$(D_T)_1 = h_1 = 0.5\ m$$

$$(D_i)_1 = \frac{0.5}{3} = 0.167\ m$$

The production-scale fermenter:

$$Volume = 6\ m^3 = \pi\left(\frac{h}{2}\right)^2 \times h$$

$$h_2 = 1.97\ m$$

$$(D_T)_2 = h_2 = 1.97\ m$$

$$(D_i)_2 = \frac{1.97}{3} = 0.656\ m$$

Now, the impeller speed required to maintain the constant (P/V) ratio has been estimated as follows:

$$N_{i2} = \sqrt[3]{\frac{N_{i1}^3 D_{i1}^2}{D_{i2}^2}} = 185\left(\frac{0.167}{0.656}\right)^{\frac{2}{3}} = 74.3\ rpm$$

b) The first step is to calculate the gas flow number FL_g and the Froude number F_r using Equations 7.10 and 7.11.

$$FL_g = \frac{F_g}{N_i D_i^3} = \frac{\left(2\frac{m^3}{min}\right)}{(74.3\ rpm)(0.656\ m)^3} = 0.095$$

$$Fr = \frac{N_i^2 D_i}{g} = \frac{\left(\frac{74.3}{min}\times\frac{min}{60\ s}\right)^2 (0.656\ m)}{9.81\frac{m}{s^2}} = 0.10$$

The step is then followed by checking if condition from Equation 7.13 is satisfied.

$$FL_g \leq 0.2\left(\frac{0.656}{1.97}\right)^{0.5} 0.10^{0.5} = 0.036$$

Since the value of FL_g calculated using Equation 7.10 is 0.095, which is more than 0.036, the air will not be fully dispersed in the bioreactor. To ensure complete gas dispersion, one can either increase the impeller speed or reduce the gas flow rate.

From the calculation, we know that the value of FL_g must be equal to or less than 0.036 to achieve complete gas dispersion. Thus, it is possible to calculate the air flow rate that can be completely dispersed by the impeller by substituting the value of FL_g obtained from Equation 7.13 into Equation 7.10.

$$F_g = FL_g N_i D_i^3 = 0.036 \times 74.3\ rpm \times 0.656^3 = 0.755\ m^3 / min$$

The air flow rate that can be completely dispersed by the impeller is 0.755 m^3/min. To ensure the gas will be fully dispersed at impeller speed of 74.3 rpm, it is recommended to sparge the fermenter with air flow rate equal to or lower than 0.755 m^3/min.

7.6 NON-GEOMETRIC SCALE-UP

One of the disadvantages of the scale-up based on geometric similarity is that they can guarantee similarity in performance only when the geometric similarity is strictly maintained. A truly geometrically similar scale-up is not an option in many industrial processes. However, even minor changes in the deviation can have major differences in the mixing pattern (Cullen et al., 2015). In non-geometric scale-up, define the most important process variable to be kept constant and estimate the next variable.

A typical non-geometric scale-up begins with a geometrically similar scale-up and makes changes one at a time. After the geometric scale-up, the variables are changed one at a time and the effect of that change is analysed. For example, after a geometric scale-up, the height-to-diameter ratio of a vessel may be altered. This step is especially relevant if an existing vessel is continued to be used for a different volume of production. In that case, the change in volume is achieved by the change in height (by increasing or decreasing the liquid level), while the diameter remains unaltered. The non-adherence to geometric similarity will allow an additional dimension to adjust during the scale-up. However, it is important to make sure that the relevant dimensionless numbers are in the acceptable range after the scale-up.

An illustration of non-geometric scale-up has been presented in Example 7.7 (c).

7.7 SCALE-UP OF AGITATED VESSELS WITH HEAT TRANSFER

In several chemical processes, the temperature control can be the primary consideration during scale-up. However, processes such as distillation and absorption are governed by the thermodynamic laws, which are size invariant (Zolkarnik, 1991). At the same time, there are processes where the primary consideration is the temperature control when the process is performed at a large scale. In such cases, we need to incorporate an additional dimension of temperature into the design space. Therefore, the dimensionless numbers such as the Grashof number and Rayleigh number can be relevant during the scale-up process.

If a chemical process taking place in an agitated vessel is heat transfer controlled, the scale-up focuses on the rate-controlling step. For heat transfer, the driving force is the temperature difference, and the resistance is represented in terms of the inverse of heat transfer coefficients as shown in Equation 7.14.

$$Q = \frac{Driving\ force}{overall\ resistance} = \frac{(T_h - T_c)}{\left(\frac{1}{UA}\right)} \tag{7.14}$$

In Equation 7.14, T_h andT_c refers to the temperature driving force, A is the heat transfer area and U is the overall heat transfer coefficient. The overall heat transfer coefficient can be estimated from the local heat transfer coefficient using Equation 7.15.

$$\frac{1}{U} = \frac{1}{h_h} + \frac{L}{k} + \frac{1}{h_c} \tag{7.15}$$

where L is the vessel thickness, k is the thermal conductivity of the vessel material, h_h is the heat transfer coefficient on the hot side of the vessel and h_c is the heat transfer coefficient on the cold side of the vessel. It can be noted that the terms $\frac{1}{U}$, $\frac{1}{h_h}$, $\frac{L}{k}$ and$\frac{1}{h_c}$ quantifies the resistance to heat flows. In Equation 7.15, we ignore the resistance to heat transfer due to the fouling factors. Therefore, we can see that the overall resistance to heat transfer is contributed by the resistance on convection to the vessel wall, convection from the vessel wall and conduction through the wall. In most cases, one of these resistances would dominate the overall resistance. In other words, typically the numerical value of one of the resistances would be significantly higher than the rest, and thus the overall resistance can be approximated by the dominating resistance. Therefore,

$$\begin{aligned}
&\frac{1}{h_h} \gg \frac{L}{K} \text{ and } \frac{1}{h_c}; \frac{1}{U} \cong \frac{1}{h_h} \text{ Convection control by hot fluid} \\
&\frac{1}{h_c} \gg \frac{L}{K} \text{ and } \frac{1}{h_h}; \frac{1}{U} \cong \frac{1}{h_c} \text{ Convection control by cold fluid} \\
&\frac{L}{K} \gg \frac{1}{h_c} \text{ and } \frac{1}{h_h}; \frac{1}{U} \cong \frac{L}{K} \text{ Conduction control by wall}
\end{aligned} \tag{7.16}$$

The conditions described in Equation 7.16 can be used during the successful scale-up of a process dominated by heat transfer. For successful scale-up,

1) Determine the rate-controlling resistance, and use this to make decisions on scale-up criteria.

2) Check whether the same resistance remains rate-controlling at the new scale.

In most cases, the vessel materials have high thermal conductivity, which makes the conduction resistance significantly lower than the other two resistances. Therefore, the focus on the scale-up in most cases is to determine the magnitude of the h_c and h_h to develop the scale-up strategy. The factors and variables associated with the heat transfer coefficients are as follows:

1) Fluid properties
   Density, viscosity, specific heat capacity and thermal conductivity
2) System geometry
   Vessel diameter, tube diameter, impeller diameter
3) Flow characteristics
   Fluid superficial velocity, or agitator speed for stirred tanks

Therefore, the general heat transfer coefficient, h, can be represented as a function of the relevant physical properties and design variables. The typical quantities are viscosity of the fluid, density of the fluid, thermal conductivity, heat capacity, impeller speed and the impeller diameter of the vessel. For the jacket side fluid, the impeller speed and impeller diameter have to be replaced with the flow velocity of the jacket side fluid and the equivalent diameter of the jacket.

$$h = f\left(\rho, \mu, k, c_p, N, D, D_i\right) \text{ For the tank side fluid}$$

$$h = f\left(\rho, \mu, k, c_p, V_j, D, D_j\right) \text{ For the jacket side fluid}$$

The dimensions of these physical properties and design variables are shown in Table 7.3.

TABLE 7.3
Fundamental Dimensions in Heat Transfer Processes

Property/Variable	Dimensions
Heat transfer coefficient	$MT^{-3}\theta^{-1}$
Thermal conductivity	$ML^{-1}T^{-3}\theta^{-1}$
Heat capacity	$L^2T^{-2}\theta^{-1}$
Density	ML^{-3}
Viscosity	$ML^{-1}T^{-1}$
Impeller speed	T^{-1}
Diameter of the vessel	L
Equivalent diameter	L
Flow velocity	LT^{-1}
Impeller diameter	L

By following the procedure in Section 7.2, a dimensionless equation can be developed as shown in Equation 7.17.

$$\frac{hD}{k} = C\left(\frac{VD\rho}{\mu}\right)^a\left(\frac{c_p\mu}{k}\right)^b \tag{7.17}$$

In Equation 7.17, $\frac{hD}{k}$ is Nusselt number, Nu, and $\frac{c_p\mu}{k}$ is Prandtl number, Pr. The value of C depends upon type of system, type of agitator, number of baffles and baffle types, nature of fluid flow, heating/cooling and the mode of heat transfer.

Equation 7.17 can be modified for the tank side and jacket side of the vessel, as shown in Equations 7.18 and 7.19.

$$\frac{h_tD}{k} = C_1\left(\frac{ND_i^2\rho}{\mu}\right)^a\left(\frac{c_p\mu}{k}\right)^b \tag{7.18}$$

$$\frac{h_jD_j}{k} = C_2\left(\frac{VD_j\rho}{\mu}\right)^a\left(\frac{c_p\mu}{k}\right)^b \tag{7.19}$$

In Equations 7.18 and 7.19, h_t and h_j refer to the tank side and jacket side heat transfer coefficients. The values of a and b are generally $\frac{2}{3}$ and $\frac{1}{3}$ (Kayode Coker, 2001).

For scale-up, the heat transfer coefficients are estimated for the small scale from Equations 7.18 and 7.19, and the dominant heat transfer coefficient is identified. During the scale-up, the parameters are set to ensure that the dominating heat transfer coefficient is maintained constant. In most cases, the process is dominated by the tank side heat transfer coefficient, h_t, which can be written as a function of the tank dimensions and the impeller speed from Equation 7.19.

$$h_t\alpha\frac{1}{D}\left(ND_i^2\right)^a \tag{7.20}$$

Since $D \propto D_i$,

$$h_t\alpha N^a D_i^{2a-1} \tag{7.21}$$

Once the scale-up is performed, the heat transfer system needs to be updated to meet the requirements of the new vessel. For example, in reactor with an exothermic process, the heat is generated in the volume of the vessel and heat is removed from the surface of the vessel. The volume of the vessel is a function of D^3, while the area is a function of D^2. Therefore, the heat removal is only possible with an increased surface area.

Similarly, if the process is dominated by the jacket side heat transfer coefficient:

$$h_j\alpha V^a D_j^{a-1} \tag{7.22}$$

The scale-up of a typical system that involves heat transfer control has been illustrated in Example 7.6.

Example 7.6. Scale-Up of Heat Transfer-Controlled Agitated Vessel

A laboratory-scale jacketed vessel has a diameter of 25 cm, and the agitator is a paddle measuring 20 cm in diameter and revolving at 100 rpm. The jacket is fitted to the cylindrical section only. The spacing between the jacket and vessel walls is 20 mm. The jacket is fitted with a spiral baffle. The pitch between the spirals is 100 mm. The coolant velocity is 5 m/s. To scale up this vessel, how should we proceed if temperature control is the predominant consideration?

The following data may be useful:

Equations for estimating the tank side heat transfer coefficients (Kayode Coker, 2001).

Agitator Type	Range of Reynolds Number	Equation
Paddle	20–4000	$Nu = 0.415Re^{0.67}Pr^{0.33}$
Paddle	>4000	$Nu = 0.36Re^{0.67}Pr^{0.33}$
Flat-blade turbine	<400	$Nu = 0.54Re^{0.67}Pr^{0.33}$
Flat-blade turbine	>400	$Nu = 0.74Re^{0.67}Pr^{0.33}$
Flat-blade turbine	2000–700,000	$Nu = 1.1Re^{0.62}Pr^{0.33}$
Propeller	>5000	$Nu = 0.64Re^{0.67}Pr^{0.33}$

For jacket side, $Nu = 0.023\ R_e^{0.8}Pr^{0.3}$

The fluid properties of the vessel and the jacket:

The Fluid Properties of the Reactor	The Jacket Fluid Properties
Density = 900 kg/m3	Heat capacity = 4.300 kJ/kg K
Viscosity = 4 cP	Density = 950 kg/m3
Specific heat = 3.2 kJ/kg.K	Viscosity = 0.9 cP
Thermal conductivity = 0.3 W/mK	Thermal conductivity = 0.6 W/mK

Solution

To get started, we need to find the rate-determining step based on the heat transfer coefficient. For the tank side, the equation to find the heat transfer coefficient depends on both the agitator type and Reynolds number.

$$\text{Reynolds number, } Re = \frac{ND_i^2\rho}{\mu} = 15000$$

For the paddle-type agitator, the relevant equation is:

$$Nu = 0.36Re^{0.67}Pr^{0.33}$$

Prandtl number, $Pr = \frac{C_p\mu}{k} = 42.67$

$$Nu = \frac{h_t D}{k} = 0.36 \times 15000^{0.67} \times 42.67^{0.33}$$

Therefore, h_t = 936 W/m²K.

For the jacket side:

$$Nu = \frac{h_j D_j}{k} = 0.023 \times Re^{0.8} Pr^{0.3}$$

where

$$Re = \frac{D_j V_j \rho}{\mu}$$

D_j is the equivalent diameter of the jacket. The equivalent diameter can be estimated by the equation:

$$D_j = \frac{4 \times \textit{cross sectional area}}{\textit{wetted perimeter}}$$

For the jacket fitted with a spiral baffle, the cross-sectional area is *ab* and the wetted perimeter is 2 (*a*+*b*) as shown in Figure 7.3. Therefore, D_j is estimated as 0.03 m.

$$Re = \frac{D_j V_j \rho}{\mu} = \frac{0.03 \times 5 \times 950}{0.0009} = 158333$$

$$Pr = \frac{C_p\mu}{k} = \frac{4300 \times 0.0009}{0.6} = 6.45$$

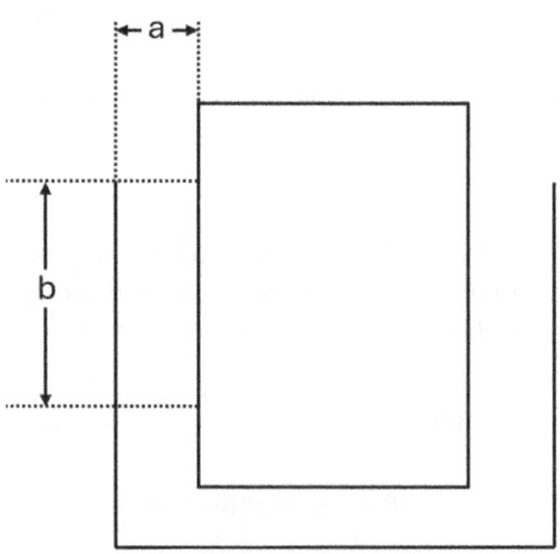

FIGURE 7.3 Equivalent diameter representation.

$$h_j = 0.023 \times \frac{k}{D_j} Re^{0.8} Pr^{0.3} = 11622\ W/m^2K$$

$$h_j \gg h_t$$

It can be seen that the tank side heat transfer coefficient dominates the overall heat transfer coefficient. Therefore, the key resistance to heat transfer is at the tank side. So, the scale-up should be performed by maintaining the same tank-side heat transfer coefficient.

$$h_{t1} = h_{t2}$$

$$h_t = \frac{k}{D_t} C_2 \left(\frac{ND_i^2 \rho}{\mu}\right)^{0.67} \left(\frac{C_p \mu}{k}\right)^{0.33}$$

Considering the physical properties remain constant during scale-up and $D_t D_t \propto D_i$:

$$h_t \propto \frac{1}{D_t}\left(ND_i^2\right)^{0.67}$$

$$h_t \propto N^{0.67} D_i^{0.34}$$

$$N_2 = N_1 \sqrt[\left(\frac{1}{0.67}\right)]{\frac{D_{i1}^{0.34}}{D_{i2}^{0.34}}}$$

To do the scale-up, the dimensions of the bigger vessel are estimated based on geometric similarity, and the impeller diameter is estimated from the equation provided.

7.8 SCALE-UP OF MASS TRANSFER-DOMINATED SYSTEMS

In the mass transfer-dominated system, the focus is on the movement of components from one medium to the other. In several systems, the focus is on the movement of mass between gas and liquid systems, and the driving force for this movement is the concentration difference. For example, if one of the components of a gaseous mixture has a preferential solubility in water, mass transfer will occur between the vapour and liquid phases. In such situations, the solubility difference in gas or liquid of different components is the key driving force. Similar to the convective heat transfer case, the molar flux can be represented in terms of the rate of mass transfer and driving force as shown in Equation 7.23.

$$J = \frac{Driving\ force}{overall\ resistance} = \frac{(C_1 - C_2)}{\left(\frac{1}{K_L}\right)} \tag{7.23}$$

where J is the molar flux, c_1 and c_2 are the component concentrations in the two mediums, and K_L is the overall mass transfer coefficient.

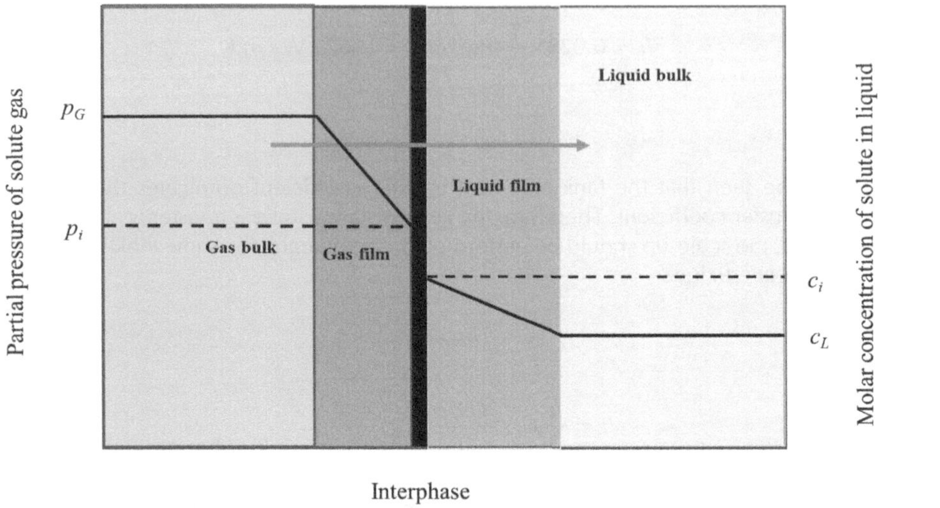

FIGURE 7.4 The two-film theory in mass transfer.

Whitman's two-film theory (Whitman, 1923) has been used to model the mass transfer of a gas between the gas and liquid interphase. In a gas-liquid contact operation, the mass transfer will occur to bring it to equilibrium. This can be visualized using two hypothetical films at the gas-liquid interphase, as shown in Figure 7.4.

According to the two-film theory, the main bodies of the gas and liquid mediums have uniform concentrations. However, the partial pressure (representation of gas concentration) falls at the interphase, whereas the liquid concentration increases at the gas-liquid interphase. These concentrations are at equilibrium. The overall mass transfer coefficients are defined to relate the interphase concentrations to the bulk liquid and gas concentrations as shown in Equation 7.24.

$$J = K_G\left(p_G - p_i\right) = K_L\left(c_i - c_L\right) \tag{7.24}$$

where K_G and K_L are the overall mass transfer coefficients based on gas-phase concentration and liquid-phase concentration. However, it is impossible to measure the interphase concentrations and partial pressures directly. Therefore, the overall mass transfer coefficients are represented in terms of the individual mass transfer, as shown in Equations 7.25 and 7.26.

$$\frac{1}{K_L} = \frac{1}{mk_G} + \frac{1}{k_L} \tag{7.25}$$

$$\frac{1}{K_G} = \frac{1}{k_G} + \frac{m}{k_L} \tag{7.26}$$

where k_G and k_L are the individual mass transfer coefficients, which quantify the rate of flow in the individual (liquid and vapour) phases, and m is the Henry's law constant.

For highly soluble gases, the value of m would be small, which makes the mass transfer resistance in the gas phase more prominent.

Gas film control: $\frac{1}{mk_G} \gg \frac{1}{k_L}$ giving $\frac{1}{K_L} \cong \frac{1}{mk_G}$

For sparingly soluble gases such as oxygen in water, the value of m would be significantly large, which makes the liquid phase resistance more prominent.

Liquid film control: $\frac{1}{k_L} \gg \frac{1}{mk_G}$ giving $\frac{1}{K_L} \cong \frac{1}{k_L}$

For successful scale-up

1. Determine the rate-controlling resistance and use this to make decisions about scale-up criteria.
2. Check that the same resistance remains rate-controlling at the new scale.

There are several correlations available to estimate the mass transfer coefficients. The scale-up procedure discussed in the previous sections can be applied to perform the scale-up by keeping the rate-controlling resistance constant.

Example 7.7. Scale-Up with Mass Transfer Considerations

A company has decided to use an existing agitated vessel to produce 10,000 L of a new pharmaceutical product via fermentation. The key design consideration is oxygen supply via aeration for such reactions. The initial proposal is to carry out this reaction (fermentation) with an impeller speed of 100 rpm. The impeller diameter is 25% of the vessel diameter, and the vessel has a height-to-diameter ratio of 3. Before finalizing the design, it was decided to test the operation in a 5 L lab scale vessel that duplicates exactly the same heterogeneity in environment that exists at the large scale. The following correlation can be used to relate the geometries of the tank and impellers, the physical properties of the liquid and the power input:

$$\frac{k_L a.D_t}{D_L} = 0.368\left(\frac{\rho N D_i^2}{\mu}\right)^{1.38}\left(\frac{\mu}{\rho D_L}\right)^{0.5}\left(\frac{\mu V_S}{\sigma}\right)^{0.5}\left(\frac{N^2 D_i}{g}\right)^{0.367}\left(\frac{N D_i}{V_s}\right)^{0.167}\left(\frac{D_i}{D_t}\right)^{0.25}\left(\frac{P}{\rho N^3 D_i^5}\right)^{0.75}$$

where $k_L a$ volumetric mass transfer coefficient, D_t is the tank diameter, D_i is the impeller diameter, D_L is the diffusivity of gas into the liquid, N is the speed of the impeller, V_S is the superficial gas velocity, P is power input, μ is the viscosity, ρ is the liquid density and σ is the surface tension.

a) Determine the dimensions of the lab-scale mixing tank and the speed of the impeller that duplicates the key design consideration. This can be achieved by keeping the same volumetric mass transfer coefficient ($k_L a$) as in the large-scale vessel.
b) If the rate of oxygen supply is the same for both vessels and geometric similarity is maintained, how would the impeller Reynolds number and power input differ between these two vessels?

c) After the scale-up was done, it was decided to increase the volume in the vessel to 12,000 L. Note that the vessel used is still the same. In that case, what should be the impeller speed of the vessel if we need to maintain the same volumetric mass transfer coefficient (k_La) as in the large-scale vessel? Note that we will not be maintaining the geometric homogeneity during scale-up.

Solution

a) Assume mixing tank to be cylindrical:

$$V = \frac{\pi}{4} D_t^2 H$$

Let existing agitated vessel be V_1 and lab scale vessel be V_2.

Since $H = 3D_t$ and $D_i = 25\%\ D_t$:

$$V_1 = \frac{\pi}{4} D_{t,1}^{\ 2} \times 3D_{t,1} = 10m^3$$

$$D_{t,1} = 1.619m$$

$$H_1 = 4.857m$$

$$D_{i,1} = 0.405m$$

$$V_2 = \frac{\pi}{4} D_{t,2}^{\ 2} \times 3D_{t,2} = 0.005m^3$$

$$D_{t,2} = 0.129m$$

$$H_2 = 0.386m$$

$$D_{i,2} = 0.03m$$

maintain constant volumetric mass transfer coefficient (k_La):

$$k_La = 0.368\left(\frac{\rho N D_i^2}{\mu}\right)^{1.38}\left(\frac{\mu}{\rho D_L}\right)^{0.5}\left(\frac{\mu V_S}{\sigma}\right)^{0.5}\left(\frac{N^2 D_i}{g}\right)^{0.367}\left(\frac{N D_i}{V_s}\right)^{0.167}\left(\frac{D_i}{D_t}\right)^{0.25}\left(\frac{P}{\rho N^3 D_i^5}\right)^{0.75}\left(\frac{D_L}{D_t}\right)$$

$$k_La \propto N^{2.281} D_i^{2.294}$$

$$\therefore N_1^{2.281} D_{i,1}^{\ 2.294} = N_2^{2.281} D_{i,2}^{\ 2.294}$$

$$N_2 = \sqrt[2.281]{100^{2.281} \times \left[\frac{0.405}{0.03}\right]^{2.294}}$$

$$N_2 = 1278rpm$$

b)

$$R_e = \frac{N D_i^2 \rho}{\mu}$$

$$R_e \propto ND_i^2$$

$$\frac{R_{e,2}}{R_{e,1}} = \frac{N_2 D_{i,2}^{\ 2}}{N_1 D_{i,1}^{\ 2}} = \frac{1278 \times 0.03^2}{100 \times 0.405^2}$$

$$R_{e,2} = 0.08\, R_{e,1}$$

The Reynolds number or lab scale is decreased by 0.08 times the existing agitated vessel.

$$P \propto N^3 D_i^5$$

$$\frac{P_2}{P_1} = \frac{N_2^{\ 3} D_{i,2}^{\ 5}}{N_1^{\ 3} D_{i,1}^{\ 5}} = \frac{1278^3 \times 0.03^5}{100^3 \times 0.405^5}$$

$$P_2 = \frac{1}{152} P_1$$

Power input is decreased by 152 times the existing agitated vessel.

c)

Geometrical similarity is not maintained:

$$V_2 = \frac{\pi}{4} D_{t,2}^{\ 2} \times H_2 = 12m^3$$

$$H_2 = 5.82m$$

$$k_L a \propto N^{2.281} D_i^{2.294} C_4^{\ 0.75}$$

where $C_4 = \frac{D_i}{H}$:

$$\therefore k_L a \propto N^{2.281} D_i^{3.044} H^{-0.75}$$

maintain constant volumetric mass transfer coefficient ($k_L a$):

$$N_1^{\ 2.281} D_{i,1}^{\ 3.044} H_1^{\ -0.75} = N_2^{\ 2.281} D_{i,2}^{\ 3.044} H_2^{\ -0.75}$$

$$N_2 = \sqrt[2.281]{100^{2.281} \times \left[\frac{0.405}{0.405}\right]^{3.044} \times \left[\frac{5.82}{4.857}\right]^{0.75}}$$

$$N_2 = 106 rpm$$

7.9 CONCLUSIONS

Chemical product industries often need to scale up processes from the laboratory to larger scales. This chapter covers the fundamental principles involved in scaling up processes across different size scales. It introduces the concepts of dimensional similarity and geometric similarity with the help of examples. Additionally, it addresses the fundamental approaches used in scaling up processes where heat and mass transfer effects are dominant. The chapter also introduces the need for and challenges associated with non-geometric scaling of processes, using an example for illustration. While these topics require more detailed treatment in specific cases, this chapter provides the essential background needed to address scale-up problems in the chemical product industry.

REFERENCES

Cullen, P. J., Romañach, R., Abatzaglou, N., Rielly, C. D., Rielly, C. D., & Abatzoglou, N. (eds.). (2015). *Pharmaceutical Blending and Mixing*. Chichester: Wiley.

Kayode Coker, A. (2001). *Fluid Mixing in Reactors- Chapter in Modeling of Chemical Kinetics and Reactor Design* (pp. 552–662). Gulf Professional Publishing.

Nienow, A. W. (1990). Gas Dispersion Performance in Fermenter Operation. *Chemical Engineering Progress*, *86*(2), 61–71.

Nienow, A. W., Warmoeskerken, M. M. C. G., Smith, J. M., & Konno, M. (1985). On the Flooding/Loading Transition and the Complete Dispersal Condition in Aerated Vessels Agitated by a Rushton-Turbine. In *Proceeding of the 5th European Conference on Mixing, Wurzburg, Germany* (pp. 143–154). BHRA Fluid Eng, Cranfield.

Warmoeskerken, M. M. C. G., & Smith, J. M. (1985). Flooding of Disc Turbines in Gas-Liquid Dispersions: A New Description of the Phenomenon. *Chemical Engineering Science*, *40*(11), 2063–2071. https://doi.org/10.1016/0009-2509(85)87023–8

Whitman, W. G. (1923). The Two-Film Theory of Gas Absorption. *Chemical and Metallurgical Engineering*, *29*, 146–148.

Zolkarnik, M. (1991). *Dimensional Analysis and Scale-Up in Chemical Engineering* (1st ed., pp. 5–22). Berlin, Germany: Springer.

Zolkarnik, M. (1998). Problems in the Application of Dimensional Analysis and Scale-Up of Mixing Operations. *Chemical Engineering Science*, *53*(17), 3023–3030.

7.10 PROBLEMS

1. During liquid makeup production, colour pigments (i.e. solids having identical particle size) are added to the product in a mixing tank. In the pilot plant, this mixing tank has a capacity of 3 L. The tank operates at 6700 rpm and has an impeller diameter of 0.035 m. The impeller diameter (D_i) is 25% of the vessel diameter (T). It has been decided to scale up this mixing tank to higher capacity. Full-scale production is conducted in a geometrically similar vessel and has an impeller diameter of 0.12 m.
   a) Determine the dimensions of the large mixing tank and the speed of the impeller after scale-up. The scale-up basis is as follows:
      i) Constant power/unit volume
      ii) Constant Reynolds number
      iii) Constant impeller tip speed
      iv) Constant torque per volume
   b) What are some of the assumptions involved in this scale-up that may affect the expected performance?
   c) How would the Reynolds number change if the scale-up is performed based on constant P/V ratio?

   The following information is useful (for other problems too):

   Impeller tip speed, $v = N \times r_i$

r_i is the radius of the impeller

Torque, $t = k_g \rho N^2 D_i^5$

k_g is a constant and ρ is the density of the mixture.

Power, $P = C_4 N^3 D_i^5 N_p$

$$C_4 = \frac{4}{\pi}\left(\frac{D_i}{T}\right)^2\left(\frac{D_i}{H}\right)$$

N_p is a function of the geometry of the impeller, T is the diameter of the vessel and H is the height of the vessel.

2. A protein suspension with a particle diameter of 3.1 μm has been reacted with sodium borohydride in the presence of ethanol at -10°C in a 30-L lab scale reactor. It has been decided to scale up this reaction to 15 m^3 capacity. The lab scale reactor has a height-to-diameter ratio of 3. The impeller diameter (D_i) is 25% of the vessel diameter (D_t). Agitator speed (N) is 350 rpm, and three Rushton impellers are used.
   a) Determine the dimensions of the large reactor if geometrical similarity is maintained.
   b) Determine the agitator speed for constant P/V (power/volume) ratio.
   c) How would the Reynolds number change if the scale-up is performed based on constant P/V ratio?
3. A steroid suspension with the particle diameter of 2.6 μm has been reacted with n-butyl lithium in the presence of THF solvent at -20°C in a 10-L lab-scale reactor. It has been decided to scale up this reaction to 10,000-L capacity. The lab scale reactor has a height-to-diameter ratio of 3. The impeller diameter (D_i) is 30% of the vessel diameter. Agitator speed (N) is 500 rpm, and three Rushton impellers are used. Determine the dimensions of the large reactor and agitator speed for:
   a) Constant P/V (power/volume) ratio
   b) Constant impeller tip speed
   c) Constant Reynolds number
4. A stirred-tank reactor is designed to produce ice cream with size of 10 m^3. The dimensions of the production tank are: D_t (diameter of the tank) = 2 m, D_i (the impeller diameter) =1.4 m and N (agitator speed) = 100 rpm. It was decided to do a trial run on a laboratory-scale stirred-tank reactor with a size of 0.1 m^3 before installing the industrial-scale production tank.
   a) Determine the dimensions of the small tank (D_i, D_t, H-height of the vessel) by using geometric similarity.
   b) Determine the required rotational speed of the impeller in the small tank if a constant P/V (power/volume) ratio must be maintained. Also, estimate the change to the impeller Reynolds number (ratio of the impeller Reynolds numbers of big and small vessels) during this scale-up.
   c) Figure 7.5 shows the effect of shear rate (S^{-1}) on the viscosity (Pa.S) of ice cream at temperatures near its melting point. Explain the rheological

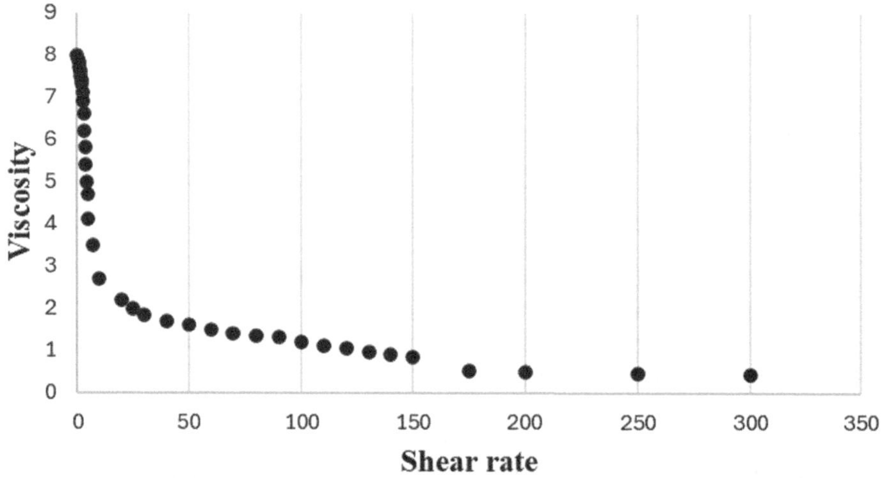

FIGURE 7.5 Viscosity of ice cream versus shear rate.

behaviour of ice cream and its impact on the reliability of scale-up results obtained in part (b).

5. A company is developing a massage cream, a type of cold cream normally used as a cleansing agent for facial skin. The product development team is responsible for identifying the key target properties of the cream and then doing the scale-up studies.
   a) It is decided to produce 10 m^3 of the cream in one batch in a cylindrical stirred-tank vessel. The dimensions of the production tank are: D_t (diameter of the tank) = 2 m and Di (the impeller diameter) = 1.5 m. It was decided to do a trial run on a laboratory-scale stirred-tank reactor with a size of 0.1 m^3 before installing the industrial-scale production tank. Determine the dimensions of the small tank (D_i, D_t, H-height) by using geometric similarity if both the vessels are only filled up to 75% of maximum capacity. You may ignore the volume of the end caps of the vessel.
   b) The optimal performance in the lab scale was obtained when an impeller speed (N) of 200 rpm is used. Determine the required rotational speed of the impeller in the large tank if a constant Reynolds number must be maintained. Mention the key assumption involved in this calculation, and explain why that assumption may not be realistic.
   c) Due to the change in demand, it was decided to reduce the production per batch of the cream to 8 m3. However, the production is to be continued in the same vessel. What should be the new rotational speed to maintain the same Reynolds number as in the previous case? How would this change affect the power-to-volume (P/V) ratio of the agitator?

6. During the production of a pharmaceutical product, it is necessary to maintain the temperature of the reaction mixture at 50°C. The predominant problem of the reactor is the removal of heat. The reaction had been carried out in a pilot plant jacketed agitated vessel with a capacity of 15 L. The jacket has an equivalent diameter of 5 cm and a coolant velocity of 2.5 m/s. The vessel has a height-to-diameter ratio of 3, and the impeller diameter (D_i) is 80% of the vessel diameter (D_t). The agitator is a flat-blade turbine and is revolving at 500 rpm. It has been decided to scale up this mixing tank to a 1200-L geometrically similar vessel.
   The fluid properties of the reactor
   Density = 900 kg/m³
   Viscosity = 0.0025 Pa·s
   Specific heat = 3100 J/kg·K
   Thermal conductivity = 0.15 W/m·K
   The jacket fluid properties
   Heat capacity = 4500 J/kg·K
   Density = 980 kg/m³
   Viscosity = 0.0003 Pa·s
   Thermal conductivity = 0.700 W/m·K
   Expressions to find heat transfer coefficients:
   For the tank side $Nu = C_1 (Re)^a (Pr)^b$
   For jacket side $Nu = C_2 (Re)^c (Pr)^d \left(\frac{\mu_j}{\mu_{wj}} \right)^{0.14}$
   where $C_I = 1.1$, $C_2 = 0.027$, $a = 0.62$, $b = 0.33$, $c = 0.8$, $d = 0.3$.
   a) Determine the dimensions of the lab-scale and full-scale vessels.
   b) Using the given data, formulate the scale-up strategy for this vessel.
   c) Carry out the scale-up based on the formulated scale-up strategy in part (b). Check how the power requirement of the impeller changes during scale-up.
7. During the production of an agrochemical product, it is necessary to maintain the temperature of the reaction mixture at 60°C. The predominant problem of the reactor is the removal of heat. The reaction had been carried out in a pilot plant jacketed agitated vessel with a capacity of 10 L. The jacket has an equivalent diameter of 4 cm and a coolant velocity of 3 m/s. The vessel has a height-to-diameter ratio of 3, and the impeller diameter (D_i) is 75% of the vessel diameter (D_t). The agitator is a flat-blade turbine and is revolving at 200 rev/min. It has been decided to scale up this mixing tank to a 1000-L geometrically similar vessel.
   The fluid properties of the reactor
   Density = 860 kg/m3
   Viscosity = 0.003 Pa.s
   Specific heat = 2900 J/kg.K

Thermal conductivity = 0.17 W/mK
The jacket fluid properties
Heat capacity = 4300 J/kg K
Density = 965 kg/m3
Viscosity = 0.0002 Pa.s
Thermal conductivity = 0.684 W/mK
Expressions to find heat transfer coefficients:
For the tank side $Nu = C_1 (Re)^a (Pr)^b$

For jacket side $Nu = C_2 (Re)^c (Pr)^d \left(\frac{\mu_j}{\mu_{wj}} \right)^{0.14}$

where $C_1 = 1.1$, $C_2 = 0.027$, $a = 0.62$, $b = 0.33$, $c = 0.8$, $d = 0.3$.

a) Determine the dimensions of the lab-scale and full-scale vessels.
b) Using the given data, formulate the scale-up strategy for this vessel.
c) Carry out the scale-up based on the formulated scale-up strategy in part (b). Check how the power requirement of the impeller changes during scale-up.

8. A company is scaling up the production of a rubber chemical (molecules that enhance the properties of rubber), which is produced in the laboratory as a thick slurry. The reaction has to be carried out at a fixed temperature of 50°C. Heat control is the most important consideration to be made during the scale-up. The reaction had been carried out in a pilot plant jacketed agitated vessel with a capacity of 20 L (0.02 m^3) and a diameter of 20 cm. The jacket has an equivalent diameter of 4 cm and a coolant velocity of 3 m/s. The impeller diameter (D_i) is 70% of the vessel diameter (D_t). The agitator is a flat-blade turbine and is revolving at 500 rev/min. The expected batch size is 5 m^3. The reaction will be carried out in a geometrically similar vessel.

   Some useful equations:

   For the tank side $Nu = C_1 (Re)^a (Pr)^b$

   For jacket side $Nu = C_2 (Re)^c (Pr)^d \left(\frac{\mu_j}{\mu_{wj}} \right)^{0.14}$

   where $C_1 = 1.1$, $C_2 = 0.027$, $a = 0.62$, $b = 0.33$, $c = 0.8$, $d = 0.3$.

   a) Determine the dimensions of the lab-scale and full-scale vessels.
   b) The heat transfer coefficients for the tank side and jacket side were estimated to be 1200 Wm^{-2}K^{-1}and 19,300 Wm^{-2}K^{-1}. Formulate the scale-up strategy and carry out the scale-up.
   c) Estimate the change of Reynolds number during scale-up. What is the significance of this change?
   d) What are the possible reasons that may lead to differences in the expected performance of the reactor after scale-up?

9. Sophorolipids are potential bio-surfactants with a structure consisting of a hydrophobic fatty acid tail of 18 carbon atoms. They are biodegradable and possess low eco-toxicity. It can be produced by an aerobic fermentation process, and the optimal conditions to produce sophorolipids have been estimated in the pilot plant scale. For this system, the volumetric mass transfer coefficient, k_L, can be estimated using the following dimensionless correlation.

$$\frac{k_L D_t^2}{D_L} = 0.06\left(\frac{\rho N D_t^2}{\mu}\right)^{1.5}\left(\frac{\mu}{\rho D_L}\right)^{0.5}\left(\frac{\mu V_S}{\sigma}\right)^{0.6}\left(\frac{N^2 D_t}{g}\right)^{0.19}\left(\frac{N D_t}{V_s}\right)^{0.32}$$

where D_t is the tank diameter, D_L is the diffusivity of gas in the liquid, N is the speed of the impeller, V_s is the superficial gas velocity, μ is the viscosity, ρ is the liquid density andσ is the surface tension.

a) In aerobic bioprocesses, it is necessary to keep a constant supply of oxygen. The pilot plant production was done in a 100 L cylindrical vessel. It has been observed that the optimum mass transfer rate was obtained when the superficial gas velocity was kept at 0.02 m/s and the impeller speed was maintained at 100 rpm. When the fermentation was started, the density of the feed was 940 kg/m^3, viscosity was 0.01 Pa.s and surface tension 72 mN/m. After the pilot plant studies, it was decided to process 1000 L of the feed in one batch in a cylindrical stirred-tank vessel. The height-to-diameter ratio is 3, and the diameter of the impeller is 80% of the vessel diameter. Determine the dimensions of the large vessel (diameter, height and impeller diameter) if geometric similarity is maintained. You may ignore the volume of the end caps of the vessel.

b) During scale-up, it is targeted to maintain the same volumetric mass transfer coefficient, k_L, obtained at the lab scale. In addition, it is also required to maintain a constant specific power input (P/V). Estimate the impeller speed and superficial gas velocity needed to attain these targets. Also, estimate the effect on Reynolds number during scale-up.

c) What are the possible reasons that may lead to differences in the expected performance of the reactor after scale-up? What possible steps can be taken to minimize these errors?

10. To support the increasing local demand for gluconic acid, a company has decided to increase its existing bioreactor scale according to the scale ratio of 1:750. The existing bioreactor has a capacity of 25 L, with a diameter-to-height ratio of 0.25. The bioreactor is stirred with a Rushton turbine, the diameter of which is 35% of the vessel diameter. The stirring speed is set at 450 rpm. The impeller off-bottom clearance to liquid height ratio is kept at 0.25. Both existing and scaled-up bioreactors are assumed to be

geometrically similar and to exhibit turbulent flow. The scaled-up bioreactor is sparged with an air flow rate of 1 m^3/min.

a) Since the cell cultures involved are shear sensitive, determine the impeller speed (rpm) in the scaled-up bioreactor if a constant impeller tip velocity must be maintained during scale-up.
b) Analyse if the impeller will be flooded. If flooding occurs, what will be the recommended gas flow rate to avoid flooding?
c) What will be the recommended gas flow rate in m^3/min to achieve complete gas dispersion?

8 Batch Processes

In batch processes, the operating conditions, compositions and mass of the operating units vary with time, even for a single product. Therefore, the process is planned based on the *recipe* given to the manufacturing unit by the product development team. The recipe is a systematic description of the materials used in each unit, their quantities, operating conditions, variations and the duration of each step. By carefully planning the operations in each unit, it is possible to maximize the production capacity of a batch processing unit. A higher production capacity usually translates to higher profit. Because of that, the ability to produce the maximum number of batches in the shortest possible time has significant value in a batch processing unit. Therefore, in this chapter, different types of batch processes, some of the challenges involved in batch processing, and various strategies followed by the industry to maximize the productivity of batch processing have been covered.

8.1 TYPES OF BATCH PROCESSES

The different ways of carrying out processes to manufacture chemicals at an industrial scale can be classified as (i) continuous processes, (ii) batch processes, (iii) fed-batch processes and (iv) batch-product removal (Felder et al., 2018). Options iii and iv are collectively classified as semi-batch processes. In a fed-batch process, some or all of the raw materials are charged continuously while the product is collected only after the process is complete. An example of a fed-batch process is the chlorination of acetyl chloride used for the industrial production of trichloroacetyl chloride. In this process, acetyl chloride is charged into the reactor with other raw materials and catalysts, and chlorination is done continuously until all the raw materials are reacted. Once the reaction is complete, the entire batch of trichloroacetyl chloride can be collected as the final product. In the batch-product removal process, the charging of raw materials is done in batch mode, and the product is collected continuously. For example, in many processes, a large amount of raw material is produced in batch mode and stored in huge storage tanks. These raw materials can be sent continuously to the rest of the processes. In such cases, as long as the production rate of the raw material is faster than the rate at which it needs to be supplied for the rest of the process, the process can be run continuously.

8.2 SAFETY CONCERNS IN BATCH PROCESSES

The safety concerns in batch processes are significantly higher than in continuous processes, and the number of accidents per process plant is much higher in batch-processing plants. Some of the unique characteristics of the batch-processing plants that contribute to the safety issues are the unsteady state nature of the batch processing, difficulty to stop the process, sub-optimal equipment, process conditions and fast scale-up.

DOI: 10.1201/9781032708911-8

One of the main reasons for this difference is the unsteady state nature of batch operations. The operating conditions, composition of the vessel content, volume, production of gases and the physical properties change as a function of time. In many cases, there may be intermediate charging or removal of materials from the vessels that can also contribute to the risk factors. The controllability of the processes is more challenging when the conditions change as a function of time.

The difficulty of controlling a process once a new batch start is another reason for the safety concerns in batch processes. Unlike continuous processes, the entire batch is charged into the units in a new batch. Therefore, in case of an emergency, it is difficult to stop the process. In some process plants, there may be additional units to transfer the batch from a malfunctioning unit, whereas in many process plants, all the equipment are utilized for the process. While this is an efficient utilization of resources from a cost standpoint, this may be an issue in case of a process malfunction.

In many instances, the reactions may be conducted near runaway conditions to increase the conversion and yield. The nature of chemical reactions is also much more complex in such processes. Because of that, minor variations in the process conditions may lead to unexpected reactions, which may lead to accidents. For example, several chemicals used in such processes are reactive with water, and the presence of water in the system after the cleaning of the vessel may lead to undesirable side reactions. Another reason for the unexpected reactions under different conditions is the relatively fast development time for many of the speciality chemicals. Because of the competitive nature of these types of chemicals, many products are rushed to the market without rigorous laboratory testing under various conditions. On many occasions, the selection of raw materials for new processes may be chosen based on the local availability, convenience and cost and without due regard to the suitability for the processes.

Unlike commodity chemicals, speciality chemicals have a short lifetime, and because of that, novel processes are introduced much more frequently than the continuous processes. In addition, batch processes use generic equipment instead of highly specialized and optimized equipment used in the commodity industry. This is due to the short lifetime of the products, and the equipment needs to be able to be used for a variety of reactions. Because of this, several operations may be conducted with poorly designed equipment for that specific operation and in many situations, would not have the safety features to prevent the scenarios that may arise during upsets or process disturbances. In multi-product production systems, the equipment is shared by different processes that involve different chemicals. This also contributes to the sub-optimal utilization of batch equipment, which may be completely incompatible with the process chemistry.

Moreover, batch processes involve many more manual operations than continuous processes, which makes them prone to human errors. The nature of batch processes also allows companies to employ unskilled labourers to conduct routine operations such as batch charging, cleaning and packaging. Because of this, there is a higher likelihood of incorrect operations or charging of different raw materials into the reactors, which may have catastrophic effects. A significant proportion of accidents (often preventable) happen because of human errors, and the majority of such mistakes are due to the ignorance of the process details by the operators involved. The operating personnel also need to spend their time near the process units, which may also lead to longer exposure to hazardous chemicals.

Scale-up from the laboratory scale also poses additional challenges to the batch process systems. Most of the process variables are size dependent, and the scale-up from laboratory scale may have unintentional impacts on some of those variables, as discussed in Chapter 7. Often, scale-up has been performed directly from laboratory scale to the plant scale without full consideration of the heat transfer and mass transfer effects as described in Chapter 7. This may also have a significant impact on the relief capacity of the vessel. If the scale-up is performed outside the safe operating envelope of the equipment, safety may not be guaranteed.

Some of the possible approaches to reduce the safety hazards are as follows (American Institute of Chemical Engineers, 1999):

1. Choose an inherently safer chemical reaction pathway.
2. Choose an inherently safer process.
3. Perform chemical reactivity testing.
4. Perform pilot-scale studies before full-scale production.
5. The pressure relief system must also consider the extreme conditions and possible upsets.
6. Educate the operating personnel about the process and the risks involved in various chemicals that are used in the process.
7. Periodic training must be made mandatory, and proper records must be kept. In addition, adequate training must be provided during the introduction of new processes/products and during the start-up and shutdown periods.
8. Perform sizing calculations before new processes to ensure that the existing units can hold the raw materials and products.
9. Implement proper labelling system in the plant.
10. Perform regular maintenance on standby units, and ensure that they are in working condition all the while.

8.3 BATCH PROCESS SCHEDULING – SINGLE PRODUCT

Effective process scheduling is one of the most important steps to maximize productivity in a typical batch processing plant. This is because the chemicals produced in batch processing plants and their demands often change, and it is necessary to make use of the available equipment to meet the demands. In a batch process production unit, products are produced in different batches. For each product, a recipe of tasks is required to convert the raw materials into the final products. However, the duration of each task varies based on the specific requirements. As a result, the utilization of equipment in a batch process system differs, and maximizing process efficiency necessitates arranging the processing in a manner that minimizes idle time for various equipment. Therefore, sequencing and scheduling are two major areas of focus for chemical engineers working in batch processing units. Sequencing refers to determining the order in which tasks are completed, while batch scheduling involves establishing the schedule for conducting unit operations in each equipment (Turton et al., 2012).

Typically, a batch is considered to be started when the raw materials are charged into the first unit in a series of tasks. The time required to complete the production of one batch of product is called the cycle time. It is to be noted that typically, there are several subtasks to be completed in an equipment before it is available for the next batch. The cycle time typically considers all the subtasks involved in the main task expected to be carried out in the given unit. The batch process scheduling targets minimize the cycle time with the available resources to achieve maximum production.

In several process plants, multiple products may be produced simultaneously, and the equipment may be shared between different products. The simplest type of scheduling approach is when the process plant produces only one major product. There are several ways to do the scheduling of batch processes. To illustrate these approaches, the sequence of operations used in the production of an insecticide has been described in Example 8.1.

Example 8.1. Insecticide Production

A pesticide has been produced through a batch process, during which the following batch recipe was applied:

The process starts with a reaction in which two raw materials and a catalyst need to be charged into the reactor, R–101. Since the catalyst is solid, the operation involves manual charging through a chute. Charging usually takes 1 hour to complete. After charging, the raw materials need to be heated to 102°C, which takes 1.5 hours to complete. It will take 4 hours to complete the reaction.

Once the reaction is complete, the reactor needs to be cooled to 50°C, which takes 1 hour. After this, the catalyst is removed by sending it through leaf filters (F-101) and then transferred to a second reactor (R-102), which can be completed in 1.5 hours. After the transfer, the filters have to be opened to remove the spent catalyst and cleaned before the next batch. The cleaning takes 1 hour to complete.

In the second reactor, a decomposition reaction takes place, which can be completed in 3 hours. One of the unreacted raw materials needs to be removed before proceeding to the next stage. Therefore, the content of the reactor is transferred to a batch distillation column, D–201. The transfer takes 30 minutes to complete, and the distillation can be completed in 2.5 hours. After recovering the unreacted raw material, the batch is transferred to another reactor, R-201 for an alkaline hydrolysis reaction, which can be completed in 8 hours.

The product of this reaction is then recovered by filtration in a rotary vacuum drum filter. (RVDF). This process takes 2 hours. The recovered product then undergoes a condensation reaction in another reactor, R-301, which can be completed in 3.5 hours to produce the crude insecticide. The crude product is then washed with water and separated using gravity separation. The whole process takes about 2 hours.

The final purification is done by dissolving the product in a solvent, followed by crystallization, which takes 7 hours to complete in a crystallizer, X–301. The purified product is then transferred to a storage vessel (T-301) for filling, which takes 1.5 hours to complete. A simplified block diagram of this process is given in Figure 8.1.

Estimate the time to complete each task in this process.

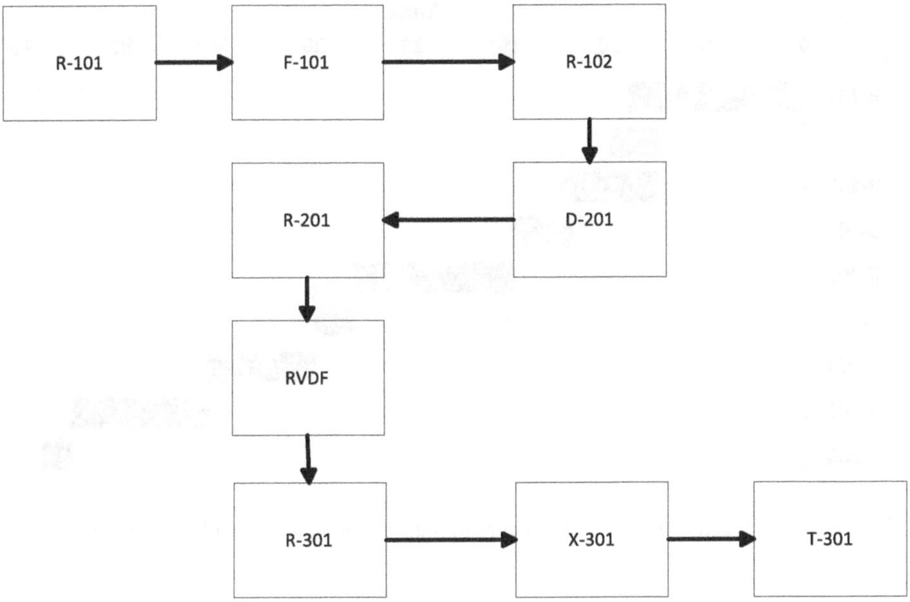

FIGURE 8.1 Insecticide production process block diagram.

Solution

The first task is reaction in **R-101**.

Since the reactor is occupied when the charging starts until the reactor is completely emptied, the time to complete the task is the sum of subtasks. In this case, the subtasks are charging, heating, reaction, cooling and transfer. It is to be noted that the cleaning time for the filter need not be included in the cycle time of this task. Therefore, the cycle time for R-101 will be $1+1.5+4+1+1.5=9$h.

F-101: $1+1.5=2.5$ h
R-102: The cycle time will be 5 hours including the transfer time.
D-101: $0.5+2.5=3$ h. Note that the transfer time to the next reactor is included in the reported cycle time.
R-201: 8 h
RVDF: 2 h
R301: $2+3.5=5.5$ h
X-301: 7 h
T-301: 1.5 h

The simplest approach for scheduling is a non-overlapping scheme in which a new batch will start only after the completion of the previous batch. In that case, the cycle time would be the sum of the individual processing time in each step. However, it should be noted that there may be an overlap of operations, typically during the transfer of batches from one unit to the other. For example, when the batch is transferred from R-101 to R-102 through F-101, all three units are

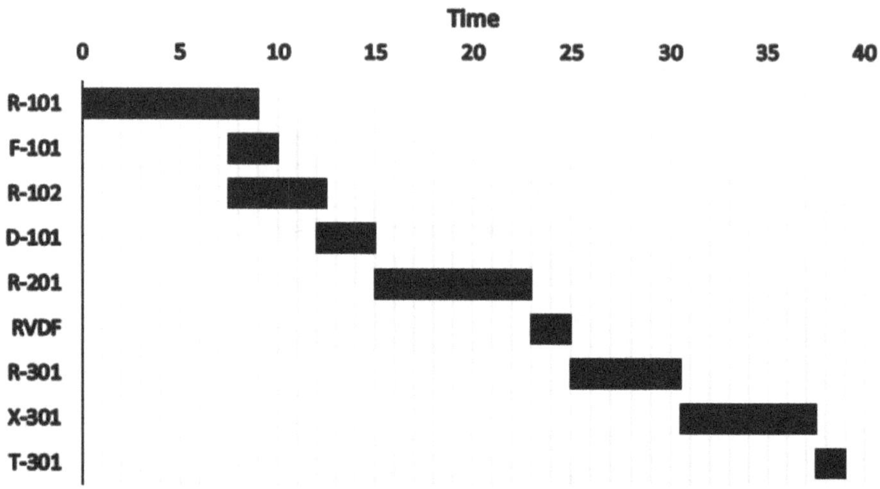

FIGURE 8.2 Gantt chart of the non-overlapping scheme for insecticide production.

simultaneously occupied. Therefore, on many occasions, it will be convenient to use a Gantt chart to visualize the production schedule. A Gantt chart is a graphical representation of the periods of time during which each equipment in a production system is utilized. The Gantt chart for a non-overlapping scheme of the process described in Example 8.1 is shown in Figure 8.2.

To simplify the scheduling, the transfer time is counted only in the unit with the longest cycle time. This way, priority can be given to the equipment that is occupied for the longer duration. In addition, the cycle time is not counted for the unit once the batch is transferred from that unit. In this example, though the filter cleaning takes 1 hour to complete, this is not part of the cycle time because the filter cleaning can be done after the batch is removed, and thus it does not affect the cycle time. Therefore, the cycle times of F-101, R-102 and D-101 are changed to 0 h (since the filter cleaning is not related to the process flow), 3.5 h and 2.5 h, respectively. With these changes, the Gantt chart for this process can be represented, as shown in Figure 8.3.

The same processes are repeated for the subsequent batches. If the number of completed batches is k, the cycle time for nonoverlapping scheme (CT_{NO}) can be estimated using Equation 8.1.

$$CT_{NO} = k\sum\nolimits_{j=1}^{o} ct_j \tag{8.1}$$

where ct_j is the time needed to complete the step j out of a total of n steps in the whole process. So, for the process described in Example 8.1, the cycle time would be 39 h. The second batch would be started after the completion of the first batch.

Even though the non-overlapping scheduling makes scheduling easier, this is obviously not efficient since none of the equipment is in operation most of the time. To improve equipment utilization and, consequently, reduce cycle time, it is recommended to implement an overlapping scheme for batch process scheduling.

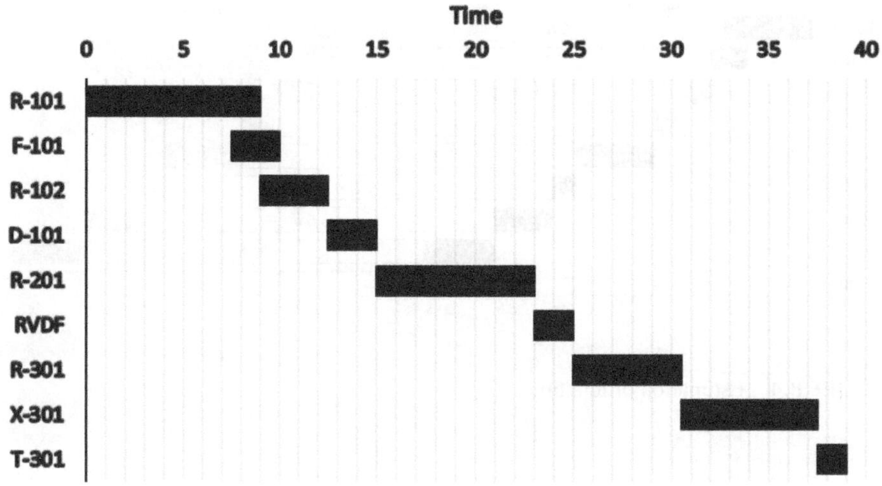

FIGURE 8.3 Modified Gantt chart of insecticide production.

The purpose of an overlapping scheme is to increase equipment utilization, thereby producing a greater number of batches. To achieve this, batches would start before the completion of the previous ones, as long as the equipment is available for processing. As a result, there would be more than one batch processed simultaneously.

The overlapping scheduling is performed by shifting the start time of the subsequent batches. This can be done on the Gantt chart by identifying the idle time of equipment in the process between batches, as shown in Figure 8.4.

In the next step, the start time of the subsequent batches is moved to the left side until two units touch each other. In reality, this shift indicates that the second batch begins as soon as equipment becomes available. When two equipment items touch each other on the Gantt chart, it indicates there is no waiting time for the equipment. In this example, the minimum idle time is between two batches of R–101. To minimize the waiting time, the second batch of R-101 can be started right after the first batch of R-101, and the rest of the operations follow in the same order as batch 1. This has been illustrated in Figure 8.5.

It can be observed that, following the first batch, each subsequent batch is completed every 9 hours, which corresponds to the time required to complete operation R–101. However, the time to complete the first batch is still the same as the non-overlapping sequence. Therefore, the total cycle time to complete k batches for an overlapping sequence (CT_O) can be estimated using Equation 8.2.

$$CT_O = \sum_{j=1}^{n} ct_j + (k-1)ct_{max} \tag{8.2}$$

where ct_{max} is the longest cycle time among all the operations in the whole process.

In order to schedule the batch operations, the term 'cycle term' has been commonly used. Cycle time refers to the average time required to complete one batch. From Equation 8.2, an expression for cycle time can be derived as shown in Equation 8.3.

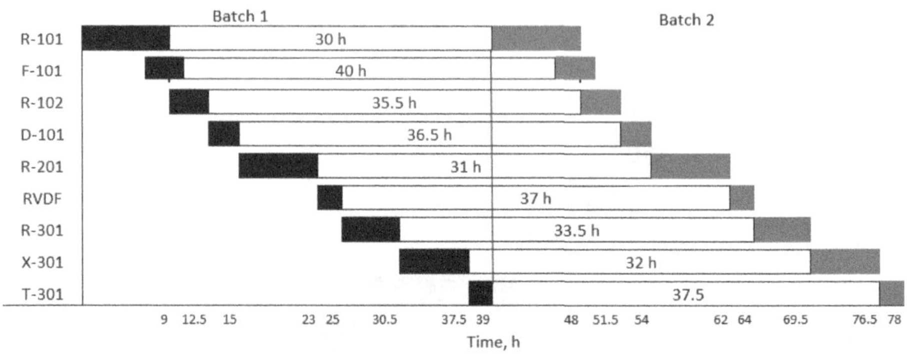

FIGURE 8.4 Estimation of idle time.

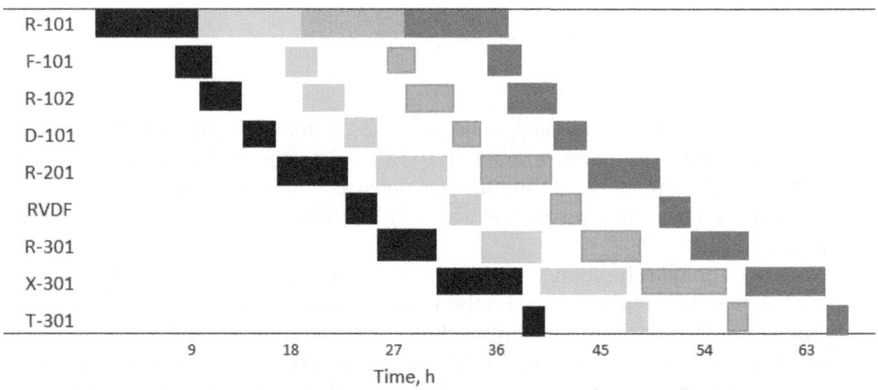

FIGURE 8.5 Overlapping sequence.

$$T_{cycle} = \frac{\sum_{j=1}^{n} ct_j + (k-1)ct_{max}}{k} \tag{8.3}$$

However, if a large number of batches are produced, the additional time needed for the first batch becomes less significant, and the cycle time can be approximated as the longest cycle time among all the operations in the whole process (ct_{max}).

Example 8.2. Cycle Time for Single-Product Production Plants

Estimate the cycle time for the process described in Example 8.1 if

i) 1000 batches of the insecticide are produced
ii) 5 batches of the insecticide are produced

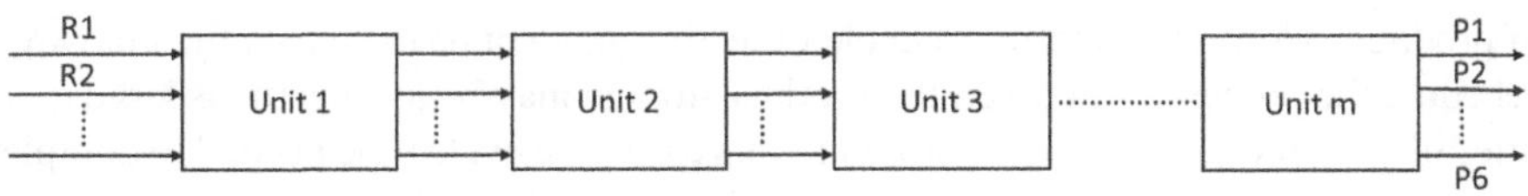

FIGURE 8.6 Flowshop plant illustration.

Solution

i) Since the number of batches is large, the average cycle time would be the time needed to complete the longest process. In this process, R-101 has the longest operation time, which is 9 h.
ii) Since the production target is 5 batches, the additional time needed for the first batch should not be ignored.

$$T_{cycle} = \frac{\sum_{j=1}^{n} ct_j + (k-1)ct_{max}}{k} = \frac{39 + 4 \times 9}{5} = 15\text{h}$$

8.4 BATCH PROCESS SCHEDULING – MULTIPLE PRODUCTS

In several types of batch process industries, a single plant may produce multiple products simultaneously. Additionally, on many occasions, the processes to produce multiple products also share the same equipment to improve cycle time. While the sharing of equipment significantly improves the flexibility of operations, it may also complicate the scheduling of processes. There are two types of models that can be used to describe batch operations, which are flowshop plants and jobshop plants (Biegler et al., 1997).

8.4.1 Flowshop Plants

In a flowshop plants, multiple products are produced by the same equipment in the same sequence. However, the duration of operations may not be the same for the equipment. This type of process can be represented as shown in Figure 8.6.

Here, six different products are produced in a plant that makes use of '*m*' different units. This plant has been considered a flowshop plant because all six products require the use of units 1 to '*m*' in the same order. However, the time spent on each of the products in each unit may be different. For example, if unit 1 refers to a reaction, it is possible that product 1 requires a reaction time of 6 hours, product 2 requires 4 hours and so on.

The scheduling approach for the flowshop plants depends on the way in which the way the products are produced. The scheduling is easier when only one product is made during each campaign. This type of scheduling is common in plants where products are seasonal or the products are delivered together. It is also important to run the production in a single product campaign when the changeover from one product to the other is labour-intensive or when the company is concerned about the contamination of products. However, in terms of cycle time, this may not be the most efficient way to schedule the process. Moreover, the product storage may be a concern, as large quantities of each product need to be stored in this case. The cycle time needed to complete the production of all the products would be the sum of the individual cycle times of the multiple products produced from this plant, as shown in Equation 8.4.

$$T_{cycle-fs} = \sum_{p=1}^{s} \left(T_{cycle}\right)_p \tag{8.4}$$

where $T_{cycle-fs}$ is the cycle time of the flowshop plant, s is the number of different products and $\left(T_{cycle}\right)_p$ is the cycle time of product p. For the scheduling of multi-product campaigns must be identified from the Gantt chart. An illustration of various scheduling approaches for flowshop plants has been presented in Example 8.3.

Example 8.3. Flowshop Plant Scheduling

A plant produces three products A, B, C, using four equipment E1, E2, E3 and E4 for one year. It is required to produce equal number of batches of these three products. The duration of the unit operations for the different products are shown in Table 8.1. Estimate the cycle times for the single product campaign, the campaign ABCABC . . . . and the campaign ACBACB . . . . Also, estimate the number of batches this plant can produce in one year through different campaigns.

Solution

Since the duration of production is one year, the number of batches is large. Therefore, it is acceptable use the simplified form of the cycle time calculation.

For single product campaign:

Cycle time for A = Max(t_i) = 7h
Cycle time for B = 8h
Cycle time for C = 7h
Total cycle time = 22 h

$$\text{Number of batches in one year} = \frac{\text{Number of hours in one year}}{\text{Cycle time}} = \frac{8000}{22} \approx 363$$

The Gantt charts of ABC campaign and ACB campaign are shown in Figures 8.7 and 8.8.

TABLE 8.1
Equipment Usage for Different Products

Product	Time in E1 (h)	Time in E2 (h)	Time in E3 (h)	Time in E4 (h)
A	7	2	5	4
B	5	8	2	5
C	4	6	7	2

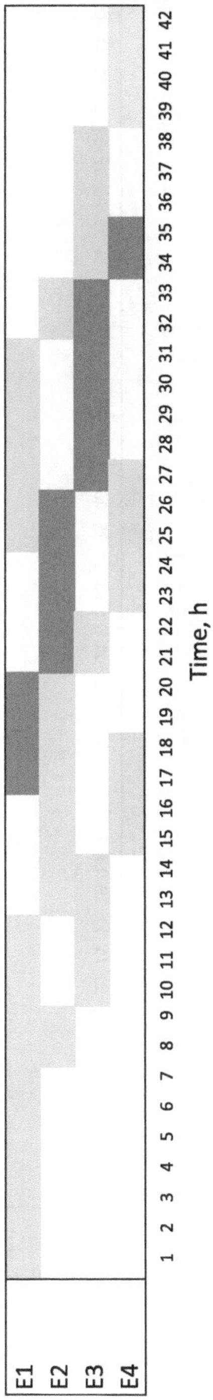

FIGURE 8.7 Gantt chart for ABC campaign.

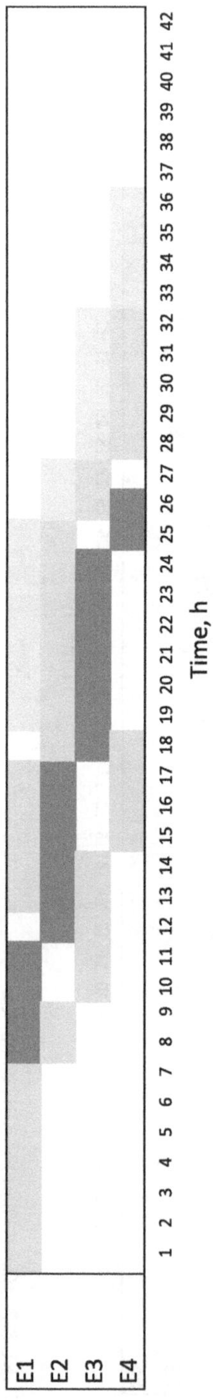

FIGURE 8.8 Gantt chart for ACB campaign.

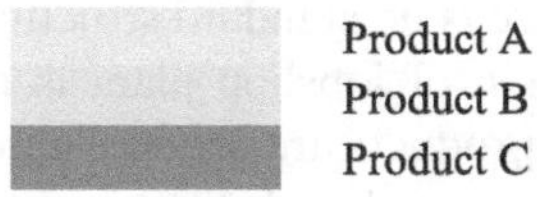

From Figure 8.7, it can be seen that the cycle time for the ABCABC . . . campaign is 42–18 = 24 hours. Therefore, the number of batches per year will be approximately 333. Similarly, the cycle time for the ACBACB . . . campaign is 36–18 = 18 hours, and the number of batches that can be produced per year would be approximately 444. Therefore, the ACBACB . . . campaign can produce approximately 81 more batches than the single-product campaign in a year, whereas following the ABCABC . . . campaign will lead to a reduction of 30 batches per year compared to the single-product campaign. This illustrates that selecting the appropriate scheduling is critical to make optimal use of the equipment and maximize production.

It is worth investigating the difference in cycle times based on different campaigns, even though the products use the same equipment in the same order. In all three products, the bottlenecks are different equipment. Because of that, the cycle time for the single product campaign is based only on the largest cycle times in the various processes. So, during each campaign, the equipment with the longest cycle time is occupied all the time. Since the bottlenecks are different for a multiproduct campaign, no equipment is occupied all the time, and there is waiting time between the use of the equipment. However, in the ACBACB . . . campaign, it can be seen that the waiting time for equipment is not large, and the overall utilization of all the equipment is larger than the single product campaign. This is possible because some equipment with longer operation times can commence operations simultaneously with another batch in a different unit. Conversely, in ABCABC . . . campaigns, more bottlenecks exist, leading to longer waiting times for each piece of equipment compared to both the single-product campaign and the ACBACB . . . campaign. This inefficiency arises due to the presence of more bottlenecks and extended waiting times in each piece of equipment.

8.4.2 Jobshop Plants

Jobshop plants represent one of the most complex yet prevalent types of batch process plants. In these plants, different products do not necessarily use the same equipment or follow the same order of operations in equipment usage. However, some equipment is shared across different processes. Due to these variations, scheduling multiproduct campaigns can be particularly challenging for jobshop plants. There are several potential campaign scenarios, including single-product campaigns, with some being more efficient and others performing less effectively than single-product campaigns.

An example of a jobshop plant is presented in Figure 8.9.

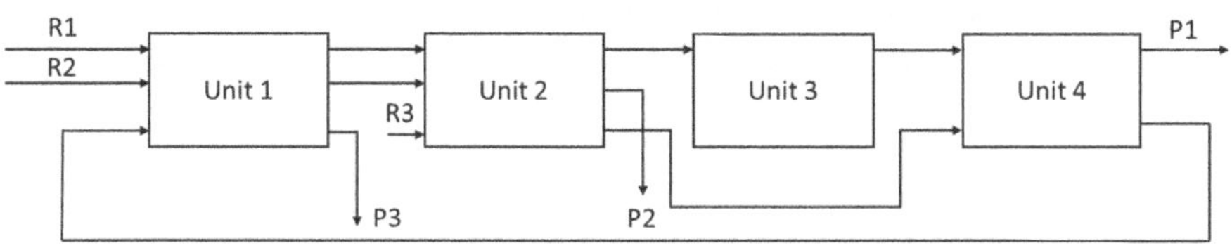

FIGURE 8.9 Example of a jobshop plant.

Unlike flowshop plants, there is no standard structure for jobshop plants. Because of that, a general representation of jobshop plant is not available. In the example presented in Figure 8.8, three products are produced in a single plant that makes use of four pieces of equipment. For product 1, the process starts from units and passes through units 2, 3 and 4 to produce the product, P1. For product 2, only two pieces of equipment are needed. Product 3 is produced in a significantly different manner. The process starts with unit 2 and then proceeds to unit 4 and finally completed in unit 1. In this case, both the pieces of equipment were involved, and the sequence are different from the rest. It can be seen that a jobshop plant can have several variations of this sequence. For example, even if product 3 follows the same sequence and equipment as product 1, it is still considered as a jobshop plant because product 2 does not utilize the same set of equipment as P1 and P3.

An illustration of various scheduling approaches for flowshop plants has been presented in Example 8.4.

Example 8.4. Jobshop Plant Scheduling

Consider the process described in Figure 8.9. A plant produces three products, namely P1, P2 and P3, using four equipment, Unit1, Unit2, Unit3 and E4, for one year. It is required to produce equal number of batches of these three products. The durations of the unit operations for the different products are shown in Table 8.2. It is to be noted that not all equipment is used in the production of the products P2 and P3, and the sequence of equipment utilization is different for P3.

Estimate the cycle times for the single product campaign and the campaign ABCABC . . . . Also estimate the number of batches this plant can produce in one year through different campaigns.

Solution

Since the duration of production is one year, the number of batches is large. Therefore, it is acceptable use the simplified form of the cycle time calculation.

For single product campaign:

Cycle time for A = Max(t_i) = 7h
Cycle time for B = 8h
Cycle time for C = 6h
Total cycle time = 21 h

TABLE 8.2
Equipment Usage for Different Products

Product	Time in Unit 1 (h)	Time in Unit 2 (h)	Time in Unit 3 (h)	Time in Unit 4 (h)
A	7	2	5	4
B	5	8		
C	4	6		2

$$\text{Number of batches in one year} = \frac{\text{Number of hours in one year}}{\text{Cycle time}} = \frac{8000}{21} \approx 381$$

The Gantt chart of ABC campaign is shown in Figure 8.10.

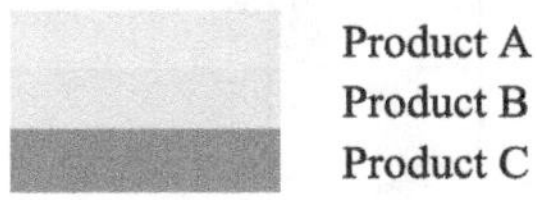

From Figure 8.10, the cycle time for one complete set of a batch can be estimated as 25h. Therefore, the number of batches in one year will be approximately 320. Therefore, in this case, the multiproduct campaign is significantly less efficient than the single-product campaign. However, in many cases, the multiproduct campaigns perform better compared to single-product campaigns. Therefore, the selection of the right campaign is a critical decision in the scheduling of job-shop plants.

In the previous example, it can be seen that the reason for the longer cycle time of this scheduling is that no equipment is utilized all the time, and the unutilized periods are lengthy. To reduce the unutilized period in unit 2, it is possible to schedule the process as ABABAB . . . CCC . . . sequence, as shown in Figure 8.11.

From Figure 8.11, it can be seen that the cycle time for ABAB . . . sequence is 13 h and CCC . . . sequence is 6 h. Therefore, the total cycle time will be 19 h. The number of batches per year will be approximately 421, which is more efficient than the single-product campaign. Therefore, by identifying the appropriate sequence, it is possible to increase the productivity of the plant significantly. However, it is to be noted that there are other factors such as cleaning time, storage space and product demand that need to be considered before making a final decision on scheduling.

8.5 STRATEGIES FOR CYCLE TIME IMPROVEMENT

On many occasions, the company's priority is to meet product demands even at a higher capital cost. This is particularly true for seasonal products when there can be high demand during peak periods. Therefore, a trade-off must be made between higher profits through increased production and the increased capital cost to produce more. Different strategies are employed to increase production in batch processing plants. The basic principle is to identify the bottleneck process and target process improvement by analysing the effect of changing the bottleneck. If the changes are worth the investment, it is prudent to make changes to the process. Three of the most common strategies for improving the cycle time and resource utilization of batch processing plants are using parallel units, intermediate storage and mixing of batches.

8.5.1 Parallel Units

This strategy is most commonly used in single-product campaigns, though it is also useful in multi-product campaigns. In several processes, one unit may have a significantly longer cycle time than the rest of the units. For example, a reactor with a longer time to complete the reaction or a crystallizer in which the controlled cooling

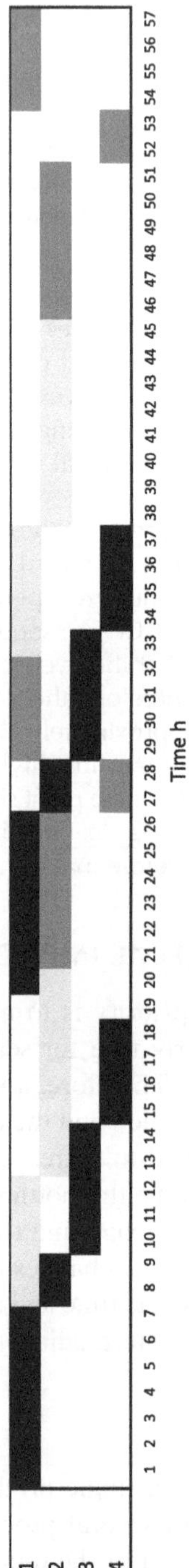

FIGURE 8.10 Gantt chart for ABC campaign of jobshop plant.

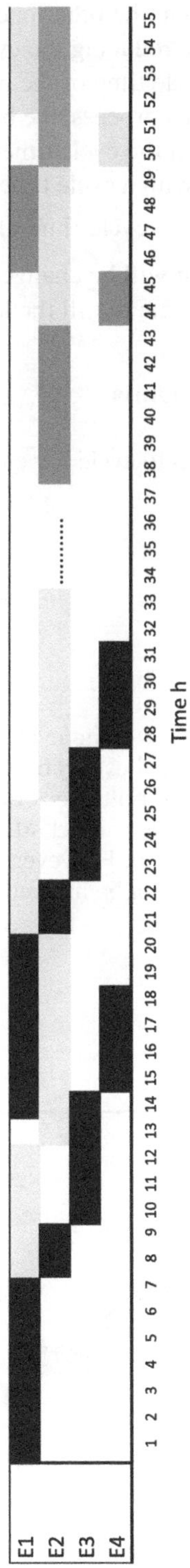

FIGURE 8.11 Gantt chart for ABABAB . . . CCCC campaign of jobshop plant.

consumes significantly more time than the other operations, thus delaying the completion of batches. In such situations, reducing the cycle time of the bottleneck units will lead to drastic changes to the cycle time of the process.

When we employ parallel units in a process, the batches sent to the different units of the same kind will alternate so that the cycle time for each unit will drop. If there are s parallel units of an equipment with a cycle time of t_i, the effective cycle time of that would be changed to $\frac{t_i}{s}$. Since the cycle time of the unit with the longest cycle time has changed, the bottleneck unit will be changed to the unit with the next highest cycle time (unless the new cycle time is still the longest).

Example 8.5. Effect of Parallel Units

In Example 8.1, what is the effect on to cycle if there is a parallel unit available for

i) R-101
ii) Crystallizer
iii) RVDF

Solution

i) With an additional reactor for R-101, the cycle time of R-101 will change to 4.5 h. This is because alternate batches can be sent to R-101a (parallel unit for R-101). So, R-101 is no longer the bottleneck unit. The new bottleneck unit is R-201 with a cycle time of 8 h. The Gantt chart is shown in Figure 8.12.
ii) Crystallizer cycle time will be 3.5 h. However, since R-101 is the bottleneck with 9 h of cycle time, there will not be any reduction to the cycle time of the process.
iii) Same as ii.

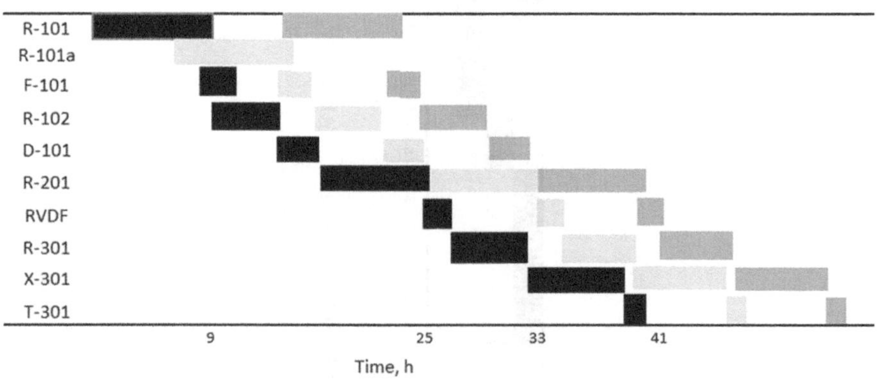

FIGURE 8.12 Gantt chart for process with parallel units.

8.5.2 Intermediate Storage

In single-product campaigns, it is possible to operate the bottleneck unit continuously, and the cycle time of the process is the longest cycle time among all the operations in the whole process. However, in multi-product campaigns, it is possible that none of the equipment is utilized completely because different products have different bottleneck operations. Therefore, one of the strategies to improve the cycle time of such processes is to make batches available for the unit with the longest cycle time. By doing this, the unit with the longest cycle time can be operated continuously and thereby reduce the cycle time of the process.

In the batch processing units with intermediate storage, the batches can be stored in storage tanks after some of the intermediate processes. The availability of intermediate storage tanks depends on a few factors.

i) Whether the improvement in cycle time is significant
ii) Whether the intermediate product can be safely stored
iii) Whether the storing cause delays in further processing

The maximum saving in cycle time can be achieved if unlimited intermediate storage (*uis*) is available. In that case, the cycle time of each equipment in all the processes are summed up. The highest among those values is the bottleneck unit. So, storage tanks need to be used to ensure that the unit with the highest cycle time operates continuously.

Example 8.6. Effect of Unlimited Intermediate Storage in Cycle Time

Revisit the process described in Example 8.4. If ABCABCABC . . . campaign is followed, no equipment is utilized completely. If unlimited intermediate storage is available, how can the cycle time be improved? Show the scheduling and utilization of storage tanks.

Solution

If unlimited storage is available, it is possible to run the unit with the longest cycle time continuously. The cycle times of the different units are shown in Table 8.3.

TABLE 8.3
Cycle Time of Different Units

Product	Time in Unit 1 (h)	Time in Unit 2 (h)	Time in Unit 3 (h)	Time in Unit 4 (h)
A	7	2	5	4
B	5	8		
C	4	6		2
Total Cycle Time	16	16	5	6

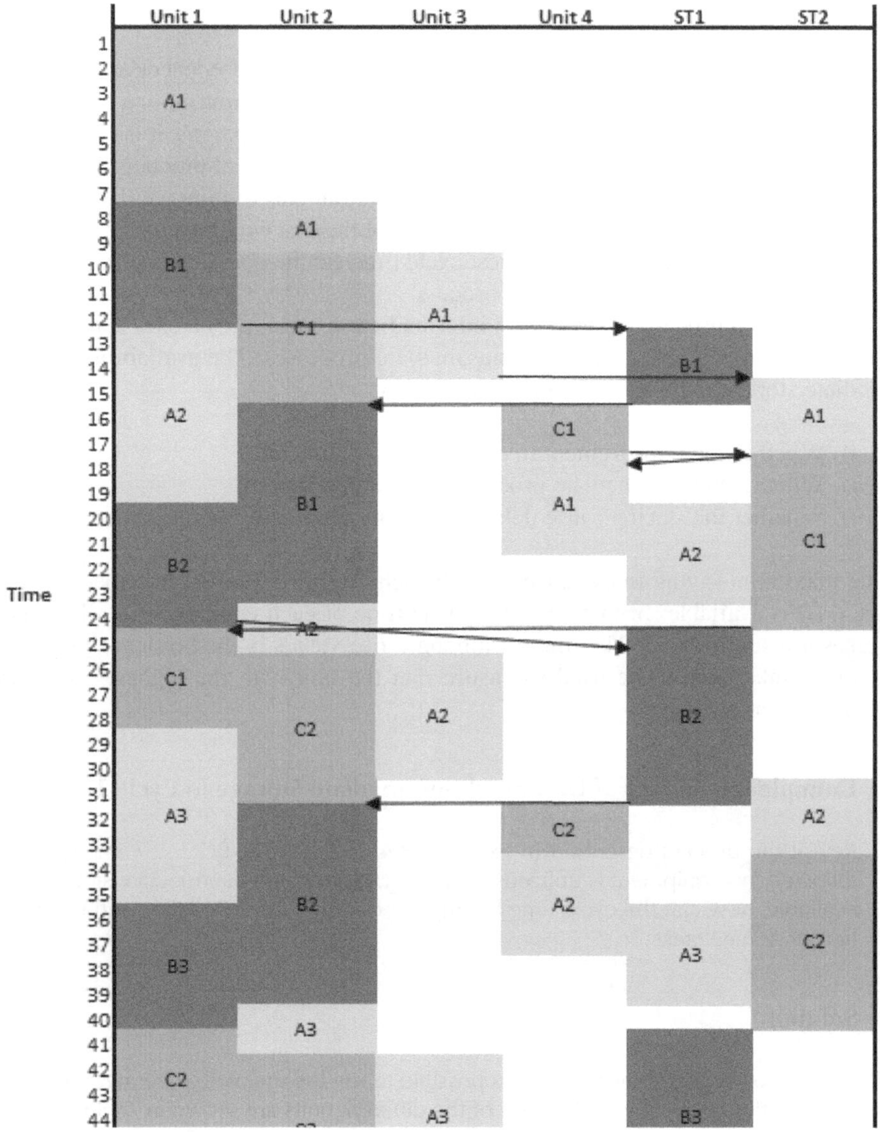

FIGURE 8.13 Jobshop plant with unlimited storage space.

Since the longest cycle time is 16 h, it is possible to run either unit 1 or unit 2 continuously. In that case, cycle time of the process will be 16 h. Since the current cycle time of ABCABC . . . sequence is 25 h, this is a significant reduction in the cycle time, and the number of batches per year will increase from 320 to 500.

The Gantt chart for this process is shown in Figure 8.13. It can be seen that by using two storage tanks, the bottleneck units can be operated continuously. The transfer of materials has been indicated in the diagram. It can be seen that there are several rounds of material transfer between units and storage tanks during the batch operation, which can make the operation complex. There will be higher capital and operating costs due to the complexity of operations, storage tanks, piping and manual labour involved in these steps.

8.5.3 Mixing of Batches

One of the strategies followed by the industrial practitioners is mixing different batches after intermediate stages. This is needed for several reasons. In some cases, there may be a significant reduction in the batch size after an intermediate stage because of the formation and removal of byproducts or wastes. In such cases, it is advantageous to mix more than one batch to utilize the available vessels in the plant. For example, if a significant amount of solvent is removed during distillation or filtration, the batches can be stored until there is sufficient material to fit in the next processing vessel. In another situation, the final product may have been produced with lower purity at an intermediate stage, and the final stages are needed for further purification of the product. In such situations, it is recommended to minimize the simultaneous handling of multiple batches because it may not provide any advantage but require more units to keep them separate.

8.6 CONCLUSIONS

Commercial production of chemical products is almost always achieved through batch processes. The efficient utilization of resources is one of the key steps in attaining maximum productivity and thus higher profits. This chapter covered the types of batch processes and typical challenges faced in the batch production plants. Various scheduling strategies typically used in single-product and multiproduct production campaigns have been explored in detail. Complex multiproduct campaigns such as flowshop and jobshop plants have been introduced, and the impact of efficient scheduling techniques has been demonstrated. Commonly used strategies to improve the productivity of these plants have also been covered in this chapter. The trade-off between the higher production, cost of production and complexity have also been discussed.

REFERENCES

American Institute of Chemical Engineers. (1999). *Guidelines for Process Safety in Batch Reaction Systems*. New York: Center for Chemical Process Safety of the American Institute of Chemical Engineers.

Biegler, L. T., Grossmann, I. E., & Westerberg, A. W. (1997). *Systematic Methods for Chemical Process Design*. Upper Saddle River, NJ: Prentice Hall PTR.

Felder, R., Rousseau, R., & Bullard, L. (2018). *Elementary Principles of Chemical Processes* (4th ed.). Wiley.

Turton, R., Bailie, R. C., Whiting, W. B., Shaeiwitz, J. A., & Bhattacharyya, D. (2012). *Analysis, Synthesis, and Design of Chemical Processes* (4th ed.). Upper Saddle River, NJ: Pearson Education International.

8.7 PROBLEMS

1. Consider the batch processes to produce products A, B and C as shown in Table Q8.1.

TABLE Q8.1
Equipment Processing Times (in Hours) for Processes A, B and C

Process	Mixer	Reactor	Crystallizer	Dryer
A	2	2	10	1.5
B	–	3	9	2.5
C	1	2.5	8	3.5

   a) Construct the Gantt chart for overlapping multiproduct campaign with sequence ACBACBACB ...... Estimate the exact number of batches of each product that can be produced in one month by assuming 500 h of operation.
   b) It is desired to run the reactor all the time so that the production can be increased. For this purpose, it is decided to buy a few parallel equipment in this process. Determine the type and number of parallel units required so that the reactor can be operated all the time. Show the Gantt chart for the new schedule. Also, how much is the improvement in cycle time?
   c) If all the processes do not have crystallizer, estimate the cycle time for the sequence ACBACB ... if the scheduling policy is unlimited intermediate storage. Let the first batch entering process A be 'n'. Show the scheduling with Gantt chart by clearly showing the batch numbers for all batches in different units. Also determine the number of storage tanks required.
2. Consider the batch processes to produce products A, B and C, as shown in Table Q8.2.

TABLE Q8.2
Equipment Processing Times (in Hours) for Processes A, B and C

Process	Reactor	Extractor	Separator
A	3	6	3
B	2	–	4
C	5	7	4

   a) Construct the Gantt chart for overlapping multiproduct campaign with sequence ABCABCABC ...... Estimate the exact number of batches of each product that can be produced in one month by assuming 500 h of operation.
   b) It is decided to buy additional equipment or storage tank so that the cycle time can be improved in this process. If the cost of having any additional

equipment or storage tank is the same, which additional equipment/storage tank can give maximum benefit in cycle time? Justify your answer with proper illustration.

c) A process modification has been suggested for product C in which the extraction step is to be conducted before the reaction step. Because of the change in process, the extraction can be completed in 6 hours and reaction time is reduced to 3 hours. If the multiproduct campaign sequence is still strictly ABCABC . . ., what is the effect in cycle time because of this change? Also propose a strategy that can significantly improve the cycle time without the purchase of additional equipment or storage tanks.

3. The recipe to manufacture products A, B and C is as follows:

A: 15 kg of liquid x and 10 kg of liquid y is to be sent to a reactor to produce A. The reaction produces 8 kg of A in 2 hours, and the density of the mixture is 900 kg/m3. The mixture is then pumped through a filter for catalyst removal to a distillation column, and it takes 1 hour to complete the transfer. The distillation process begins after the transfer is complete. It takes 4 hours to separate 7 kg of product A from the unreacted raw materials.

B: 20 kg of liquid x is heated in a reactor to produce 14 kg of B in 8 hours. Density of the mixture is 800 kg/m3. The mixture is to be sent to a crystallizer to recover 10 kg of B in 6 hours. After the crystallization process, the mixture needs to pass through a filter to recover 9 kg of B. Filtration takes 1 hour to complete.

C: 20 kg of a liquid mixture (density 800 kg/m3) of y and z is sent to a distillation column to remove most of the y in 4 hours. The remaining 12 kg of the mixture is heated in the presence of a catalyst in a reactor for 2 hours to produce 10 kg of C. The density of the mixture is 700 kg/m3. The mixture of C and unreacted y and z is sent to a crystallizer to separate 9 kg of C from the mixture in 6 hours. After the crystallization process, the mixture needs to pass through a filter to recover 8 kg of C. Filtration takes 2 hours to complete.

a) Construct the Gantt chart for overlapping multiproduct campaign with sequence ABCABCABC . . . . . . Estimate the number of batches of each product that can be produced in a typical month. Assume 500 h of operation. Compare this sequencing with a single product campaign to produce equal number of batches of A, B and C. Comment on the possible reasons for the better performance of the single product/multiproduct campaign based on the schedule.

b) How much will be the saving in cycle time for the following cases in the multiproduct (ABC . . .) campaign.

i) Availability of an additional reactor
ii) Availability of an additional distillation column
iii) Availability of an additional crystallizer
iv) Availability of a storage tank

4. The products A, B and C are produced in lab scale by the following procedure.
   A: 10 kg of liquid x and 20 kg of liquid y are sent to a reactor to produce A. The reaction produces 9 kg of A in 3 hours, and the density of the mixture is 1000 kg/m^3. The mixture is then sent to a distillation column to separate 8 kg of A from the mixture in 2 hours.
   B: 25 kg of liquid x is heated with a catalyst in a reactor to produce 20 kg of B in 4 hours. Density of the mixture is 900 kg/m^3. The mixture is then sent to a crystallizer, and 18 kg of B is produced in 5 hours.
   C: 30 kg of a liquid mixture (density 800 kg/m^3) of y and z is sent to a distillation column to remove most of the y in 4 hours. The remaining 20 kg of the mixture is heated in the presence of a catalyst in a reactor. The density of the mixture is 850 kg/m^3. 12 kg of C is produced in 4 hours. The mixture of C and unreacted y and z is sent to a crystallizer to separate 11 kg of C from the mixture in 6 hours.
   a) Construct the Gantt chart for an overlapping multiproduct campaign with the sequence ABCABCABC...... Compare this sequencing with a single product campaign to produce equal number of batches of A, B and C. Comment on the possible reasons for the better performance of the single product/multiproduct campaign based on the schedule.
   b) In the multiproduct campaign done in part a, how can the cycle time be improved with the help of storage tanks? Show the scheduling and transfer of materials between the tanks and the units and explain the flow.
5. A company has received an order to produce the products A, B and C in one week. The demand for products A, B and C are 20, 25 and 30 tonnes, respectively. The recipe to produce these products is given next:
   A: 10 kg of liquid X is sent to a reactor to produce crude A. The reaction produces 6 kg of A in 6 hours. The mixture is then sent to a batch extraction system. It takes 3 hours to extract 5 kg of product A from the unreacted raw materials. In the next step, 4.5 kg of pure A is separated using a distillation column in 3 hours.
   B: 20 kg of liquid Y and 15 kg of X are heated in a reactor to produce 20 kg of B in 3 hours. The mixture is to be sent to a distillation column to recover 10 kg of B in 4 hours. After the distillation, product B is passed through an adsorption tower to remove traces of water. This process is completed in 2 hours.
   C: 20 kg of C and Z is sent to a batch extraction system to remove most of the Z in 6 hours. The mixture of C and unrecovered Z is sent to a distillation column to separate 12 kg of C from the mixture in 3 hours.
   a) If the plan is to produce equal number of batches of A, B and C, compare the number of batches that can be produced through
      i) A single product campaign
      ii) A multiproduct campaign with sequence ABCABCABC......
      Gantt chart needs to be constructed only for the multiproduct campaign.
   b) Construct the Gantt chart of the multiproduct campaign ABCABC... that gives the best cycle time if unlimited storage space is available.

9 Common Separation Processes in Chemical Product Industry

The separation process that dominates the chemical product industry is different from those commonly used in the commodity chemical industry. Fractional distillation is the most common separation technique used in the commodity industry. However, because of the small quantities of chemicals used in the product industry, continuous operations are not that common. In addition, the majority of the processes involve solid handling or dealing with thermally sensitive materials in the product industry. Because of that, the role of distillation is typically limited to a smaller fraction of the operations, such as solvent recovery using batch distillation, conducted in such processes. While fractional distillation is not commonly used in the major separation stages of speciality chemicals, steam distillation is commonly used in the recovery of essential oils. Therefore, it is important to have an understanding of this process in the product industry.

The cost of separation is a major component in the price of the final product for many speciality chemicals because they are produced in low concentrations and need to be concentrated to extremely high purities. The separation cost estimations for fine chemical production can be approximated using the Sherwood plots, as covered in Chapter 6. For pharmaceutical products, the level of impurities needs to be lowered to nearly zero due to the effect of such impurities on efficiency and their side effects. Therefore, a series of separation steps needs to be operated at different concentrations.

Most separation methods rely on the presence of a significant difference in some physical property to achieve a good separation between a target component and the rest of the mixture. With bio-oil mixtures that contain several components, the probability that many components have highly similar or overlapping parameters, such as boiling point, melting point, density and miscibility in organic and aqueous phases, is high. To deal with such mixtures, the RIPP (Recovery, Isolate, Purify, Polish) scheme is commonly employed in the pharmaceutical sector, as seen in the extraction of phytosterols (high-value biomolecules consumed for different pharmacological reasons) (Cussler and Moggridge, 2011). In the recovery stage, the foreign substances are removed through filtration. The second stage is the isolation of the target component from the rest of the feed. Since the feed is typically supplied at low concentrations, the goal at this stage is to increase the concentration of the target component. The most common approach used at this stage is extraction. While extraction cannot usually target a single component at high purity, it can eliminate most of the impurities. The next stage is the purification of the concentrated solution separated

DOI: 10.1201/9781032708911-9

through extraction. The most common approaches are adsorption and ion exchange processes. The final polishing is done through crystallization, where products can be concentrated to extremely high purities. After crystallization, the products are washed and dried before packaging.

While there are several separation tools used in the chemical product industry, the focus of this chapter is on the most commonly used separation tools, which are steam distillation, extraction, adsorption, crystallization and spray drying. The rest of the sub-sections will provide an overview of these tools.

9.1 STEAM DISTILLATION

Steam distillation is the most widely used separation technique in the essential oil industry to recover fragrant oils of vegetable origin. This process helps to conduct distillation at significantly lower temperatures than the boiling points of those chemicals. This is an extremely valuable method to separate chemicals with high boiling points and components that decompose below their boiling points. Chemical products such as perfumes, essential oils such as lavender oil, eucalyptus oil and citrus oil, are typically purified by this method. This can also be used in separating petroleum fractions and waxes at relatively low temperatures.

In steam distillation, the target products must be completely immiscible with water, which is usually the case with essential oils. Steam is generally directly passed through the mixture from which the target components are to be separated. The vessel in which this process is carried out is called still. Steam distillation is usually carried out by passing steam into the mixture through a series of perforations on coils that are placed inside the still. However, it is also possible to heat the mixture externally. The vapour product from this process contains the target compounds and water. The vapour is condensed and collected in receivers, and they form two immiscible layers (aqueous layer and organic layer of targeted chemicals) from which the targeted compounds can be separated by gravity separation. A simplified representation of a steam distillation system is shown in Figure 9.1.

9.1.1 THERMODYNAMICS OF STEAM DISTILLATION

Since steam distillation is used to separate the oils that are completely immiscible with water, the components will exert their own individual vapour pressures at the temperature at which the system is kept. The phase diagram of the system under constant pressure is shown in Figure 9.2.

In Figure 9.2, a binary system of two components A and B is presented where the vapour pressure of A is significantly higher than that of B. Let A be steam and B be the component targeted for separation. Consider the condition at the point M in the diagram where the vapour pressures of both A and B are below the pressure at which they are kept. If the mixture is heated, both A and B exert their own vapour pressure because they are completely immiscible. Therefore, unlike a solution where the original components lose their identity, the constituents in the mixture preserve their individual identity. If this mixture is heated to temperature T_1, the sum of their vapour pressures reaches the pressure at which the mixture is kept, and the boiling begins.

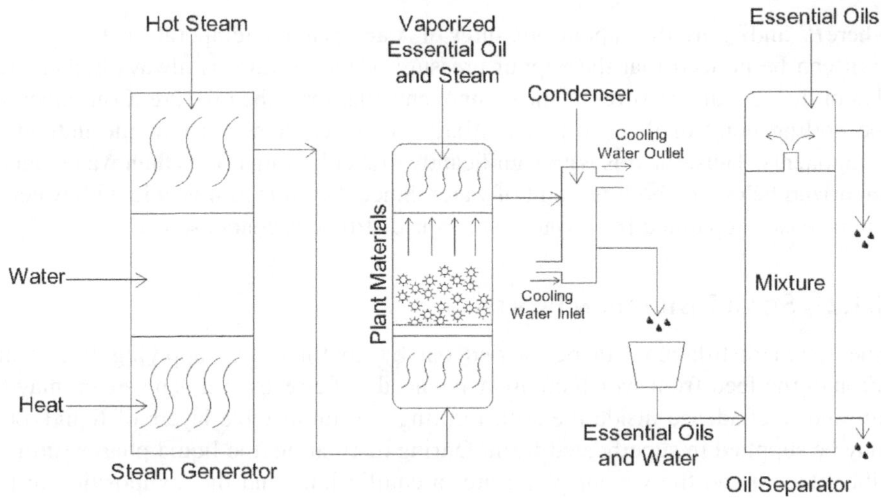

FIGURE 9.1 Steam distillation system.

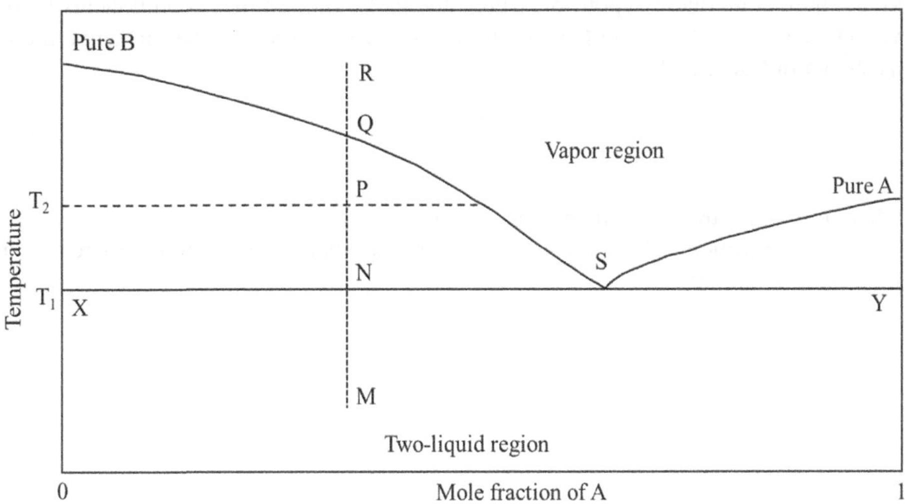

FIGURE 9.2 Phase diagram of immiscible systems.

At this point, the pure liquids and one vapour phase co-exist in equilibrium. The composition of the vapour phase, which is in equilibrium with the liquids, is shown by the point *S*. The temperature at which the vaporization begins can be estimated based on the sum of vapour pressures of the pure liquids, as shown in Equation 9.1.

$$P = P_A^S + P_B^S \tag{9.1}$$

where P_A^S and P_B^S are the vapour pressures of A and B at the temperature T_1.

It can be noticed that the vapour pressure of the mixture is always higher than that of the vapour pressures of the components that form the mixture. Consequently, the boiling point of the two-phase mixture is lower than either of the individual components. Thus, the oils with significantly higher boiling points than water can be vaporized below the boiling point of water. Since the oil is immiscible with water, it can be easily separated from water when the distillate is condensed.

9.1.2 Steam Distillation Process

The steam distillation can be carried out by continuously supplying live steam through the feed from which components need to be recovered. The steam may be allowed to condense inside the still, forming two immiscible layers of liquid, or it may be supplied in superheated form. During distillation, the liquid phases (immiscible phases) and the vapour phase are in equilibrium, and the composition of the vapour is represented by point S in Figure 9.2. The distillation temperature can be obtained by plotting the pressure against temperature. The temperature at which the sum of the vapour pressures reaches the system's pressure is the distillation temperature. Since both the steam and the component to be recovered exert their own individual pure component vapour pressures, the molar ratio of these components in the vapour will be in the ratio of their vapour pressures at the distillation temperature, as shown in Equation 9.2.

$$\frac{n_A}{n_B} = \frac{P_A^S}{P_B^S} \tag{9.2}$$

where n_A and n_B are the number of moles of A and B.

The compositions of steam and the target components can be estimated from Equations 9.3 and 9.4.

$$y_A = \frac{n_A}{n_A + n_B} = \frac{P_A^S}{P} \tag{9.3}$$

$$y_B = \frac{n_B}{n_A + n_B} = \frac{P_B^S}{P} \tag{9.4}$$

The amount of steam needed to recover the target component can be estimated from Equation 9.5.

$$w_A = w_B \left(\frac{P_A^S}{P_B^S} \right) \times \frac{M_A}{M_B} \tag{9.5}$$

where w_A is the mass of steam required to recover w_B kg of B, M_A and M_B are the molecular weights of A and B.

In many situations, especially for the recovery of plant-based substances such as essential oils and fragrances, the target component comprises several chemicals.

Steam distillation typically targets to remove all these chemicals together. Since these chemicals have chemically similar structure and miscible with each other, we can apply Raoult's law to estimate the vapour pressure exerted by that mixture.

$$P_B^S = P_{B1}^S x_1 + P_{B2}^S x_2 + P_{B3}^S x_3 + \ldots \tag{9.6}$$

In Equation 9.6, $P_{B1}^S, P_{B2}^S, P_{B3}^S \ldots$ are the vapour pressures of various components B_1, B_2, B_3 in the oil and x_1, x_2 are their mole fractions in the oil.

Equation 9.5 is based on the assumption that there will be perfect mixing between steam and the target component. However, in the real systems, the partial pressure exerted by the material in the vapour will be less than the expected pressure. The term vaporization efficiency, E_v, is used to quantify this deviation and has been defined as shown in Equation 9.7.

$$E_v = \frac{P_{B,\,actual}^S}{P_B^S} \tag{9.7}$$

Equation 9.5 is now modified to estimate the actual steam requirement to recover the target component.

$$\left(w_A\right)_{actual} = w_B \left(\frac{P_A^S}{E_v P_B^S}\right) \times \frac{M_A}{M_B} \tag{9.8}$$

In addition to the steam needed for the recovery of the components, steam is also needed to heat the feed from the supply conditions to the operating conditions. In addition, steam also needs to provide the latent heat required for the vaporization of the feed. If C_{pF} and λ_F are the specific heat and latent heat of vaporization of the feed, F is the feed, T_F is the feed temperature and the distillation temperature is T, the total requirement can be estimated from Equation 9.9.

$$\left(W_A\right)_{total} = \frac{1}{\lambda_S}\left[FC_{pF}\left(T - T_F\right) + w_B \lambda_F\right] + w_B \left(\frac{P_A^S}{E_v P_B^S}\right) \times \frac{M_A}{M_B} \tag{9.9}$$

In Equation 9.9, λ_S is the latent heat of vaporization of steam.

It is to be noted that the components separated using steam distillation usually have significantly higher boiling point than water and thus exert extremely low vapour pressure near the boiling point of water. Because of this, the steam required to carry the vapour is usually significantly higher than the steam needed for heating.

Example 9.1. Steam Distillation

1. A mixture of two essential oils (A and B) is to be recovered from a number of non-volatile impurities. The oils are heat sensitive and will decompose well below their boiling points. A pilot scale unit is charged with 1 mole of crude oil mixture with a molar composition of 75% A and 25% B at 50°C. The oils

TABLE 9.1
Vapour Pressure Data

Temperature, K	353	373
Vapour pressure of water, kPa	47.98	101.3
Vapour pressure of oil A, kPa	2.68	5.34
Vapour pressure of oil B, kPa	2.64	5.30
Vapour pressure of impurity, kPa	0.0001	0.0002

TABLE 9.2
Latent Heat and Heat Capacity Data

	Latent Heat of Vaporization (kJ/mol)	Heat Capacity (J/mol K)
Water	125 (at 105 kPa)	75
Oil A	100	230
Oil B	105	300

form an ideal mixture that obeys Raoult's law and are immiscible with water. It is proposed to recover the oils by allowing them to mix with steam directly at a total pressure of 95 kPa. The vapour pressure data are given in Table 9.1. Assume that the impurities do not affect the vaporization characteristics. Also, assume that the vapour pressure varies linearly with temperature. Based on this, calculate the following:

a) The equilibrium temperature.
b) The composition of the resulting vapour.
c) The amount of steam required to recover all the oils if the steam is supplied at 105 kPa. Temperature of saturation of steam at 105 kPa = 375 K. Enthalpy of saturated steam at 95 kPa=148 KJ/mol and 149 kJ/mol at 105 kPa. The data on latent heat and heat capacity are given in Table 9.2.

Solution

a) Vapour Pressure (V.P.) of mixture

$$(\mathrm{A+B})\ \text{at}\ 353\mathrm{K} = 0.75 \times 2.68 + 0.25 \times 2.64 = 2.67\,kPa$$

$$\text{V.P. of A+B at } 373\mathrm{K} = 5.34 \times 0.75 + 5.3 \times 0.25 = 5.33\,kPa$$

$$\text{Total V.P. at } 353\mathrm{K} = 47.98 + 2.67 = 50.65\,kPa$$

$$\text{Total V.P. at } 373\mathrm{K} = 106.63\,kPa$$

$$\text{Increase in } \frac{\text{V.P.}}{1°\text{C}} = \frac{106.63 - 50.65}{20} = 2.799\ kPa$$

Required increase in Temperature = ?

$$\text{V.P.} = 50.65 - 95\ kPa$$

$$\therefore \text{Increase in Temperature} = \frac{95 - 50.65}{2.799} = 15.85\ k\text{Pa}$$

$$\therefore \text{Operating Temperature} = 368.85\ K$$

b) $$\text{At } 368.85\text{K},\ P_A^{H_2O} = 47.98 + \left(\frac{101.3 - 47.98}{20}\right) \times 15.85 = 90.24\ kPa$$

$$\therefore \text{Mole fraction of } H_2O = \frac{90.24}{95} = 0.95$$

$$\text{Vapour composition} = 95\%\ H_2O, 5\%\ oil\ (3.75\%\ A \text{ and } 1.25\%\ B)$$

c) $$\text{Amount of water needed per mole of oil} = \frac{95}{5} = 19\,moles$$

$$\text{Sensible heat} = 0.75 \times 230 \frac{\text{J}}{mol.K} \times 45.85 = 11348J = 11.35kJ$$

$$\text{Latent heat} = 0.75 \times 100 + 0.25 \times 105 = 101.25kJ$$

$$\text{Heat released by 19 mole steam} = 19 \times (148.5 - 148) = 9.5\ kJ$$

$$\text{Heat needed} = 11.35 + 101.252 - 9.5 = 103.1\ kJ$$

$$\text{Heat of vaporization of steam at } 105\ kPa = 125 \frac{kJ}{mol}$$

$$\text{Heat given by 1mol of steam} = 125 + \frac{75}{1000} \times 1(375 - 368.85) = 125.46\ kJ$$

$$\therefore \text{Steam needed for heating} = \frac{103.1kJ}{125.46} = 0.82mol$$

$$\therefore \text{Total steam needed} = 19.82mol = 356.8kg$$

9.2 EXTRACTION

In the chemical product industry, the main role of extraction is to increase the concentration of the target component before further purification. Typically, the extraction of target components is carried out in batch units. It is to be noted that extraction can rarely be highly selective. Therefore, the focus at this stage is to minimize losses and also to minimize the extraction of non-targeted components present in the mixture. Four aspects merit discussion are as follows:

1) Choice of solvent.
2) How much separation in one batch.
3) How much losses.
4) How to purify by repeated extractions.

9.2.1 Solvent Selection

In the chemical process industries, the successful performance of a solvent-based operation has a great influence on the feasibility of a downstream chemical product manufactured through that process. In turn, the successful selection of solvent plays a key role in the successful performance of an extraction process. The feasibility of employing a solvent in a given application is dominated by various considerations of efficiency, including both technical (e.g. performance) and economic factors (e.g. solvent cost) (Clarke et al., 2018). An ideal solvent should provide the best performance for the process.

The key criteria for solvent selection are capacity, selectivity and solvent loss (Gmehling and Schedemann, 2014). However, capacity and selectivity are often antagonistic properties. This has been observed as solvents with a high selectivity often exhibit a low capacity and vice versa. The properties of a mixture of two solvents that exhibit opposite behaviour may be enhanced compared to those of pure solvents (Gmehling and Schedemann, 2014). Both selectivity and capacity of the pure solvent can be estimated through calculation of the infinite-dilution activity coefficient of the solute in the solvent. The inverse relationship between infinite-dilution activity coefficient and capacity (solvent power) is shown in Equation 9.10.

$$Sp_i = \frac{1}{\gamma_{i,s}^{\infty}} \tag{9.10}$$

where Sp_i denotes the solvent power of the pure solvent for a component i and $\gamma_{i,s}^{\infty}$ is the infinite-dilution activity coefficient of the extracted component i in the solvent. Equation 9.11 is used to calculate the selectivity (Sel_{ij}) of the pure solvent for a component i relative to component j, which is simply the ratio between infinite-dilution activity coefficient of the extracted component i and component j in the solvent.

$$Sel_{ij} = \frac{\gamma_{i,s}^{\infty}}{\gamma_{j,s}^{\infty}} \tag{9.11}$$

A good compromise needs to be made between the selectivity and solvent power to select the ideal solvent. In addition to these two key targets, the performance of a solvent is determined from several key criteria, which are listed in Table 9.3.

TABLE 9.3
Attributes for Solvent Selection

Solvent Specifications	Target Properties	Justifications on Product Specifications
Exist as liquid under atmospheric conditions	Normal boiling point Normal melting point	The solvent or solvent mixture must be in liquid phase to be used in LLE operation under atmospheric conditions. The melting point of solvent should not be too close to the extraction temperature, to avoid solid precipitation caused by a decrease in temperature.
Easy to recover the solvent	Normal boiling point/azeotrope formation	The boiling point of the solvent should be significantly different from the solute so that the solvent recovery using distillation is easy. Solvent also should not form azeotropes with the feed or the solute.
Density	High density difference	Higher density difference with the feed makes the separation easier.
Viscosity	Low	To reduce the power consumption for mixing and transfer.
Interfacial tension between the feed	Should not be too low or too high	Low interfacial tension between the feed and the solvent would enhance the mass transfer between the solute and the solvent. However, too low interfacial tension between the solvent and feed phases will make their separation harder.
Possess favourable safety and health characteristics	Total penalty score of the inherent safety and health sub-indexes.	To decrease the risks for the personnel who would handle and be exposed to the solvent. The details of these indices are given in Chapter 5.

9.2.2 Extraction Thermodynamics

Consider a dilute product 1 in water which needs to be recovered using a solvent. Since the solute has a higher solubility in the solvent compared to water, it will move from water into the solvent until an equilibrium has been reached.

Once the equilibrium has been reached between feed and solvent, their chemical potentials must be the same:

$$\mu_1(\text{in solvent}) = \mu_1(\text{in feed}) \tag{9.12}$$

$$\mu_1(in\,solvent) = \mu_1^o(\text{in solvent}) + RT\,ln\,cx_1(in\,solvent) \tag{9.13}$$

where μ_1 is the chemical potential, μ_1^o is the chemical potential in a standard state, c is total concentration, x_l and y_l are mole fractions in solvent and feed, respectively.

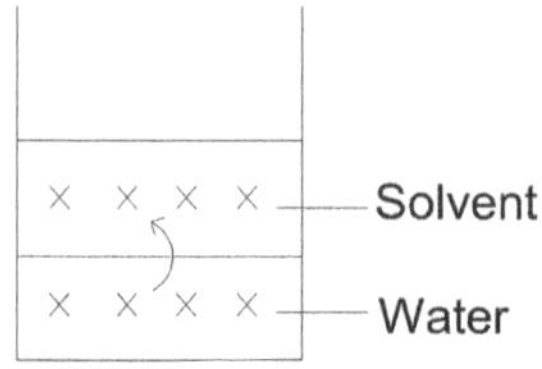

FIGURE 9.3 Extraction equilibrium.

$$\mu_1^o\left(\text{in solvent}\right)+RT\,ln\,cx_1=\mu_1^o\left(\text{in feed}\right)+RT\,ln\,cy_1 \tag{9.14}$$

Since the objective of extraction is to achieve a higher concentration of the solute in the solvent compared to the feed, the term 'partition coefficient' (m) has been introduced. The partition coefficient is the ratio of the solute concentrations in the solvent and the feed:

$$\text{m}=\frac{y_1}{x_1}=\exp\left[\frac{\mu_1^\text{o}(\text{in solvent})-\mu_1^\text{o}(\text{in feed}}{\text{RT}}\right] \tag{9.15}$$

For efficient extraction, the value of partition coefficient must be small. However, Equation 9.15 may not be directly used to find the value of m as the chemical potential in a standard state may not be available easily. To obtain a useful expression to find the value of m, consider a saturated solution of the solute in the feed. In that case:

$$\mu_1(as\ \text{solid})=\mu_1^o(\text{in feed})+RT\,ln\,cy_1(\text{saturated feed}) \tag{9.16}$$

Similarly for saturated solvent:

$$\mu_1(as\ \text{solid})=\mu_1^o(\text{in solvent})+RT\,ln\,cx_1(\text{saturated feed}) \tag{9.17}$$

Solving for the standard state values and inserting into (9.17):

$$\text{m}=\frac{\text{y}_1(\text{in feed})}{\text{x}_1(\text{in solvent})}=\frac{\text{y}_1(\text{saturated feed})}{\text{x}_1(\text{saturated solvent})} \tag{9.18}$$

In Equation 9.18, m is defined in terms of the saturation concentrations in the feed (usually water) and the solvent. Since saturation concentrations are easier to get, this equation provides a way to get the partition coefficient for any feed concentrations.

9.2.3 Separation Efficiency

Consider a process where H moles of feed with a solute concentration of y_0. To recover the solute, L moles of pure solvent have been used. The concentration of the solute in the feed is changed to y_1, and the solvent concentration is increased to x_1.

From mass balance:

$$Hy_0 = Hy_1 + L\,x_1 \tag{9.19}$$

$$y_1 = mx_1$$

The fraction extracted f is then:

$$f = \frac{L\,x_1}{H\,y_0} = \frac{Lx_1}{Hy_1 + Lx_1} = \frac{Lx_1}{Lx_1 + mHx_1} = \frac{1}{1 + (mH/L)} \tag{9.20}$$

Thus, a small value of m gives a large f.

The concentrations of feed and the solvent phases can also be estimated as shown in Equations 9.21 and 9.22.

$$x_1 = \frac{fHy_0}{L} = \frac{Hy_0}{\left(1 + \frac{mH}{L}\right)L} = \frac{y_0}{\frac{L}{H} + m} \tag{9.21}$$

$$y_1 = mx_1 = \frac{y_0}{1 + \frac{L}{Hm}} \tag{9.22}$$

From Equation 9.22, the loss of the solute can be estimated. If the solute is an expensive material, it is worth conducting multiple stages of extraction to recover most of the solute. The solute feed concentration after two rounds of extraction is estimated from Equation 9.23.

$$y_2 = \frac{y_1}{\left(1 + \frac{L}{Hm}\right)} = \frac{y_0}{\left(1 + \frac{L}{Hm}\right)^2} \tag{9.23}$$

So, for a multistage extraction unit, the losses can be estimated from Equation 9.24.

$$\text{Losses} = Hy_n = \frac{Hy_0}{\left(1 + \frac{L}{mH}\right)^n} \tag{9.24}$$

where *n* is the number of extraction stages.

Example 9.2. Extraction of Chemicals from Dilute Solution

After the fermentation process, the pharmaceutical product is extracted from the fermentation broth using an organic solvent in a multi-stage extraction unit. The distribution coefficient (the ratio of mole fractions of pharmaceutical product in solvent to water) of product at pH=4 is 25. The volume of aqueous phase is 6000 L. If only two extraction units are available to recover 98% of product from the aqueous phase, determine the required amount of the organic solvent in the extraction unit. You can set the same amount of solvent in both units.

Solution

Fractional extraction after two rounds of extraction:

$$f = 1 - \frac{1}{\left(1 + \frac{L}{mH}\right)^N}$$

In this case, $f = 0.98$, $m = 1/25 = 0.04$, $N = 2$, $H = 6000 l$

$L = 1457\ l$

9.3 ADSORPTION

Adsorption is a mass transfer process that occurs when a gas or liquid solute is selectively transferred onto the surface of a solid, forming a molecular or atomic film. The solid material is referred to as the adsorbent, and the solute adsorbed on the adsorbent is called the adsorbate. The adsorbent is usually in the form of small particles arranged in a fixed bed. As the fluid passes through this bed, the solid particles adsorb components from the fluid. Once the bed approaches saturation, the flow is stopped, and the bed undergoes regeneration by desorbing the adsorbate through thermal or other methods. The adsorbate is thus recovered, and the solid adsorbent is ready for another cycle of adsorption. The process is illustrated in Figure 9.4.

The reverse of adsorption is called desorption. Regeneration or desorption is the process where a substance detaches from a surface. This process allows valuable adsorbates to be recovered and enables the adsorbent to be reused in subsequent cycles. To desorb an adsorbate, one can either raise the temperature of adsorbent or lower the partial pressure of the adsorbate if it is in gas phase, or its concentration if it is in the liquid phase. Since the adsorbent will be subjected to varying operating conditions, such as transitioning from low temperature to high temperature, it must move from one location to another. Nonetheless, moving adsorbents without damaging particles and equipment is impractical. Therefore, the adsorption system must have two zones (or fixed beds) that allow for periodic changes in operating conditions in each zone. Practically, this requires at least two beds arranged in parallel but out of phase to achieve a pseudo steady-state condition. Most adsorption processes operate in a semi-batch mode, where adsorbent beds are periodically loaded and unloaded.

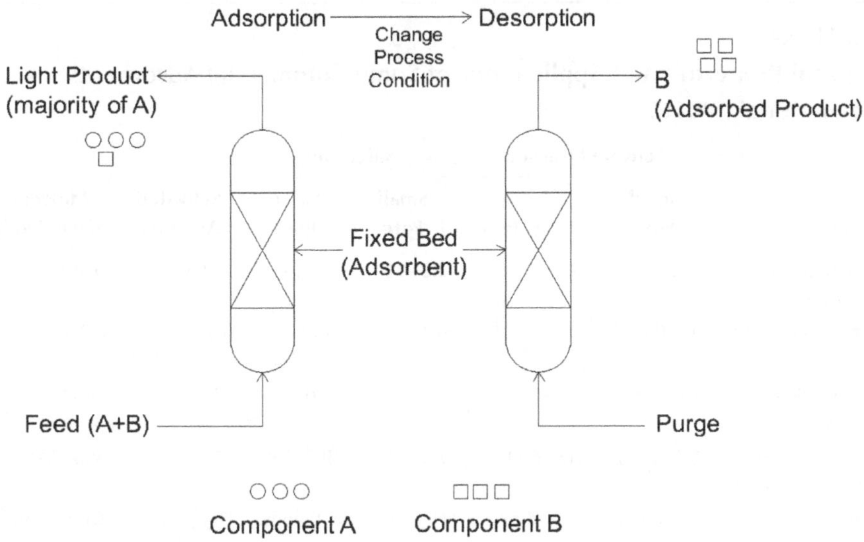

FIGURE 9.4 An example of cyclic fixed-bed adsorption system operates in batch mode. A feed with components A and B, represented by circles and squares, respectively, enters the fixed-bed adsorber. Component B, the targeted product is adsorbed on the adsorbent surface when the bed is operating in adsorption mode. By changing the process operating condition, i.e., increase the operating temperature, the saturated bed is regenerated, and the adsorbate, component B, is recovered.

Today, adsorption processes are widely utilized in many real-world applications. They have become the most widely used non-vapour-liquid technology for molecular separations in the natural gas, chemical, petroleum and petrochemical industries. For instance, gas-phase adsorption is employed for the dehydration of air and natural gas, the filtration of dust, particles, chemicals and odours from air and the removal of sulphur components from natural gas. Similarly, liquid-phase adsorption is used for removing organics from water, eliminating colourizing agents, and extracting sulphur compounds from organic solutions.

9.3.1 Adsorbents

Adsorbents are normally available as small cylindrical pellets, flakes, beads or granules, with diameters ranging from 20 μm to 12 mm and specific surface areas from 300 to 1200 m^2/g (Seader et al., 2016). These adsorbent particles possess a highly porous structure, characterized by numerous fine pores that can constitute up to 50% of the total particle volume. Examples of commercially used absorbents include activated carbon, silica gel, activated alumina and molecular sieve zeolites. Their physical properties and applications are shown in Table 9.4.

TABLE 9.4
Physical Properties and Applications of Some Commercial Adsorbents (Seader et al., 2016)

	Activated Carbon		Silica Gel			
Adsorbent	Small Pore	Large Pore	Small Pore	Large Pore	Activated Alumina	Molecular Sieve Zeolites
Pore diameter, d_p (nm)	1–2.5	> 3	2.2–2.6	10–15	1–7.5	0.3–1
Particle porosity, ε_p	0.4–0.6	-	0.47	0.71	0.5	0.2–0.5
Particle density, ρ_p (kg/m^3)	500–900	600–800	1090	620	1250	1400
Specific surface area, S_g (m^2/g)	400–1200	200–600	750–850	300–350	320	600–700
Typical Applications	Gas-phase adsorption	Liquid-phase application	Dry gases and liquids, to fractionate hydrocarbons		Dry gases and liquids	Drying, air separation, CO_2 removal

The size and shape of an adsorbent should be carefully selected to minimize mass transfer effects. When mass transfer effects are reduced, equilibrium can be reached more quickly. It should also reduce the pressure drop to maximize flow rate within the system. Additionally, for an adsorbent to be effective, it must exhibit the following three characteristics:

i) It must selectively concentrate one or more components from a fluid, achievable through:
   a) Selective binding of one or more components onto its surface
   b) Selective geometric exclusion of specific components (molecular sieving)
   c) Selective differences in intra-particle diffusion rates of components
ii) It must demonstrate reversibility to allow the recovery of adsorbate and the reuse of adsorbent.
iii) It should ideally exhibit a high delta loading. Delta loading is defined as the change in weight of adsorbate per unit weight of adsorbent during adsorption and desorption processes. In another words, delta loading indicates how much adsorbate can be adsorbed onto the adsorbent under a given condition, and how efficiently it can be desorbed afterwards.

Other than the aforementioned properties, bulk density, ρ_b; true solid density, ρ_s; and bed porosity, ε_b, are also important in the design of fixed bed adsorbers. The relationship between ε_b, ρ_b and ρ_p is expressed by Equation 9.25, whereas relationship between ε_p, ρ_s and ρ_p is given by Equation 9.26.

$$\varepsilon_b = 1 - \frac{\rho_b}{\rho_p} \tag{9.25}$$

$$\varepsilon_p = 1 - \frac{\rho_p}{\rho_s} \tag{9.26}$$

9.3.2 Adsorption Mechanisms

Adsorption is a spontaneous and exothermic process. Initially, the adsorbate diffuses from the bulk phase through the fluid film surrounding the particle to reach the external surface of particle (external film mass transfer). Upon reaching the external surface, adsorption occurs instantaneously, establishing equilibrium between the adsorbate in the fluid and that on the adsorbent surface. The adsorbed material then travels through the pores and is adsorbed onto the pore surfaces (active sites). Figure 9.5 illustrates the major steps involved in an adsorption process. Note that the steps are reversed during regeneration of adsorbent.

The type of adsorption is determined by the strength of interaction between the adsorbate molecules and the adsorbents. There are two types of adsorptions, that

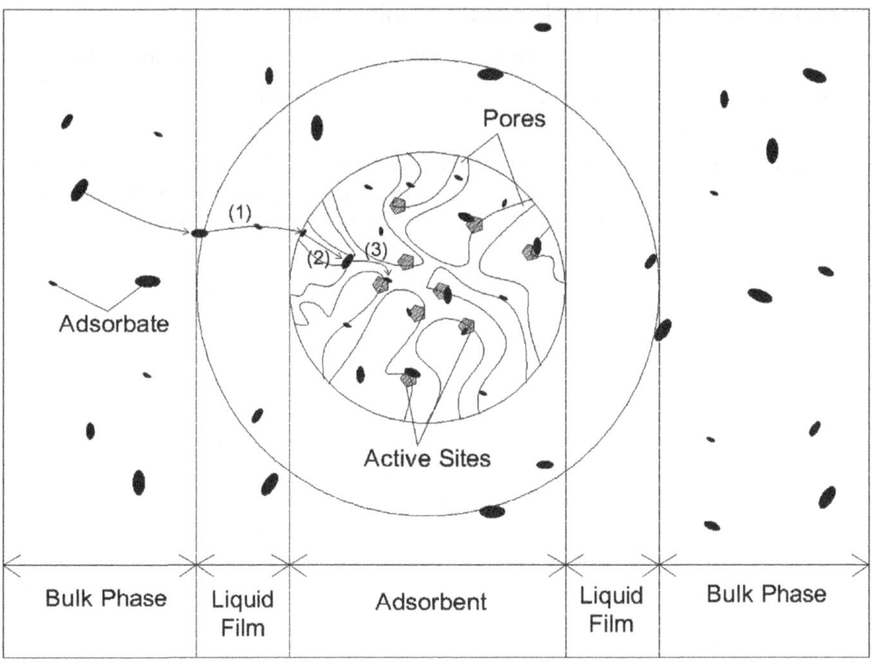

FIGURE 9.5 Steps involved in an adsorption process: (1) external film diffusion, (2) internal diffusion and (3) adsorption on active sites.

is, physisorption and chemisorption. Physisorption, also known as physical adsorption, is a type of adsorption where the adsorbate accumulates on the surface of a solid adsorbent only through weak intermolecular forces. These forces are primarily van der Waals interactions between the adsorbate and the adsorbent. In contrast, chemisorption, or chemical adsorption, is a type of adsorption where the adsorbate attaches to the surface of adsorbent through the formation of chemical bonds, such as covalent bonds. The characteristics of physisorption and chemisorption are compared in Table 9.5.

9.3.3 Adsorption Thermodynamics and Equilibrium

Under comparable conditions, different fluids are adsorbed to varying extent. Thermodynamics can be applied to adsorption to calculate the phase equilibrium between a fluid and a solid adsorbent. This phase equilibrium is crucial for characterizing the adsorption process, which involves the distribution of molecules between the fluid phase and the adsorbent surface. The adsorbate continuously attaches to and detaches from the adsorbent surface, eventually reaching a dynamic equilibrium when the rates of adsorption and desorption are equal. This equilibrium is often

TABLE 9.5
Characteristics of Physisorption and Chemisorption

	Physisorption	Chemisorption
Binding forces	Van der Waals forces	Chemical bonds
Activation energy	Low	High
Nature of adsorption	Non dissociative. Reversible process as the process is instantaneous and non-activated due to low activation energy	Often dissociative. Irreversible process as the process is slow due to high activation energy
Heat of adsorption	Small (20–40 kJ/mol)	Large (80–240 kJ/mol)
Saturation uptake	Monolayer or multilayer	Monolayer only
Specificity	Not specific with respect to adsorbent	Highly specific and occurs only when chemical bond can be formed between adsorbent and adsorbate
Effect of temperature	As temperature increases, the amount of gas physically adsorbed decreases. Thus, it is only significant at relatively low temperatures	It must be conducted at temperatures suitable for chemisorption to occur. With an increase in temperature, chemisorption initially increases and then decreases
Effect of pressure	Physisorption increases with an increase in pressure	Very high pressures are favourable. Small changes in pressure do not significantly affect chemisorption
Effect of surface area of adsorbent	Increases with an increase in surface area of adsorbent	Increases with an increase in surface area of adsorbent

described quantitatively using adsorption isotherms – mathematical models that express the relationship between the amount of adsorbate on the adsorbent and its concentration in the fluid phase at a constant temperature. The isotherms help to predict the adsorption capacity of an adsorbent under the given conditions. The amount of adsorbate adsorbed is often represented as a function of partial pressure (for gases) or concentration (for liquids) at a constant temperature.

Adsorption isotherms are semi-empirical models, where the constants of standard equations are fitted to experimental data. Some common examples include Henry's Law, Langmuir and Freundlich isotherms. These isotherms can be obtained experimentally using traditional bottle-point method. It is advisable to prepare at least 8 to 10 bottles to ensure sufficient data points for isotherm fitting. Each bottle is filled with a known volume, V, and known initial concentration of adsorbate solution, c_0 (kg adsorbate (solute) per m^3 of solution). Assume this adsorbate solution is a binary mixture. After adding an adsorbent of known weight m to the solution, the solution is stirred or shaken until equilibrium is reached. Once reaching equilibrium, the equilibrium concentration, c_{eq}, is measured. By neglecting any volume change, the amount of adsorbate adsorbed on the adsorbent (or concentration in the solid phase, or adsorbate loading) at equilibrium, expressed as q_{eq} (kg adsorbate per kg adsorbent), can be calculated by performing simple mass balance for a batch adsorption process. The process is illustrated in Figure 9.6.

Note that $q_0 = 0$ since fresh adsorbent is normally used. One can then determine q_{eq} using Equation 9.27.

$$q_{eq} = \frac{V}{m}\left(c_0 - c_{eq}\right) \tag{9.27}$$

By plotting q against c, an adsorption isotherm is obtained (McCabe et al., 1993). Examples of isotherm curves are displayed in Figure 9.7. An isotherm with an upward convex shape is desirable since higher adsorbate loading can be achieved at low fluid concentrations. In contrast, an isotherm with an upward concave shape is undesirable due to lower adsorbate loading. The horizontal line shows irreversible adsorption, where the amount of adsorbate adsorbed is independent of fluid concentration.

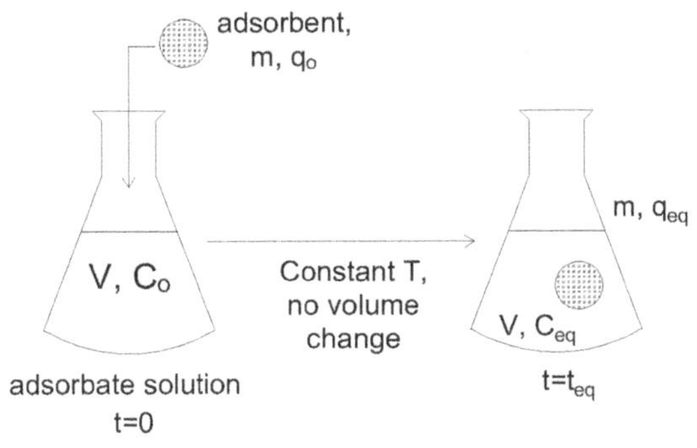

FIGURE 9.6 Experimental measurement of adsorption equilibrium data.

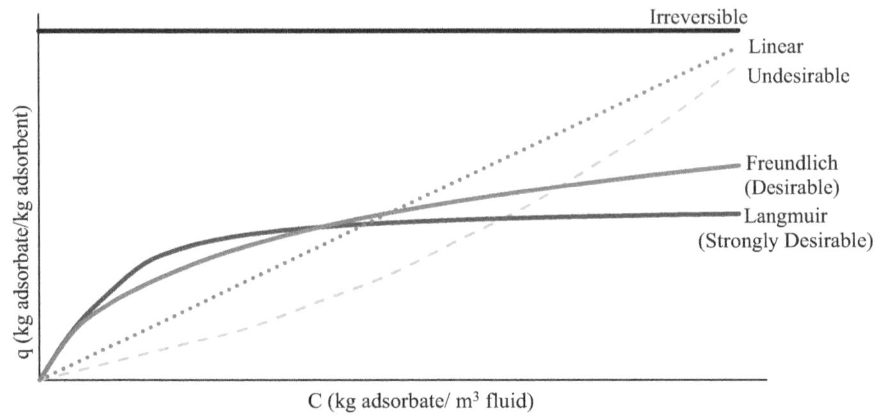

FIGURE 9.7 Adsorption isotherms.

9.3.4 Adsorption Isotherms for Single-Solute Adsorption

To date, no single adsorption isotherm equation has been able to describe all experimental curves with the same accuracy. Most existing single-solute adsorption isotherms were initially developed for gas-phase adsorption, where equilibrium adsorbate loading is a function of partial pressure. By substituting the pressure term with concentration, these isotherms can be extended to describe solute adsorption. The subsequent sections discuss several common isotherms that have been developed for single-solute adsorption, namely Henry's Law, the Langmuir, Freundlich, Temkin, Dubinin-Radushkevich (D-R) and Redlich-Peterson isotherms. They can be categorized into one-, two- and three-parameter models.

9.3.4.1 One-Parameter Isotherm

9.3.4.1.1 Henry's Law

Henry's Law is the simplest adsorption isotherm, which contains only one parameter, as depicted by Equation 9.28. It assumes a linear relationship between the amount of adsorbate adsorbed and the concentration of adsorbate in the bulk phase. Note that this isotherm is only applicable to dilute systems.

$$q_{eq} = K_H c_{eq} \tag{9.28}$$

where q_{eq} (mg/g) represents the amount of adsorbate at equilibrium, K_H (L/g) denotes Henry's adsorption constant, which is the isotherm parameter and c_{eq} (mg/L) is the concentration of adsorbate on the adsorbent at equilibrium.

9.3.4.2 Two-Parameter Isotherms

Examples of two-parameter isotherms include the Langmuir, Freundlich, Temkin and Dubinin-Radushkevich (D-R) isotherms. Among these, Langmuir and Freundlich equations are the most frequently used.

9.3.4.2.1 Langmuir Isotherm

Langmuir isotherm postulates that a single layer of adsorbate molecules is adsorbed onto specific homogeneous sites, with no interaction occurring between the adsorbed molecules (Langmuir, 1918). The adsorption is reversible and eventually reaches a state of equilibrium. The Langmuir isotherm equation is defined as:

$$q_{eq} = \frac{q_{max,L} K_L c_{eq}}{1 + K_L c_{eq}} \tag{9.29}$$

where $q_{max,L}$ (mg/g) and K_L (L/mg) are the isotherm parameters, which represent the Langmuir maximum adsorption capacity and the Langmuir constant, respectively. Hall separation factor, R_L (Hall et al., 1966), represented by Equation 9.30, is used to define the characteristic of the Langmuir isotherm. The adsorption is desirable if R_L value falls between 0 and 1, undesirable if $R_L > 1$, linear if $R_L = 1$ and irreversible if $R_L = 0$.

$$R_L = \frac{1}{1 + K_L c_0} \tag{9.30}$$

where c_0 (mg/L) is the initial concentration of adsorbate.

9.3.4.2.2 Freundlich Isotherm

In contrast to the Langmuir isotherm, the Freundlich isotherm assumes that *multi-layer adsorption* of the adsorbate occurs on a heterogeneous surface, with no interactions between the adsorbed molecules (Freundlich, 1906). It has been widely applied to heterogeneous adsorption systems and is expressed as:

$$q_{eq} = K_F c_{eq}^{1/n} \tag{9.31}$$

where K_F ((mg/g)(L/mg)$^{1/n}$) implies the Freundlich constant, while n (dimensionless) is Freundlich intensity parameter. Both constants are the isotherm parameters. The larger the K_F value, the higher the adsorbate loading.

9.3.4.2.3 Temkin Isotherm

Unlike the Langmuir and Freundlich isotherms, Temkin isotherm accounts for interactions between adsorbent and adsorbate. It assumes that heat of adsorption reduces linearly with the increase in coverage of adsorbent surface (Temkin and Pyzhev, 1940). The Temkin isotherm is given by Equation 9.32.

$$q_{eq} = \frac{RT}{B} \ln K_T c_{eq} \tag{9.32}$$

where K_T (L/mg) and B (J/mol) are isotherm parameters representing the Temkin isotherm coefficient and the constant related to adsorption heat, respectively, R (8.314 J/(mol K)) represents universal gas constant and T (K) represents solution temperature.

9.3.4.2.4 Dubinin-Radushkevich (D-R) Isotherm

D-R isotherm is often used to determine the mean free energy of adsorption, E, which aids in distinguishing between physisorption and chemisorption (Dubinin and Radushkevich, 1947). Mean free energy of adsorption represents the free energy required to transfer 1 mol of adsorbate from the bulk phase to the adsorbent surface. The process is classified as physisorption if E < 8 kJ/mol, and as chemisorption if E falls between 8 and 16 kJ/mol. Equations 9.33 to 9.35 are used to describe D-R isotherm.

$$q_{eq} = q_{max,DR} e^{-\beta_{DR}\varepsilon^2} \tag{9.33}$$

$$\varepsilon = RT \ln\left(1 + \frac{1}{c_{eq}}\right) \tag{9.34}$$

$$E = \frac{1}{\sqrt{2\beta_{DR}}} \tag{9.35}$$

where $q_{max,DR}$ (mg/g) represents the D-R maximum adsorption capacity, β_{DR} (mol²/kJ²) represents D-R coefficient related to mean free energy of adsorption, E (kJ/mol) and ε (kJ/mol) is the Polanyi potential. $q_{max,DR}$ and β_{DR} are the isotherm parameters.

9.3.4.3 Three-Parameter Isotherm

9.3.4.3.1 Redlich-Peterson (R-P) Isotherm

The Redlich-Peterson (R-P) isotherm, derived from a combination of the Langmuir and Freundlich isotherms, is the most common three-parameter model. It applies to both homogeneous and heterogeneous systems (Redlich and Peterson, 1959), and its equation is denoted as Equation 9.36. According to this equation, the numerator demonstrates a linear dependency on concentration, while the denominator illustrates how adsorption changes exponentially with concentration.

$$q_{eq} = \frac{K_{RP} c_{eq}}{1 + \alpha_{RP} c_{eq}^{\beta_{RP}}} \tag{9.36}$$

where K_{RP} (L/g) and α_{RP} (L/mg) represent constants of R-P isotherm, and β_{RP} is the exponent that can be varied between 0 and 1. These three constants are the isotherm parameters. R-P model approaches Langmuir model when $\beta_{RP} = 1$, and it approaches Freundlich isotherm when $\alpha_{RP} c_{eq}^{\beta_{RP}} \gg 1$.

9.3.5 Adsorption Isotherm Models Fitting

To determine the adsorption behaviour and mechanism of a system, it is essential to identify which model the adsorption isotherm fits best. The traditional method for fitting isotherms to experimental data involves linear regression analysis. This process begins by linearizing the isotherm equations, followed by plotting the

TABLE 9.6
Linear Forms and Plots of Various Isotherm Models

Isotherm	Linear Form	Plot
Henry's Law	$q_{eq} = K_H c_{eq}$	q_{eq} vs. c_{eq}
Langmuir	$\frac{1}{q_{eq}} = \frac{1}{q_{max,L} K_L}\left(\frac{1}{c_{eq}}\right) + \frac{1}{q_{max,L}}$	$1/q_{eq}$ vs. $1/c_{eq}$
Freundlich	$\ln q_{eq} = \ln K_F + \frac{1}{n}\ln c_{eq}$	$\ln q_{eq}$ vs. $\ln c_{eq}$
Temkin	$q_{eq} = \frac{RT}{B}\ln K_T + \frac{RT}{B}\ln c_{eq}$	q_{eq} vs. $\ln c_{eq}$
Dubinin-Radushkevich (D-R)	$\ln q_{eq} = \ln q_{max,DR} - \beta_{DR}\varepsilon^2$	$\ln q_{eq}$ vs. ε^2
Redlich-Peterson (R-P)	$\ln\left(K_{RP}\frac{c_{eq}}{q_{eq}} - 1\right) = \beta_{RP}\ln c_{eq} + \ln\alpha_{RP}$	$\ln\left(K_{RP}\frac{c_{eq}}{q_{eq}} - 1\right)$ vs. $\ln c_{eq}$

model-predicted isotherms. The isotherm parameters can then be determined from these linear plots. By comparing the goodness of fit between model-predicted isotherms, the isotherm with the highest correlation coefficient, R^2 value, represents the best-fit model. Note that the adsorption isotherms discussed in previous section are mostly expressed in non-linear mathematical expressions. The linear forms of these isotherms, along with the parameters to plot the equations, are summarized in Table 9.6.

To date, there are debates regarding the use of linear regression analysis, which may introduce inherent bias through linearization. To address this limitation, non-linear regression analysis has been extensively studied and applied to optimize isotherm fitting. The non-linear regression method minimizes the error distribution between predicted isotherm and actual experimental equilibrium data. An added advantage of non-linear regression is its ability to handle isotherms that cannot be linearized. Error functions, such as not limited to R^2, sum of squared errors (SSE), average relative error deviation (ARED) and root mean square error (RSME), are commonly utilized to quantify the difference between predicted isotherm and experimental data. To reduce the error distribution between predicted isotherm and experimental equilibrium data, these error functions can be minimized or maximized depending on their definition. The 'solver add-in' tool of Microsoft Excel is often utilized for fitting and predicting various adsorption isotherm parameters by non-linear regression.

Example 9.3. Adsorption Isotherms

In a batch adsorption experiment, granular activated carbon is used to remove chromium from a solution. The equilibrium data at 30°C are presented in Table 9.7. Determine which isotherm model – Henry's Law, Langmuir, Freundlich or Temkin best fits the data through linear regression analysis.

TABLE 9.7
Concentration and Adsorption Capacity Data

c_{eq} (mg/L)	q_{eq} (mg/g)
1.281	4.680
5.177	13.706
15.180	21.205
33.477	41.631
56.481	60.880
88.357	77.911
183.117	104.221
319.670	120.083

TABLE 9.8
Linearized Isotherm Data

Henry's Law		Langmuir		Freundlich		Temkin	
c_{eq}	q_{eq}	$1/c_{eq}$	$1/q_{eq}$	$\ln c_{eq}$	$\ln q_{eq}$	$\ln c_{eq}$	q_{eq}
1.281	4.680	0.781	0.214	0.247	1.543	0.247	4.680
5.177	13.706	0.193	0.073	1.644	2.618	1.644	13.706
15.180	21.205	0.066	0.047	2.720	3.054	2.720	21.205
33.477	41.631	0.030	0.024	3.511	3.729	3.511	41.631
56.481	60.880	0.018	0.016	4.034	4.109	4.034	60.880
88.357	77.911	0.011	0.013	4.481	4.356	4.481	77.911
183.117	104.221	0.005	0.010	5.210	4.647	5.210	104.221
319.670	120.083	0.003	0.008	5.767	4.788	5.767	120.083

Solution

1. Generates tables for each isotherm containing the necessary parameters required for graph plotting as presented in Table 9.8.
2. Plot the model-predicted isotherms and identify the isotherm that gives the highest R^2 value as presented in Figure 9.8.

Based on R^2 values, Langmuir is the best-fit model, which can be used to describe the mentioned adsorption process. The isotherm parameters of Langmuir, $q_{max,L}$ and K_L, are calculated as 67.57 mg/g and 0.0572 L/mg, respectively.

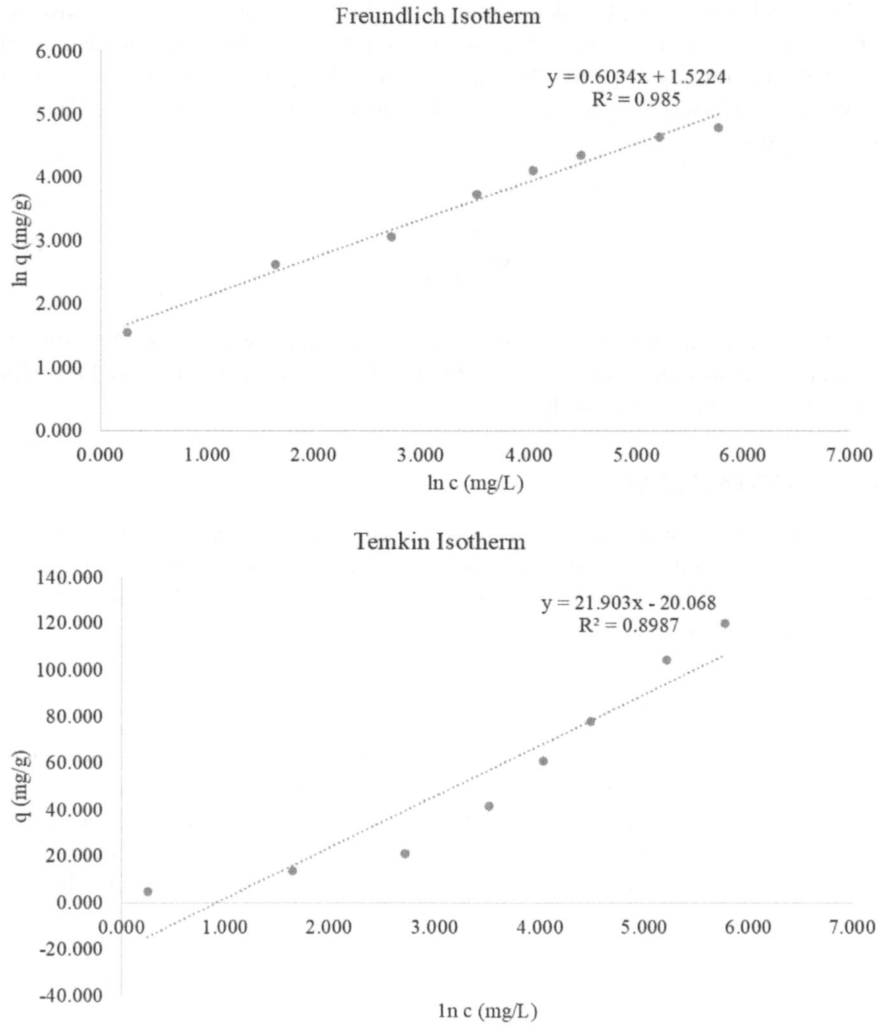

FIGURE 9.8 Model-predicted isotherm plots.

9.3.6 Adsorption Isotherms for Multi-Solute Adsorption

Wastewater typically contains more than one solute (adsorbate) that needs to be treated through adsorption. When an adsorbent is introduced into a multicomponent solution, the adsorption of adsorbate *i* occurs in the presence of other competing solutes. Thus, the interaction between these solutes must be considered. This competitive adsorption can be described using multi-solute isotherm equations or thermodynamic models. Thermodynamic models are more general in nature and offer a wider range of applications compared to multi-solute isotherms. Among thermodynamic models, the Ideal Absorbed Solution Theory (IAST) stands out for its ability to predict multi-solute adsorption using single-solute isotherm parameters (Worch, 2012).

The IAST can be used to determine the equilibrium concentration of adsorbate i in the presence of other competing organic solutes, $c_{eq,i}$, expressed in terms of the adsorbate loading, q_i. It allows the single-solute adsorption behaviour described by the Freundlich adsorption isotherm to predict multicomponent adsorption. The IAST is expressed as:

$$c_{eq,i} = \frac{q_i}{\sum_{j=1}^{N} q_j} \left(\frac{\sum_{j=1}^{N} n_j q_j}{n_i K_i} \right)^{n_i} \quad (9.37)$$

where $c_{e,i}$ is the equilibrium concentration in multicomponent system, N is the total number of solutes, K_i and $1/n_i$ represent the Freundlich capacity and Freundlich intensity parameter, respectively.

9.4 CRYSTALLIZATION

Crystallization is one of the final purification steps in the manufacturing process of several value-added chemicals, such as pharmaceutical chemicals. It is also one of the oldest separation processes used to produce various salts and sugars for a long time (Richardson et al., 2002).

9.4.1 Supersaturation

The crystallization process is based on the variation of solubility with temperature. In general, solubility increases with temperature, though there are several exceptions. The diagrams that represent the solubility of substances as a function of temperature are called solubility curves. Solubility curves for several common substances can be found in the literature, with some examples shown in Figure 9.9.

The solubility curves provide the equilibrium limit on the concentration of substances in solutions. For a substance whose solubility increases with temperature, the concentration of the substance in the solution increases with temperature. Therefore, once a solution with a higher concentration is reached at a higher temperature, it is expected that part of the dissolved substance will precipitate if the temperature is decreased according to the solubility curve. However, solutions generally do not follow the solubility curve once a higher concentration is reached through heating. This is because the system needs to provide the surface energy required to form the crystals. Even though the solution is in a thermodynamically unstable state, the excess Gibbs free energy at the supersaturated state can still be lower than the surface energy required for the formation of crystals. Therefore, the supersaturated solution would continue to exist in a metastable state. The difference between the actual concentration (c) and the concentration based on solubility (c^*) is called the degree of supersaturation, Δc.

$$\Delta c = c - c^* \quad (9.38)$$

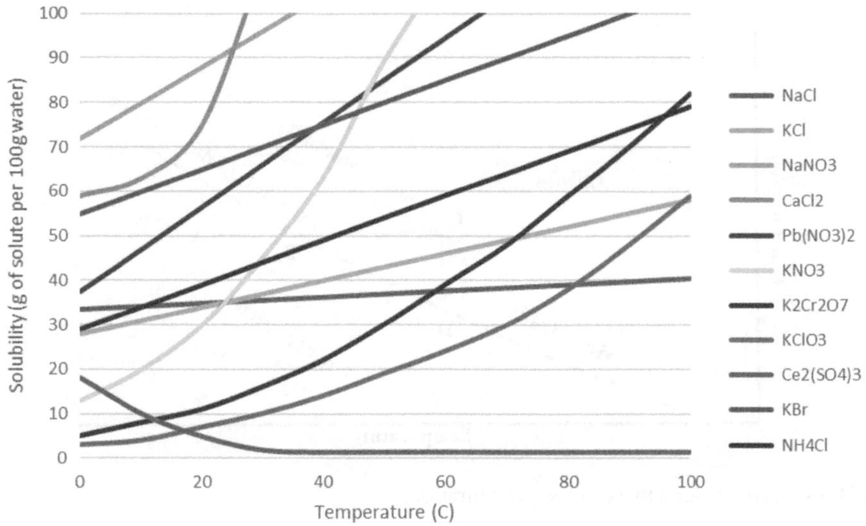

FIGURE 9.9 Solubility curves of some of the common salts.

As the temperature decreases, the degree of supersaturation increases. As the degree of supersaturation increases, the solution becomes more unstable and thus the tendency to form crystals increases. Eventually, at a certain concentration, the increase in Gibbs free energy will exceed the surface energy needed for crystal formation, and crystallization will begin. Therefore, the degree of supersaturation is the driving force for crystallization. The degree of supersaturation can be increased by the following methods:

1. Temperature reduction.
2. Solvent reduction by evaporation of solvent.
3. Addition of non-solvents that alter the solubility of substances. Typically, the non-solvents have significantly different polarity from the substance in the solution.

9.4.2 Nucleation

Nucleation is the formation of new crystals from the solution. In the industrial crystallizers, there are two forms of nucleation, which are primary and secondary nucleation. In primary nucleation, crystals are formed purely based on supersaturation. In secondary nucleation, the presence of crystals and the interaction of crystals with the environment in which the crystallization is carried out will lead to the formation of crystals.

The onset of nucleation can be explained by Mier's theory, which can be schematically presented in Figure 9.10.

According to Mier's theory of supersaturation, there is a relationship between the concentration of a solution and crystallization (Miers and Isaac, 1906). Crystallization

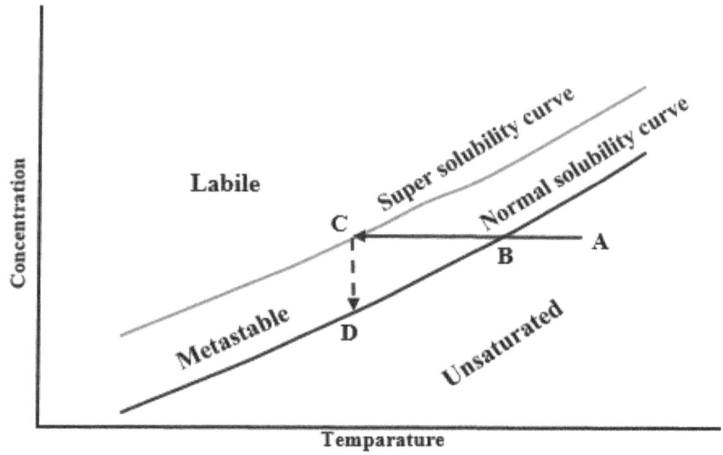

FIGURE 9.10 Mier's theory of supersaturation.

begins spontaneously from a solution once it reaches a concentration higher than the saturation concentration. The saturation concentrations and the concentrations at which crystallization begins spontaneously can be plotted as a function of temperature. When an unsaturated solution, represented by point A in Figure 9.10, is cooled, the temperature drops along the line ABC. Crystallization cannot begin if the concentration is below the normal solubility curve, where the solution is in a thermodynamically stable state. This means that it is possible to produce a solution with the concentration indicated by point A. Once the temperature reaches the point indicated by B, the saturation concentration has been reached at that temperature. Crystallization will not begin, but any subsequent cooling will result in a metastable solution. This solution is called metastable because it is not possible to reach a concentration higher than the saturated concentration; however, no spontaneous crystallization would begin at that concentration. However, this will not continue indefinitely. Once the temperature reaches the point C, crystallization begins simultaneously. Any concentration above that point is thermodynamically unstable and would result in immediate crystallization. This zone is also called the labile zone. The temperature-concentration relationship can also be seen vertically. Any reduction in concentration from point C will make the solution metastable. If the concentration is below point D, crystallization is impossible.

Mier's theory works only in idealized situations when the solution is completely free of any solid particles, such as crystals or foreign particles. Solid substances will act as nuclei on which the crystals may grow well before the super solubility curve shown in Figure 9.10. Because of that, the super solubility curve predicted by Mier's theory is seldom used in practice. Once crystallization begins due to the high degree of supersaturation, it is known as primary nucleation. Instead, a zone of super solubility is commonly used in the study of the kinetics of industrial crystallizers.

Nucleation in a crystallizer and be both primary nucleation and secondary nucleation. Primary nucleation takes place in a crystallizer based on two mechanisms, which are homogeneous and heterogeneous nucleation. Homogeneous nucleation takes place when the supersaturation is so high that the increase in Gibbs free energy exceeds the surface energy required to form crystals. This process is uninfluenced by the solids and interaction with other substances present in the system. Crystallization begins with the formation of clusters due to the random association of the solute molecules. Once the size of a cluster exceeds a critical size, it forms a crystal embryo. Once the particles in the solution form a crystal embryo, more and more particles from the solution grow on the embryo and thus increase the size of the crystal. The theoretical rate of homogeneous nucleation I estimated from Equation 9.39 (Myerson, 2002):

$$B^{o} = A\exp\left[\frac{-16\pi\sigma^{3}V_{M}^{2}}{3k^{3}T^{3}\left(ln\left(s+1\right)\right)^{2}}\right] \tag{9.39}$$

where B^o = Rate of homogeneous nucleation, A = a constant called frequency factor, Σ = interfacial tension, V_M = molar volume, k = Boltzmann constant, T = absolute temperature and s = Degree of supersaturation.

From Equation 9.39, it is evident that the nucleation rate increases rapidly with the degree of supersaturation.

Primary nucleation is rarely the dominating mechanism in the industrial crystallizers. In the industrial crytallizers, seed crystals have been introduced to start the crystal growth in the supersaturated solution to overcome the thermodynamic barrier. Seed crystals provide the active sites on which the crystals can grow, thereby eliminating the need to have a large Gibbs free energy increase that exceeds the surface energy needed to form the crystals. Since the energy required for crystal growth is significantly lower than that for homogeneous crystallization, heterogeneous crystallization is typically the dominating mechanism in the industrial crystallizers.

In an ideal situation, the number of crystals produced in an industrial crystallizer will be the number of seeds introduced into the system. However, in general, this is true only at the beginning of the process. As the crystals start to grow in the solution, several mechanisms contribute to the formation of more crystals during crystallization by breaking the existing crystals in the system. The newly formed crystals would act as seeds for more crystallization and this mechanism is known as secondary nucleation. The secondary nucleation could be the results of the excessive fluid shear which may break the crystals, collision between different crystals, collision between the crystals and the agitator or the collision of crystals to the vessel walls. The secondary nucleation due to the collision is the most common mechanism and thus dominates the crystal formation in the industrial crystallizers.

9.4.3 Crystal Growth Rate

The crystal growth rate has been modelled using different theories in which the mass transfer theory is the most established one (Wankat, 1994). According to this, the first step is the convective mass transport of the solute from the supersaturated solution to the crystal surface. In the next step, the crystals grow in layers due to the accumulation of the transferred solute particles on the crystal surface. Thus, the driving force for crystallization is the degree of supersaturation, and the rate of crystal growth can be represented using Equation 9.40.

$$\frac{dm_c}{dt} = K_L A_c \left(c - c^*\right) \tag{9.40}$$

where $\frac{dm_c}{dt}$ = rate of increase of crystal mass, A_c= area of one crystal and K_L= overall mass transfer coefficient which is a combination of mass transfer coefficients for the convective mass transfer and the growth rate.

The growth rate can also be represented in terms of the characteristic size of the crystals. The characteristic size is related to the mass, volume and area through shape factors.

$$m_c = \rho_c \varnothing_v L^3 \tag{9.41}$$

$$A_c = \varnothing_a L^2 \tag{9.42}$$

$$v_c = \varnothing_v L^3 \tag{9.43}$$

In Equations 9.41–9.43. L is the characteristic size, ϕ_v is the volume shape factor and ϕ_a is the area shape factor. By combining these equations with Equation 9.40.

$$\frac{dm_c}{dt} = \frac{d}{dt}\left(\rho_c \varnothing_v L^3\right) = K_L A_c \left(c - c^*\right) \tag{9.44}$$

$$3\rho_c \varnothing_v L^2 \frac{dL}{dt} = K_L \varnothing_a L^2 \left(c - c^*\right) \tag{9.45}$$

$$G = \frac{dL}{dt} = k_D \left(\frac{c - c^*}{c^*}\right) \tag{9.46}$$

where G is the growth rate of the crystal and $k_D = \frac{c^* K_L \varnothing_a}{3\rho_c \varnothing_v}$.

Therefore, the growth rate is independent of the size of the crystal and depends only on the degree of supersaturation. Based on the experimental values of saturated concentrations, the growth rate can be determined. The growth rate will allow us to estimate the time needed to reach the crystals with desired size.

9.4.4 Cooling Rate

In a crystallizer, the cooling rate has to be controlled in such a way that a continuous growth rate can be maintained for the crystals. If the growth rate is too slow, the process gets delayed, and if the growth rate is too fast, it leads to the accidental capture of impurities into the crystal lattice which reduces the purity. From Equation 9.46, it can be seen that, to maintain the constant growth rate, the difference between the saturated concentration and the actual concentration needs to be maintained constant at as the cooling takes place. Since the solubilities are strong functions of temperatures, cooling rate has to be controlled in such a way that the difference between the saturated concentration and the actual concentration would remain constant.

9.4.5 Scale-Up Challenges

In an ideal situation, the crystallization in an industrial crystallizer can be controlled based on the number of seeding crystals and the cooling rate. While homogeneous crystallization does occur, the quantity of crystals formed is typically insignificant in most cases. It is to be noted that, in the older crystallizers with wear and tear, there may be cracks in which the solution can penetrate and thus get significantly more cooling than the bulk solution. In such cases, there can be more formation of the fine crystals produced during the crystallization process. However, the major challenge for the crystallization process is secondary nucleation because this process occurs randomly, and it is not possible to predict the formation of additional crystals due to random collisions. In addition, the size of the new crystals varies significantly between systems. This is because the breaking of crystals differs in different cases. There will also be a large number of fine crystals produced due to secondary nucleation. Because of this, the crystals present in the crystallizer at any time are of different sizes. The representation of this distribution is represented using *crystal size distribution* (CSD).

In order to prevent losses of crystals during the filtration of the crystals, it is preferred to have a narrow range of sizes for the crystals. However, the different mechanisms of crystal formation mentioned in the previous paragraph make it impossible to predict the crystal size distribution in an actual equipment. The formation of such fine crystals also leads to significant losses at the crystallization stage.

9.5 SPRAY DRYING

Spray drying systems are commonly used in the chemical product industry to produce a diverse range of products in powdered form, which allows for easier handling, storage and transportation. Examples of spray-dried products include emulsion polymers, powdered detergents, fertilizers, pesticides, antibiotics, catalysts, coffee creamer and milk powder. Spray drying is desirable when slurry or suspension feeds containing solid particles cannot undergo mechanical drying. It operates as a single-step continuous process in convection mode, where a solution rapidly dries into a dry powder form as hot gas passes through.

Owing to its short residence time, it is capable of handling both heat-sensitive and non-heat-sensitive fluids as long as the materials are pumpable. Its ability to handle large production rates and a wide range of operating temperatures makes it highly attractive, as it can be easily tailored and designed to meet specific capacity requirements. The shape, particle size, bulk density and moisture content of the final product can be effectively controlled through the use of spray dryer. However, there are limitations to using spray dryers, such as low thermal efficiency, high installation costs and potential fire hazards or product degradation resulting from buildup in the drying chamber (Filková and Mujumdar, 2020). Additionally, it consumes more energy as compared to other types of dryers for the same amount of water evaporation (Filková and Mujumdar, 2020).

Despite these drawbacks, spray drying remains indispensable in both the chemical and pharmaceutical industries. It is crucial for manufacturing speciality chemicals like catalysts, pigments and emulsion polymers, where specific particle size and surface morphology play critical roles in performance. It is also widely used to encapsulate bioactive compounds, drugs and food ingredients to enhance stability and prolong shelf life. The following sections will delve into its basic working principle, mass and energy balances, thermal efficiency and estimation of residence time in the drying chamber.

9.5.1 Working Principle

There are normally five steps in spray drying system:

a) Feed concentration
b) Feed Atomization
c) Contact of droplet with the hot gas
d) Moisture evaporation
e) Particle separation from exit air

Liquid feed is often concentrated before pumping it into spray drying system. This is to increase the solid concentration in the feed to save energy required to evaporate the moisture. The concentrated feed is then pumped to the spray drying chamber, where it contains atomizer to atomize liquid feed into droplets. There are two major types of atomizers, namely rotary atomizer and nozzle atomizer. The droplets are then immediately exposed to the hot gas (usually hot air), where the heat rapidly evaporates the moisture within the droplets, transforming them into solid particles. The solid particles fall to the base of the drying chamber and are entrained in the drying gas, which is then passed through cyclone separator to separate them from the exit gas. The solid particles are collected at the bottom of cyclone and fine particles

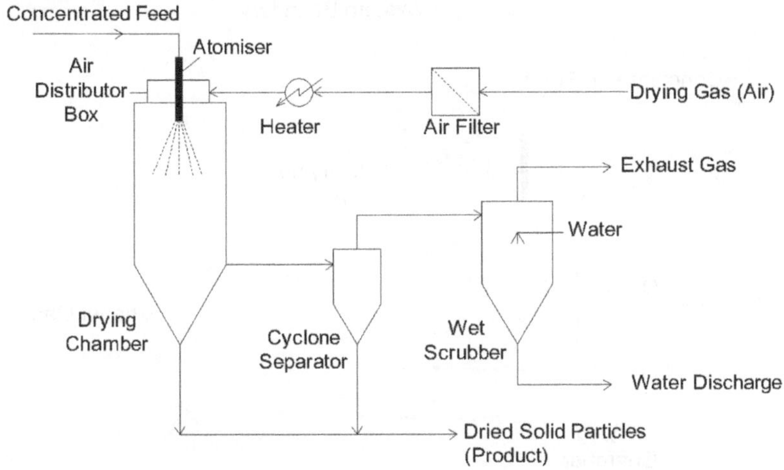

FIGURE 9.11 Conventional open-cycle co-current flow spray drying system.

that remain trapped in the gas will be passed through wet scrubber for further separation. The control of various process parameters, such as inlet air temperature, feed rate and nozzle design, allows for precise control over the properties of the final product. Figure 9.10 illustrates the conventional open-cycle co-current flow spray drying system.

9.5.2 Mass and Energy Balances

To design a spray drying system, it is essential to perform mass and energy balance around the system. Figure 9.11 illustrates the critical data required to calculate mass and energy balance around a single-stage spray dryer. It is assumed that the drying operation occurs at steady state, that is, no accumulation, and it uses air as drying gas. Figure 9.12 illustrates the mass and energy balance around the spray dryer.

The critical data of each stream are defined as follows:

Concentrated Feed Stream

$\dot{m}_F$ = mass flow rate of solid in the feed (kg dry solid/h)

W_{F1} = moisture content of the feed (kg H_2O/ kg dry solid)

T_{F1} = inlet feed temperature (°C)

q_{F1} = specific enthalpy of feed (humid solid) entering the dryer (kJ/kg dry solid)

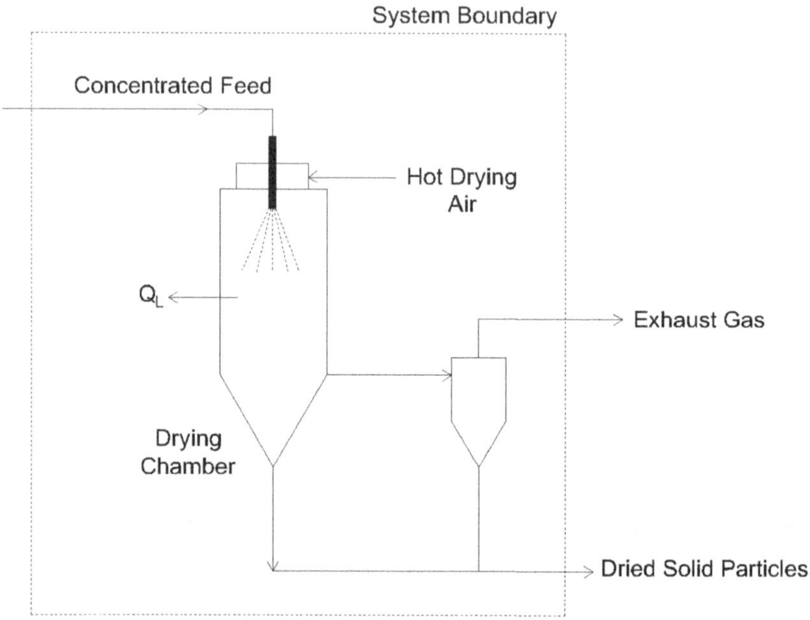

FIGURE 9.12 Spray dryer data for mass and energy balance calculation.

Hot Drying Air

$\dot{m}_a$ = mass flow rate of drying air (kg dry air/h)

H_1 = absolute humidity of inlet air (kg H_2O/ kg dry air)

T_{a1} = temperature of drying air entering the dryer (°C)

q_{a1} = specific enthalpy of drying air entering the dryer (kJ/kg dry air)

Dried Solid Particles (Product)

W_{F2} = moisture content of the product (kg H_2O/ kg dry solid)

T_{F2} = temperature of product (°C)

q_{F2} = specific enthalpy of product leaving the dryer (kJ/kg dry solid)

Exhaust Gas

H_2 = absolute humidity of outlet air (kg H_2O/ kg dry air)

T_{a2} = temperature of air leaving the dryer (°C)

q_{a2} = specific enthalpy of air leaving the dryer (kJ/kg dry air)

Q_L = heat loss from the dryer

9.5.2.1 Mass Balance

Since it is assumed that the system is at steady state and there is no accumulation, mass entering and leaving the spray dryer must be equal. Mass balance can be performed on water component (moisture):

Moisture in the feed + Moisture in the drying air = Moisture in the product
+ Moisture in the exhaust gas (9.47)

$$\dot{m}_F\left(W_{F1}\right)+\dot{m}_a\left(H_1\right)=\dot{m}_F\left(W_{F2}\right)+\dot{m}_a\left(H_2\right) \tag{9.48}$$

9.5.2.2 Energy Balance

Similarly, energy balance can be calculated using the following equations:

Total heat entering the dryer = total heat leaving the dryer
+ heat loss from the dryer (9.49)

$$\dot{m}_F\left(q_{F1}\right)+\dot{m}_a\left(q_{a1}\right)=\dot{m}_F\left(q_{F2}\right)+\dot{m}_a\left(q_{a2}\right)+Q_L \tag{9.50}$$

where $\dot{m}_F\left(q_{F1}\right)$ and $\dot{m}_a\left(q_{a1}\right)$ represent the enthalpy of feed and air entering the dryer, respectively, whereas $\dot{m}_F\left(q_{F2}\right)$ and $\dot{m}_a\left(q_{a2}\right)$ show the enthalpy of product and air leaving the dryer, respectively.

Note that heat loss is generally minimal in well-insulated drying chambers. However, heat losses are significantly higher in drying chambers that are not insulated or are equipped with cooling air jackets to maintain cool dryer walls, which is normally used when the product is hygroscopic and heat sensitive (Masters, 1976). For design purposes, a 10% heat loss through the dryer walls is often taken into account.

The specific enthalpy of feed entering the dryer, q_{F1}, can be obtained by the summation of both the specific enthalpy of dry solid and its moisture (Masters, 1976), illustrated as Equation 9.51.

$$q_{F1}=c_{P_{DS}}\left(\Delta T_F\right)+c_{P_W}\left(W_{F1}\right)\left(\Delta T_F\right) \tag{9.51}$$

where c_{p_DS} is the specific heat capacity of dry solid (kJ/(kg dry solid °C)), c_{p_W} is the specific heat capacity of moisture (liquid) (kJ/(kg H_2O °C)) and ΔT_F is the difference between feed temperature and the temperature of reference, which is the freezing point of water, that is, 0°C.

Using a similar approach, the specific enthalpy of product leaving the dryer, q_{F2} can be estimated through Equation 9.52.

$$q_{F2}=c_{P_{DS}}\left(\Delta T_P\right)+c_{P_W}\left(W_{F2}\right)\left(\Delta T_P\right) \tag{9.52}$$

where ΔT_P is the difference between product temperature and the reference temperature (°C).

Equation 9.53 is the general equation used to estimate the specific enthalpy of drying air, q_a.

$$q_a = c_s\left(\Delta T\right) + h_{v_W}\left(H\right) \tag{9.53}$$

where h_{v_W} is the latent heat of vaporization of water at 0°C (approximately 2500.9 (kJ/kg H_2O)), H is the absolute humidity of air and c_s is the specific heat capacity of humid air (kJ/(kg dry air °C)), which is defined as the amount of heat required to increase the temperature of one unit mass of air and its vapour by 1°C at constant temperature. Thus, it can be estimated using Equation 9.54.

$$c_s = c_{p_a} + c_{P_v}\left(H\right) \tag{9.54}$$

where c_{p_a} is the specific heat capacity of dry air (approximately 1.005 kJ/(kg dry air °C)), and c_{P_v} is the specific heat capacity of water vapour (approximately 1.884 kJ/(kg H_2O °C)). Hence, the specific enthalpy of drying air entering and leaving the dryer, q_{a1} and q_{a2}, can be calculated using Equations 9.55 and 9.56, respectively.

$$q_{a1} = \left[1.005 + 1.884\left(H_1\right)\right] \times \Delta T_{a,F} + 2500.9\left(H_1\right) \tag{9.55}$$

$$q_{a2} = \left[1.005 + 1.884\left(H_2\right)\right] \times \Delta T_{a,P} + 2500.9\left(H_2\right) \qquad 9.56)$$

where $\Delta T_{a,F}$ is the difference between T_{a1} and the temperature of reference, which is the freezing point of water, that is, 0°C; $\Delta T_{a,P}$ is the difference between T_{a2} and the temperature of reference, which is the freezing point of water, that is, 0°C.

Example 9.4. Spray Dryer

A co-current flow spray dryer is used to produce 850 kg/hr of dried pigment containing 3 wt% moisture. The concentrated feed that contains 40 wt% solid is pumped into the drying chamber at 30°C. The dried pigment leaves the dryer at 55°C. Ambient air at 32°C and 40% relative humidity is heated to 180°C prior to entering the chamber. Air exits the dryer at 90°C. Assume the heat loss is 80,000 kJ/h and the specific heat capacity of the dry pigment and liquid moisture is 1.5 kJ/kg K and 4.2 kJ/kg K, respectively. Calculate the air flow rate required for this process and its absolute humidity at the dryer outlet.

Solution

Mass and energy balances should be performed to determine the unknown parameters.

Mass balance:

Using Equation 9.48: $\dot{m}_F\left(w_{F_1}\right) + \dot{m}_a\left(H_1\right) = \dot{m}_F\left(w_{F_2}\right) + \dot{m}_a\left(H_2\right)$

Total concentrated feed flow rate $= \frac{850 \times 0.97}{0.4} = 2061.25 kg/h$

Mass flow rate of solid $(\dot{m}_F) = 850 \times 0.97 = 824.5 kg\, dry\, solid/h$

Moisture content of the feed $(w_{F_1}) = \frac{0.6}{0.4} = 1.5\, H_2O/kg\, dry\, solid$

Moisture content of the dried pigment (product)

$(w_{F_2}) = \frac{3}{97} = 0.0309\, kg\, H_2O/kg\, dry\, solid$

Note that absolute humidity of inlet air (H_1) has to be determined from psychometric chart.

$At\ 32°C\ and\ 40\%\ RH, H_1 = \frac{0.012\, kg\, H_2O}{kg\, dry\, air}$

Although air is heated to 180°C, it still carried the same moisture loading of 0.012 kg $H_2O\, /\, kg\, dry\, air$

$$\therefore \frac{824.5\, kg\, dry\, solid}{h}\left(\frac{1.5 kg H_2O}{kg\, dry\, solid}\right) + \dot{m}_a\left(\frac{0.012 kg H_2O}{kg\, dry\, air}\right)$$

$$= \frac{824.5\, kg\, dry\, solid}{h}\left(\frac{0.0309 kg H_2O}{kg\, dry\, solid}\right) + \dot{m}_a(H_2)$$

$\dot{m}_a(H_2 - 0.012) = 1211.27$ [Equation (A)]

Energy Balance

Specific heat capacity of dry pigment $(c_{P_{DS}}) = 1.5\, kJ\, /\, kg.K\ ; Q_L = 80000 kJ\, /\, h$

using Equation 9.50 : $\dot{m}_F(q_{F_1}) + \dot{m}_a(q_{a_1}) = \dot{m}_F(q_{F_2}) + \dot{m}_a(q_{a2}) + Q_L$

$q_{F_1} = c_{P_{DS}}(\Delta T_p) + C_{P_w}(w_{F_2})(\Delta T_p)$ [Equation 9.51]

$$q_{F_1} = \frac{1.5 kJ}{kg\, dry\, solid.K}(303.15K - 273.15K) + \frac{4.2 kJ}{kg H_2O.K}\left(1.5\frac{kg H_2O}{kg\, dry\, solid}\right)(303.15K - 273.15K)$$

$q_{F_1} = 234 kJ\, /\, kg\, dry\, solid$

$q_{F_2} = c_{P_{DS}}(\Delta T_p) + c_{P_w}(w_{F_2})(\Delta T_p)$ [Equation 9.52]

$$q_{F_2} = \frac{1.5 kJ}{kg\, dry\, solid.K}(328.15K - 273.15K) + \frac{4.2 kJ}{kg H_2O.K}\left(0.0309\frac{kg H_2O}{kg\, dry\, solid}\right)(328.15K - 273.15K)$$

$q_{F_2} = 89.6379\, kJ\, /\, kg\, dry\, solid$

$$q_{a_1} = [1.005 + 1.884H_1] \times \Delta T_{a_1F} + 2500.9(0.012)$$

$$q_{a_1} = [1.005 + 1.884(0.012)][180 - 0] + 2500.9(0.012)$$

$$q_{a_1} = 214.98\ kJ / kg\ dry\ air$$

$$q_{a_2} = [1.005 + 1.884H_2] \times \Delta T_{a_1p} + 2500.9H_2 \quad \text{[Equation 9.56]}$$

$$q_{a_2} = [1.005 + 1.884H_2][90 - 0] + 2500.9H_2$$

$$q_{a_2} = 90.45 + 2670.46H_2\ kJ / kg\ dry\ air$$

Equation 9.50 becomes:

$$824.5(234) + \dot{m}_a(214.98) = 824.5(89.6379) + \dot{m}_a(90.45 + 2670.46H_2) + 80000$$

$$124.53\dot{m}_a - 2670.46\dot{m}_a H_2 = -39026.5. \quad \text{[Equation (B)]}$$

Combine Equation (A) and (B):
From Equation (A):

$$\dot{m}_a H_2 = 1211.27 + 0.012\dot{m}_a \quad \text{[Equation (C)]}$$

Substitute Equation (C) to Equation (B):

$$124.53\dot{m}_a - 2670.46(1211.27 + 0.012\dot{m}_a) = -39026.5$$

$$92.48\dot{m}_a = 34554.7\ kg / hr$$

$$H_2 = \frac{1211.27 + 0.012(34554.7)}{34554.7}$$

$$H_2 = 0.047\ kg\ H_2O / kg\ dry\ solid$$

9.5.3 Efficiency of Spray Dryer

The performance of a spray dryer is dictated by its thermal efficiency, which is defined as the ratio of the amount of heat energy used for evaporating the moisture to the total heat supplied to the spray dryer (Masters, 1976). Its thermal efficiency depends on the operating temperatures. Assuming that the spray drying process is truly adiabatic, where heat loss is negligible, the overall thermal efficiency, $\eta_{overall}$ can be estimated by Equation 9.57, which can be defined as the ratio of the heat used in evaporation for adiabatic operation to the total heat supplied to the dryer (Masters, 1976).

$$\eta_{overall}(\%) = \frac{T_{a1} - T_2}{T_{a1} - T_{a0}} \times 100 \quad (9.57)$$

where T_{a0} represents the atmospheric temperature of inlet air, T_{a1} represents the heated inlet air temperature and T_2 represents the outlet air temperature if the operation is adiabatic. Note that T_2 shall be read from psychometric diagram. At ideal conditions, maximum overall efficiency can be achieved when the drying air leaves the dryer at saturated state, that is, the outlet air temperature, T_2, is equal to the wet bulb temperature, T_{sat}.

Evaporative efficiency, η_{evap}, refers to the effectiveness of spray dryer in converting the feed into dry product by moisture removal. It represents the ratio of actual evaporative capacity to the evaporative capacity achieved under ideal conditions, where air exits the dryer at adiabatic saturation temperature (wet bulb temperature) and can be calculated using Equation 9.58.

$$h_{evap}(\%) = \frac{T_{a1} - T_2}{T_{a1} - T_{sat}} \times 100 \quad (9.58)$$

Example 9.5. Spray Dryer Efficiency

Ambient air at 25°C with 20% relative humidity is heated to 100°C before entering the spray drying chamber. The air exits the dryer at 60°C with 15% relative humidity. Calculate:

a) Maximum thermal efficiency
b) Overall thermal efficiency
c) Evaporative efficiency
d) Percentage heat loss

Solution

The drying process of this example is illustrated on a psychometric chart, as shown in Figure 9.13.

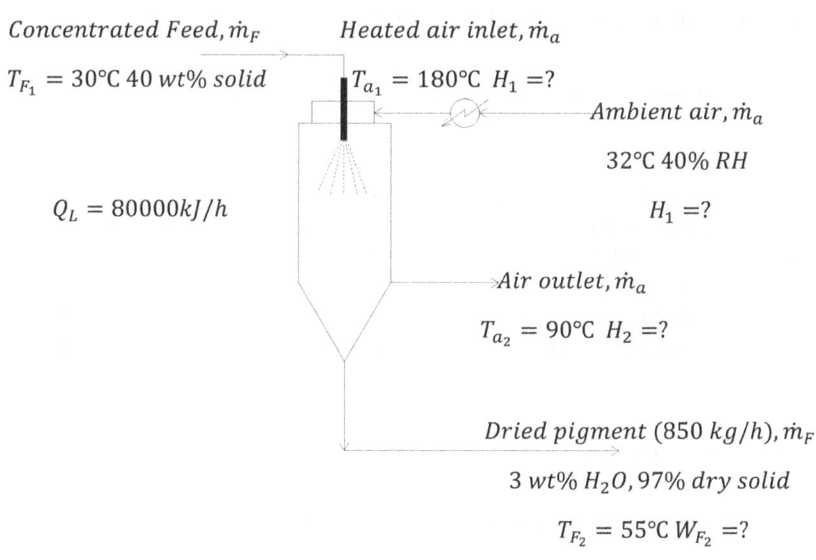

FIGURE 9.13 Schematic diagram of mass and energy balance data around spray dryer.

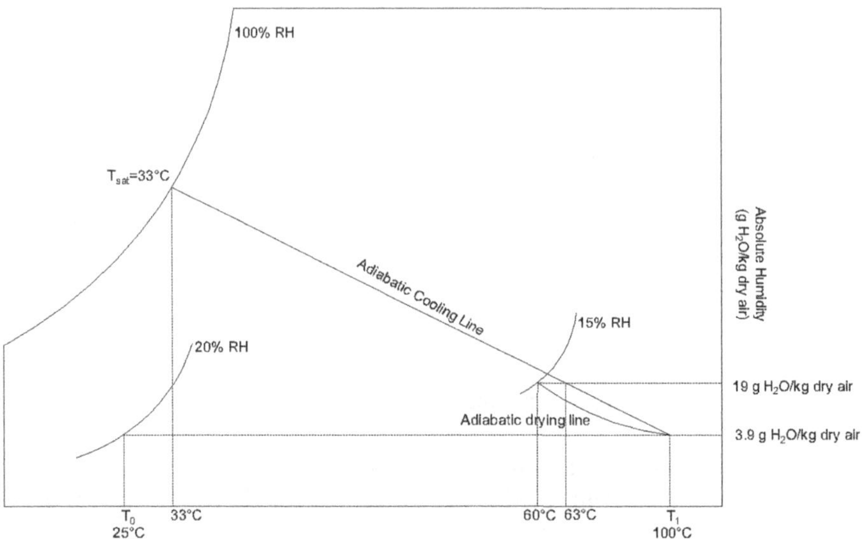

FIGURE 9.14 Drying process of example XX.

a) Maximum thermal efficiency can be calculated using Equation 9.57, with T_2 is equal to the wet bulb temperature, T_{sat}. To determine T_{sat}, a psychometric chart should be used. First, locate the point at T_{a1}, which is 25°C with 20% relative humidity on the chart. Under adiabatic cooling conditions, it follows a constant enthalpy line. T_{sat} can be identified at the intersection of the adiabatic cooling line through T_{a1} and the 100% relative humidity line. From Figure 9.14, ambient air at 25°C with 20% relative humidity carries absolute humidity of 0.0039 kg H_2O/kg dry air and T_{sat} is 33°C.

Thus, maximum thermal efficiency$(\%) = \dfrac{100-33}{100-25} \times 100 = 89.3\%$

b) Overall thermal efficiency is given by Equation 9.57, with T_2 represents the outlet air temperature at adiabatic condition. Similar to part (a), a psychometric chart should be used to determine T_2. Psychometric chart shows that the outlet air at 60°C with 15% relative humidity carries absolute humidity of 0.019 kg H_2O/kg dry air. It should be noted that under adiabatic cooling, absolute humidity remains the same and T_2 is the temperature where the absolute humidity line intersects the adiabatic cooling line through T_{a1}. From Figure 9.14, $T_2 = 63°C$.

Hence, $h_{overall}(\%) = \dfrac{100-63}{100-25} \times 100 = 49.3\%$

c) Evaporative efficiency is given by Equation 9.58.

$$h_{evap}(\%) = \frac{100-63}{100-33} \times 100 = 55.2\%$$

d) Percentage heat loss $= \dfrac{63-60}{100-60} \times 100 = 7.5\%$

9.5.4 Residence Time in Spray Drying Chamber

Residence time is one of the important factors that affects the quality of final product. One must ensure adequate residence time to complete the drying process. Hence, drying chambers must be designed to achieve two objectives: (i) to process an air volume with sufficient heat to dry the spray droplets and (ii) to ensure sufficient air residence time for the droplets to reach the desired moisture content (Masters, 1976). It should be noted that residence time of particle droplets and air in the drying chamber is different from the drying time of droplets. By assuming the minimum residence time, t_{min}, of product is equivalent to the average residence time of air, t_{min} can be estimated by dividing the volume of drying chamber by the total air volumetric flow rate at the outlet temperature (Masters, 1976). The volume of drying chambers with various shapes can be calculated using Equations 9.59 through 9.62, respectively, as detailed in Table 9.9.

The design of a spray dryer often relies on experience and data obtained from pilot plant, such as residence time, air flow rate and air conditions. Residence times of commercial spray dryers are in the range of 3 to 50 seconds. For many products dried in co-current flow spray dryers with rotary air flow, residence times between 20 and 40 seconds are normally observed (Masters, 1976). For highly heat-sensitive products, low drying air temperatures are favourable. However, this results in low driving forces and extended residence times, leading to a very tall drying chamber.

TABLE 9.9
Estimation of Volumes of Drying Chambers with Various Shapes

Shape of Drying Chamber	Formulae to Estimate Drying Chamber Volume	Equation
Rectangular	$V = Lbh$	9.59
Cone	$V = \frac{1}{12} D^2 h$	9.60
Cylinder	$V = \frac{\pi}{4} D^2 h$	9.61
Conical (with cone angle 60°)	$V = 0.7854 D^2 (h' + 0.2886 D)$ *Note that the V estimated through Eq(16) is in ft. dimension.*	9.62

where L = length, b = breadth, h = height, h' = cylindrical height and D = chamber diameter

TABLE 9.10
Recommended Residence Time Ranges for Different Products (Filková and Mujumdar, 2020)

Type of Products	Recommended Residence Time in Drying Chamber
Fine, non-heat-sensitive products, non-hygroscopic	Short (10–20 s)
Fine to coarse spray droplets	Medium (20–35 s)
Large powder (200–300 μm), heat-sensitive products	Long (> 35 s)

Given the same inlet air temperature, larger spray droplets require longer residence time as compared to finer droplets. Table 9.10 shows the recommended residence time ranges for different products.

Example 9.6. Residence Time of Spray Drying Process

In a production of skim milk, a co-current flow spray dryer is supplied with 55,000 kg/hr air. The drying chamber has a 60° cone with a diameter of 7 m and a cylindrical height of 5m. Determine the minimum residence time in the drying chamber if the air is exhausted at 100°C with an absolute humidity of 0.044 kg H_2O/kg of dry air.

Solution

Volume of conical-based spray dryer is given by Equation 9.62.

$$V = 0.7854D^2\left(h' + 0.2886D\right)$$

Given, spray dryer diameter (D)=7 m and cylindrical height (h')=5 m
Convert them into ft

$$D = 7\text{m} \times \frac{3.2808\text{ft}}{1\text{m}} = 22.97\text{ft}$$

$$h' = 5m \times \frac{3.2808ft}{1m} = 16.404ft$$

$$\therefore V = 0.7854 \times 22.97^2\left[16.404 + 0.2886 \times 22.97\right.$$

$$V = 9545ft^3$$

$$V = 9545ft^3 \times \frac{1m^3}{(3.2808)^3 ft^3}$$

$$\therefore V = 270.3\text{m}^3$$

Air flow rate = 55000kg / hr air

By assuming air is ideal gas, ideal gas law can be used to deduce volumetric flow rate of air at exhaust temperature. Since air carries some moisture, the specific humid air volume(V_h) at exhaust temperature can be calculated using the following equation:

$$V_h = \frac{22.4m^3}{kgmol}\left[\frac{1}{29kg / kmol\,Air} + \frac{\mathrm{H}}{18kg / kgmolH_2O}\right]\left[\frac{\mathrm{T}_2\,K}{273.15K}\right]\left[\frac{1atm}{P\,atm}\right]$$

Assume no pressure drop, the equation can be reduced to

$$V_h = \frac{22.4m^3}{kgmol}\left[\frac{1}{29kg / kmolAir} + \frac{H}{18kg / kgmolH_2O}\right]\left[\frac{T_2 K}{273.15K}\right]$$

$$\therefore V_h = 22.4\left(\frac{1}{29} + \frac{0.044}{18}\right)\left(\frac{373.15}{273.15}\right)$$

$$V_h = 1.13m^3 kg\,dry\,air$$

Total volumetric flow rate of air at 100°C

$$= \frac{55000kg}{hr}\left(\frac{1.13m^3}{kg}\right) \times \frac{1hr}{3600s} = 17.3m^3/s$$

Minimum residence time

$$= \frac{drying\ chamber\ volume}{total\ volumetric\ flow\ rate\ of\ air\ at\ outlet\ temperature}$$

$$= \frac{270.3m^3}{\frac{17.3m^3}{s}}$$

$$= 15.6s$$

Example 9.7. Spray Dryer Sizing

A spray dryer pilot plant has been constructed to produce a heat-sensitive pharmaceutical product. The drying chamber, which has a cylindrical shape with a diameter of 5 m, is supplied with 25,000 kg/hr of dry air. According to pilot plant data, a residence time of 50 seconds is required. The air exits the dryer at 90°C with an absolute humidity of 0.05 kg H_2O/kg dry air. Determine the height of this spray dryer.

Solution

Assume ideal gas and no pressure drop, specific humid air volume(V_h):

$$V_h = 22.4\left(\frac{1}{29}+\frac{0.05}{18}\right)\left(\frac{363.15}{273.15}\right)$$

$$V_h = 1.11 m^3/kg\ dry\ air$$

$$\text{volumetric flow rate of air} = \frac{25000kg}{hr}\left(\frac{1.11m^3}{kg}\right)\left(\frac{1hr}{3600s}\right) = 7.71m^3/s$$

$$\text{Average velocity of dry air} = \frac{7.71m^3}{s}\left(\frac{4}{p(5^2)m^2}\right) = 0.393m/s$$

$$\text{height of spray dryer} = \text{velocity} \times \text{residence time} = 0.393 \times 50 = 19.6m$$

9.6 CONCLUSIONS

The separation tools used in the chemical product industry are significantly different from those commonly used in the bulk chemical industry. In this chapter, the most commonly used separation techniques in the chemical product industries have been discussed. The fundamental principles involved in the separation processes, such as steam distillation, batch adsorption, extraction, crystallization and spray drying, have been discussed in detail with examples. Comparisons have been made against the separation approaches used in the bulk chemical industry to have a clearer understanding of the distinction between these industries.

REFERENCES

Clarke, C., Tu, W., Levers, O., Bröhl, A., & Hallett, J. (2018). Green and Sustainable Solvents in Chemical Processes. *Chemical Reviews*, *118*(2), 747–800.

Cussler, E. L., & Moggridge, G. D. (2011). Molecular Products. In *Chemical Product Design. Cambridge Series in Chemical Engineering* (pp. 311–362). Cambridge University Press.

Dubinin, M. M., & Radushkevich, L. V. (1947). Equation of the Characteristic Curve of Activated Charcoal. *Proceedings of the Academy of Sciences of the USSR: Physical Chemistry Section*, *55*(1), 331–337.

Filková, I., & Mujumdar, A. S. (2020). Industrial Spray Drying Systems. In *Handbook of Industrial Drying* (pp. 263–307). CRC Press.

Freundlich, H. M. F. (1906). Over the Adsorption in Solution. *Journal of Physical Chemistry*, *57*(385471), 1100–1107.

Gmehling, J., & Schedemann, A. (2014). Selection of Solvents or Solvent Mixtures for Liquid–Liquid Extraction Using Predictive Thermodynamic Models or Access to the Dortmund Data Bank. *Industrial & Engineering Chemistry Research*, *53*(45), 17794–17805.

Hall, K. R., Eagleton, L. C., Acrivos, A., & Vermeulen, T. (1966). Pore- and Solid-Diffusion Kinetics in Fixed-Bed Adsorption Under Constant-Pattern Conditions. *Industrial and Engineering Chemistry Fundamentals*, *5*(2), 212–223. https://doi.org/10.1021/i160018a011

Langmuir, I. (1918). The Adsorption of Gases on Plane Surfaces of Glass, Mica and Platinum. *Journal of the American Chemical Society*, *40*(9), 1361–1403.

Masters, K. (1976). *Spray Drying: An Introduction to Principles, Operational Practice, and Applications*. G. Godwin.

McCabe, W. L., Smith, J. C., & Harriott, P. (1993). Adsorption. In *Unit Operations of Chemical Engineering* (pp. 810–837). McGraw-Hill.

Miers, H. A., & Isaac, F. (1906). Refractive Indices of Crystallizing Solutions. *Journal of the Chemical Society*, *89*, 413–454.

Myerson, A. (2002). *Handbook of Industrial Crystallization*. Butterworth-Heinemann.

Redlich, O., & Peterson, D. L. (1959). A Useful Adsorption Isotherm. *The Journal of Physical Chemistry*, *63*(6), 1024. https://doi.org/10.1021/j150576a611

Richardson, J. F., Harker, J. H., & Backhurst, J. R. (2002). Chapter 15 – Crystallisation. In *Chemical Engineering Series, Chemical Engineering* (5th ed., pp. 827–900). Butterworth-Heinemann.

Seader, J. D., Henley, E. J., & Roper, D. K. (2016). Adsorption, Ion Exchange, and Chromatography. In *Separation Process Principles: With Applications Using Process Simulators* (pp. 451–517). Wiley.

Temkin, M. J., & Pyzhev, V. (1940). Recent Modifications to Langmuir Isotherms. *Acta Physiochimica URSS*, *12*, 217–225.

Wankat, P. C. (1994). *Rate Controlled Separations*. Dordrecht: Springer.

Worch, E. (2012). Adsorption Equilibrium II: Multisolute Adsorption. In *Adsorption Technology in Water Treatment: Fundamentals, Processes, and Modeling* (pp. 77–122). De Gruyter. https://doi.org/10.1515/9783110240238

9.7 PROBLEMS

1. Cinnamon fruit contains several constituents that can be separated via steam distillation to form an essential oil. The essential oil contains more than 50 components; however, the major components are cadinene (A), cadinol (B) and caryophyllene (C). The average molecular mass of the essential oil is 210 g/mol. The normalized mole fractions of A, B and C in the oil are 0.6, 0.25 and 0.15, respectively. It can be assumed that the rest of the components are only present in insignificant quantities. The components of the essential oil are immiscible with water. It can be assumed that A, B and C form an ideal mixture. The essential oil needs to be recovered from a number of non-volatile impurities in the crude cinnamon fruit. The oils are heat sensitive and will decompose well below their boiling points. A still is charged with 200 kg of fruits that contain 5% essential oil by weight. It is

TABLE 9.11
Vapour Pressure Data at 95°C

Vapour pressure of water, kPa	83.1
Vapour pressure of A, kPa	3.7
Vapour pressure of B, kPa	5.2
Vapour pressure of C, kPa	2.6

proposed to recover the oils by allowing them to mix with steam directly at a temperature of 95°C. Steam is passed at a rate of 30 kg/h. The vapour pressure data are given in Table 9.11.
Calculate the pressure to be maintained in the still and check whether all the oil can be recovered in 45 minutes by ignoring the losses. Assume a vaporization efficiency of 75%.

2. Geraniol ($C_{10}H_{18}O$) is an essential oil of commercial value. It is conventionally purified from the nonvolatile impurities by steam distillation. A batch of 2000 kg of impure geraniol is purified by steam distillation at a pressure of 101.3 kPa using saturated steam. A feed containing 85 weight % geraniol and 15% water is fed to a still at atmospheric conditions. The vaporization efficiency was found to be 75%. The vapour pressure of geraniol and water can be estimated as follows:

Geraniol $\ln P^s = 22.1 - \dfrac{7200}{T}$

Water $\ln P^s = 16.26 - \dfrac{3800}{T - 46.8}$

where T is the temperature in K, and P^s is the vapour pressure in kPa.

a) Estimate the distillation temperature and the composition of the vapour that leaves the steam distillation unit.
b) If the saturated steam supplied to the still is used for the process, estimate the amount of steam required to recover all the germinol from the crude feed.
Data:
Latent heat of vaporization of germinol = 300 kJ/kg
Heat capacity of germinol = 2 kj/kg K
Heat capacity of water = 4.2 kj/kg K
Latent heat of saturated steam at 101.3 kPa = 2257 kJ/kg

3. Crude turpentine can be purified by steam distillation using saturated steam at 105 kPa. A feed of 1000 kg of crude turpentine is charged to the still at

TABLE 9.12
Vapour Pressure Data

VP_{water} (kPa)	$VP_{Turpentine}$ (kPa)	T (K)
12.4	2.5	323
19.9	3.5	333
31.1	5	343
47.3	7.5	353
70.1	11.8	363
101.3	17.5	373

303 K. Saturated steam is supplied to the still at 130 kPa. The vaporization efficiency is 80%. The vapour pressure (VP) data is shown in Table 9.12. Determine

a) The composition of the vapour. You can assume that the vapour pressure changes linearly between the given temperature intervals.
b) The amount of steam needed for this process.

Data:

Latent heat of vaporization of turpentine = 300 kJ/kg

Heat capacity of turpentine = 2 kJ/kg K

Heat capacity of water = 4.2 kJ/kg K

Latent heat of saturated steam at 130 kPa = 2238 kJ/kg

Specific enthalpy of saturated steam at 105 kPa = 2678 kJ/kg

Specific enthalpy of saturated steam at 130 kPa = 2687 kJ/kg

Molecular weight of turpentine = 140 g/mol

Molecular weight of water = 18 g/mol

Temperature of saturation of steam at 130 kPa = 380.3 K

4. A penicillin broth fermenter (bioreactor) has 1 kg of penicillin in 150 L of water. Compare the effectiveness of one 100 L versus two 50 L extractions with amyl acetate. Solubility of penicillin in amyl acetate is 10 times more than that in water. Also, calculate the amount of penicillin that is left in the aqueous layer after two extractions. If there are three extraction units available to recover penicillin from the aqueous phase, determine the outlet composition of aqueous phase if 100 L of solvent is used in each stage.
5. Powdered activated carbon was used in a batch adsorption experiment to remove nickel from a solution. The equilibrium data at 50°C are presented in Table 9.13. Determine which isotherm model – Henry's law, Langmuir, Freundlich or

TABLE 9.13
Concentration and Adsorption Capacity Data

c_{eq} (mg/L)	q_{eq} (mg/g)
2.091	4.844
11.842	12.040
26.261	18.435
60.297	34.926
139.447	65.138
186.146	78.464
245.845	88.539
321.525	94.619

Dubinin-Radushkevich (D-R) best fits the data through linear regression analysis.

6. To produce 1500 kg/hr of catalyst with 5 wt% moisture, a feed that contains 45 wt% solid is pumped into a co-current flow spray dryer at 35°C. Atmospheric air at 35°C and 30% relative humidity is heated to 200°C prior to entering the chamber. The dried catalyst leaves the dryer at 60°C, and air leaves the dryer at 85°C. The specific heat capacity of the catalyst and liquid moisture is 1.05 kJ/kg K and 4.2 kJ/kg K, respectively. 100,000 kJ/h of heat is lost during the process. Compare the air flow rate and its absolute humidity at dryer outlet for this process with those of the process when heat loss is negligible.
7. Ambient air at 35°C with 30% relative humidity is heated to 115°C before entering the spray drying chamber. The air exits the dryer at 68°C with 15% relative humidity. Calculate:
   a) Maximum thermal efficiency
   b) Overall thermal efficiency
   c) Evaporative efficiency
   d) Percentage heat loss

10 Unconventional Processing Techniques in Chemical Product Industry

The conventional unit operations discussed in the previous chapter, such as distillation, adsorption and crystallization, are typically sufficient for producing single-species molecular products. Examples include solvents and refrigerants, which are manufactured using well-established processing techniques and produced in large quantities for economic feasibility. Even blended products like fuel blends require only simple processing methods. However, there are shortfalls for using these traditional unit operations to produce complex formulated and functional products, such as soaps, detergents and various chemical devices.

Producing advanced devices and functional products often requires innovative techniques such as physical vapour deposition, solution coating and granulation, which are beyond the scope of conventional unit operations. In these cases, unconventional processing techniques must be considered. These unconventional processing techniques can be grouped into two major classifications, including addition/formation and removal/destruction (Zhang et al., 2018). Addition/formation process involves either adding a component or a mixture of components to a product or using physical and/or chemical methods to create a specific macrostructure or microstructure in the product. Under the addition/formation classification, techniques such as coating, agglomeration and printing are prevalent. On the other hand, removal/destruction process involves breaking down a component or a mixture of components from the manufactured product. Surface removal and breakage are typical examples under the removal/destruction category. In this chapter, different types of unconventional processing techniques will be discussed.

10.1 COATING

Coating is used to enhance the properties of an object or to provide new functionalities. It refers to applying a thin continuous layer of material onto the surface of an object. It can be applied in the form of solid, liquid or gas. But most of the time, coating is applied in liquid form. They serve two main purposes: functional and decorative. Most coatings function as a protective layer to shield surfaces from biological, chemical or mechanical damage caused by contact with surrounding media. They can also function as decorative finishes by adding colour and smoothing out surface

DOI: 10.1201/9781032708911-10

irregularities. For example, paint is one of the most common types of conventional coatings used in our daily lives. Paints comprise binders, resins, solvents and pigments, which give colour, adhesion and protection to surfaces such as metal and wood against sunlight, moisture and abrasion. Besides paints, coatings also include materials such as lacquers, undercoats, sealers, enamels and varnishes. They are all developed based on the same fundamental principles and normally consist of four key components, including film-forming substances like resin and epoxy, pigments for colour, solvents and additives. When the desired properties of a final product are known, a coating can be formulated by identifying the necessary components and their required amounts to achieve the desired functionalities. Unlike paints that are often applied with a roller, brush or sprayer, some specialized coatings require more advanced techniques. The following section discusses some common types of coating techniques.

10.1.1 Types of Coating Techniques

There are different types of coating techniques available depending on the physical form of precursors, that is, solid (powder coating), liquid (solution) and gas, as well as molten or semi-molten state. Among these, the effective methods include physical vapour deposition (PVD), chemical vapour deposition (CVD), sol-gel and thermal spraying. Due to their unique deposition techniques and the ability to adjust the type of material, thickness and density, each of these methods is well-suited for various applications. The typical coating thickness ranges from 0.1 μm to 10 mm, whereas the deposition temperature varies from room temperature up to 1200°C. Some characteristics of the common coating techniques mentioned are compared in Table 10.1. The following sub-section discusses the general overview of the mentioned coating techniques.

TABLE 10.1
Comparison of Characteristics among Common Coating Techniques (Holmberg and Matthews, 2009)

Characteristics \ Coating Techniques	Chemical Vapour Deposition (CVD)	Physical Vapour Deposition (PVD)	Sol-Gel	Thermal Spraying
Coating thickness (μm)	0.5–2000	0.1–1000	1–10	50–1000
Uniformity of coating	Very good	Good	Fair/good	Variable
Deposition rate (kg/h)	Up to 1	Up to 0.5 per source	0.1–0.5	0.1–10
Substrate deposition temperature (°C)	150–1200	50–500	25–1000	100–800
Substrate material	Metals, ceramics, polymers	Metals, ceramics, polymers	Metals, ceramics, polymers	Metals

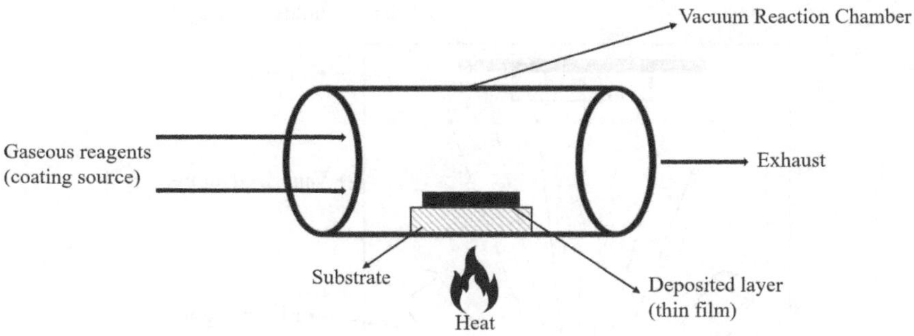

FIGURE 10.1 Schematic diagram of CVD process.

10.1.2 Chemical Vapour Deposition (CVD)

In the CVD process, gaseous reagents serve as the source of coating species. The solid substrate, which can be made of various materials such as metals, ceramics or polymers, is placed in a reaction chamber operating under vacuum conditions. To initiate the reaction, the substrate is subjected to external stimuli such as heat (typically at elevated temperatures ranging from 800–1200°C) or plasma. Source gases are introduced into the reaction chamber, where they react and deposit on the substrate, forming a coating. The by-product generated during the reaction must then be removed. The deposition that occurs during the CVD process is multi-directional, allowing for a more uniform coating across complex surfaces. An example of a CVD process subjected to heat and vacuum conditions is illustrated schematically in Figure 10.1.

10.1.3 Physical Vapour Deposition (PVD)

In contrast to CVD, where the source is introduced in the vapour phase, PVD involves forming the coating species (vapours) through evaporation or atomization from a solid source placed within the vacuum chamber. These vaporized atoms then deposit onto the substrate to form a coating. PVD is categorized based on how the source material is atomized, which can occur through either heat input or the transfer of kinetic energy. The two main categories of PVD are evaporation and sputtering. Evaporation involves the use of heat energy to vaporize the solid material source. The heat energy required for evaporation can be supplied by different means such as electron beam, inductive heating, resistive evaporation and cathodic arc evaporation (Holmberg and Matthews, 2009). Figure 10.2 shows the PVD process occurring through evaporation.

On the other hand, sputtering is a non-thermal vaporization process where the source material, known as the sputtering target, is made cathodic. The setup allows ions, typically from an inert gas such as Argon, to bombard the cathode surface. Upon ion bombardment, coating atoms are ejected and vaporized from the source by momentum transfer. These coating atoms then travel through the vacuum and finally deposit onto the substrate surface to form a thin layer of coating. Some common configurations used in sputtering PVD include diode, triode, ion beam and magnetron (Holmberg and Matthews, 2009). Figure 10.3 shows the sputtering PVD process.

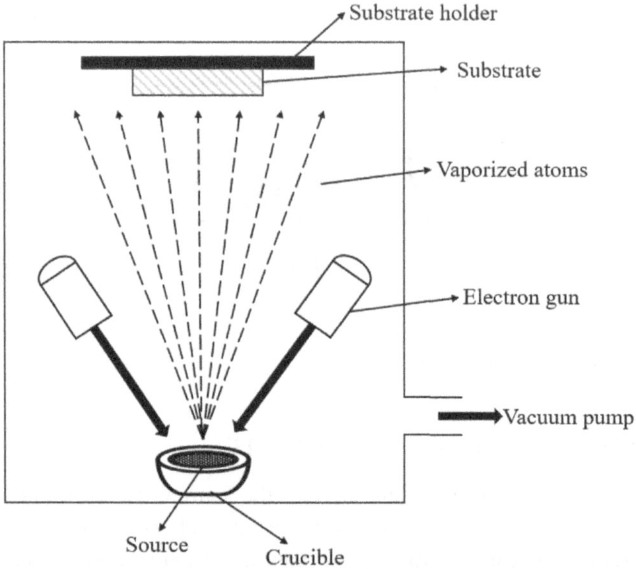

FIGURE 10.2 Schematic diagram of a PVD setup utilizing electron beam as the thermal source.

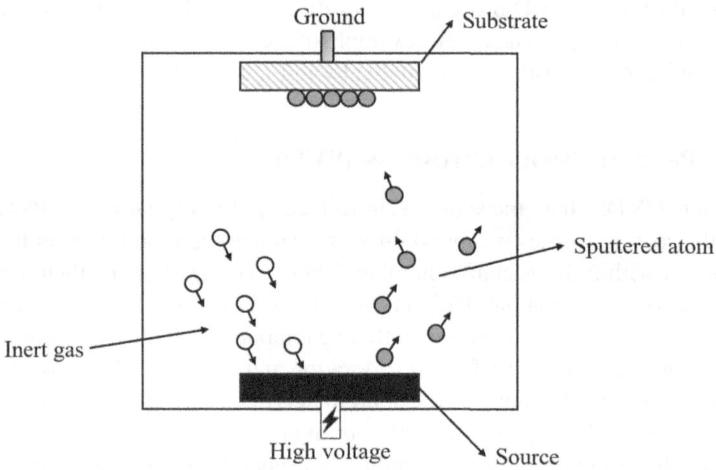

FIGURE 10.3 Sputtering PVD process.

Nonetheless, compared to CVD, PVD has some drawbacks, such as it produces less uniform coating on complex shapes, involves a more complex process and incurs higher operating costs. The less uniform coating in PVD is primarily due to its more directional deposition, which results in a non-conformal film. PVD is normally used to deposit a conductive metal oxide, typically indium tin oxide, to form a transparent conductive thin film on the glass substrate. This glass substrate, which is coated with one or more transparent conductive layers, is commonly used in manufacturing touch screen devices.

10.1.4 Sol-Gel

The sol-gel process is one of the most widely used techniques in the category of solution coating. It is commonly used to produce thin film for solar energy applications. In solution coating, the coating material, often a polymer or resin, is applied as a solution onto the substrate. This involves dissolving the precursors in a solution. The solution then forms a dry film upon drying. In sol-gel process, it involves transforming a liquid colloidal solution (sol) into a gel-like network. There are four (4) main steps involved in sol-gel process. The first step involves creating a 'sol' by dissolving metalorganic precursors like metalloid alkoxide in a solvent such as alcohol. The following step is to deposit the sol solution to produce coating on the substrate, which is normally done by dipping, spinning or spraying. The particles in sol then undergo hydrolysis and condensation reactions to form gel; this process is normally known as gelation. The step is then followed by ageing and drying to allow further condensation and the formation of a thin film on the substrate surface. The sol-gel process is schematically illustrated in Figure 10.4.

10.1.5 Thermal Spraying

Thermal spraying uses a heat source such as combustion, electricity or plasma to melt materials. The molten or semi-molten material is then sprayed onto the surface of substrate to form a protective layer. There are two types of thermal spraying, including low-energy and high-energy processes (Holmberg and Matthews, 2009).

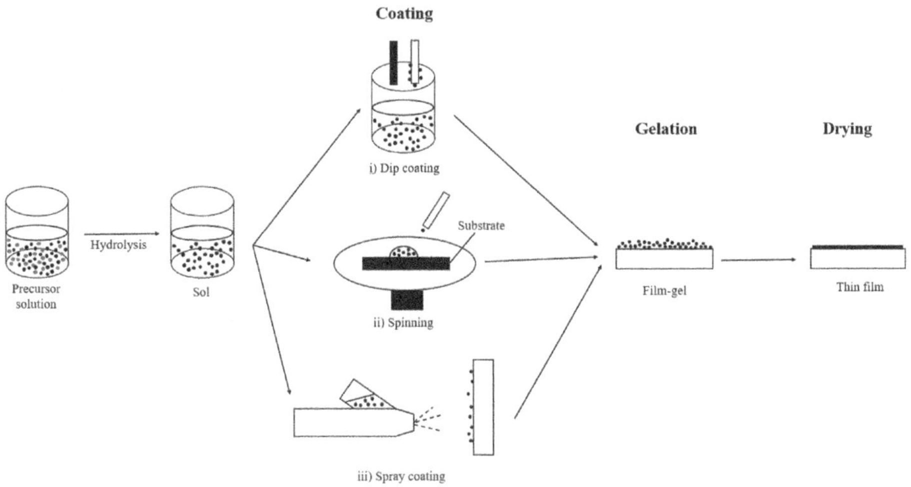

FIGURE 10.4 Schematic illustration of sol-gel process.

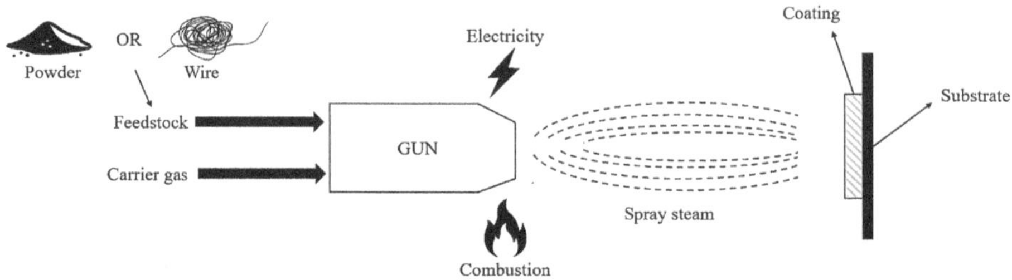

FIGURE 10.5 Thermal spraying process.

Arc and flame spraying are examples of low-energy processes, whereas high-energy processes include methods such as plasma spraying, high-velocity oxy-fuel (HVOF) spraying. The coatings produced through thermal spraying are often used to provide protection from corrosion and are considerably thicker than those produced by CVD, PVD or the sol-gel method.

The thermal spraying process is illustrated in Figure 10.5. The process begins with the coating material (feedstock), in the form of either powder or wire, entering the gun. Note that the coating materials should not decompose at or near their melting point. The feedstock is then heated to a molten or semi-molten state and accelerated towards the surface of the substrate using a carrier gas such as argon or helium. As the accelerated particles strike the substrate surface at high velocity, they flatten and form a coating layer.

10.2 AGGLOMERATION

Agglomeration is a widely used particle-size enlargement technique in chemical process industries. It is a process where solid particles adhere to one another or to surfaces through physical and/or chemical bonding. This process enhances the functionalities and characteristics of a product while addressing some common processing challenges, such as poor flowability, low bulk density, dusting and segregation. In many cases, it is challenging to handle or transport fine particles as they do not flow easily. This difficulty arises from their tendency to clump together because of surface forces acting between them. Additionally, fine particles can lead to adverse health and environmental consequences as they tend to form dust clouds that pose inhalation risks and contaminate the environment. Agglomeration helps mitigate these issues by increasing the bulk density of products, enabling them to flow more easily. When particles are held together in the form of agglomerates, their tendency to segregate and form dust is also reduced. Moreover, agglomeration allows for better control over the functional properties, porosity and solubility of the product (Green and Perry, 2008).

Agglomeration process is found across various industries. In the chemical industry, it is used to produce catalysts, detergents and bleach powders. In the food industry, it facilitates the production of instant foods and beverages such as milk powder, instant coffee or tea powder, chocolate drink powder and dry soups. In agriculture, it is utilized for producing pesticides, fertilizers and fuel pellets. Unit operations used for agglomeration process can be broadly categorized into two main types: agitation methods and compression methods (Green and Perry, 2008). Agitation agglomeration, commonly known as granulation, involves mixing or stirring formulated materials to form agglomerates. Compression or pressure agglomeration involves pressing materials into larger and denser structures. This technique can be used as the final stage in the pharmaceutical industry to produce tablets and capsules for medicinal purposes after the material has been granulated. Agglomeration can also be induced by heat, leading to a process known as sintering. The following sections will discuss the common agglomeration processes used in the industries.

10.2.1 Dry Granulation

There are two major granulation processes such as dry granulation and wet granulation. In dry granulation, the chemical composition of the product remains the same as no liquid is used in the process. Slugging/pre-compression and roller compaction are the two methods used for dry granulation. Both methods require high pressure to agglomerate the particles. However, the difference lies in the form of how powder is processed. In the slugging process, a heavy-duty tablet press is used to produce slugs or large tablets, whereas the powder is compressed between two rolls to form a continuous sheet or plates of materials in the roller compaction process. A typical dry granulation process is illustrated in Figure 10.6.

Note that dry granulation is employed when the product is sensitive to heat and/or moisture as wet granulation could result in product degradation. In general, roller compaction has become more popular and has replaced slugging in most industries as slugging suffers from some limitations, such as poor process control and lower production capacity.

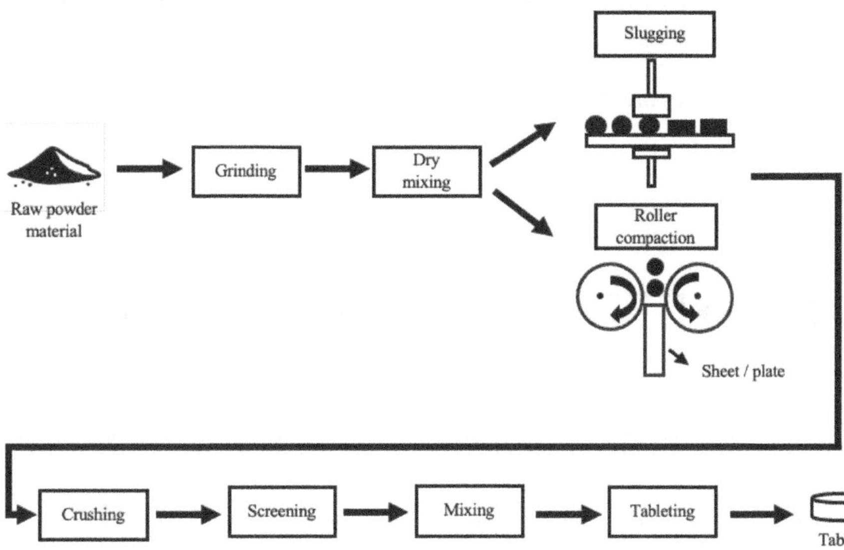

FIGURE 10.6 Dry granulation process.

10.2.2 Wet Granulation

Wet granulation involves mixing bulk material with a liquid binder or solvent to aid agglomeration process by forming a wet mass through adhesion. The liquid binder can be added to the powder material through various methods such as spraying, pouring and melting. This process is typically performed in high-shear mixers, drums, disc granulators or fluidized bed granulators. Compared to dry granulation, wet granulation requires more time, energy and labour due to the necessary drying step to reduce moisture content within the wet granules. The common steps involved in wet granulation process are depicted in Figure 10.7. Despite being a longer, more complex and expensive process, wet granulation remains the preferred choice because the attributes of the finished granule product can be customized.

Wet granulation is governed by three sets of rate processes (Green and Perry, 2008):

a) Wetting and nucleation
   In this stage, the liquid binder is brought into contact with the dry powder materials, resulting in the formation of liquid bridges between particles. The wetted particles adhere to each other and form small aggregates or nuclei with a loose structure.
b) Consolidation and growth
   This occurs when two granules collide with each other or when they collide with the walls of the equipment, leading to compaction and growth. This step dictates the density and porosity of the final granule product.
c) Attrition and breakage
   This is the phase where some granules may experience attrition or break into smaller particles due to mechanical forces or compaction within the granulator. This leads to a reduction in granule size, which subsequently influences the final size distribution of the granules.

These rate processes are illustrated schematically in Figure 10.8.

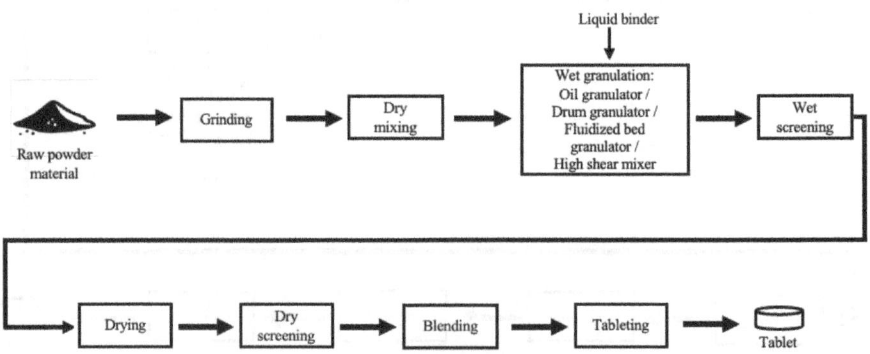

FIGURE 10.7 Steps involved in wet granulation process.

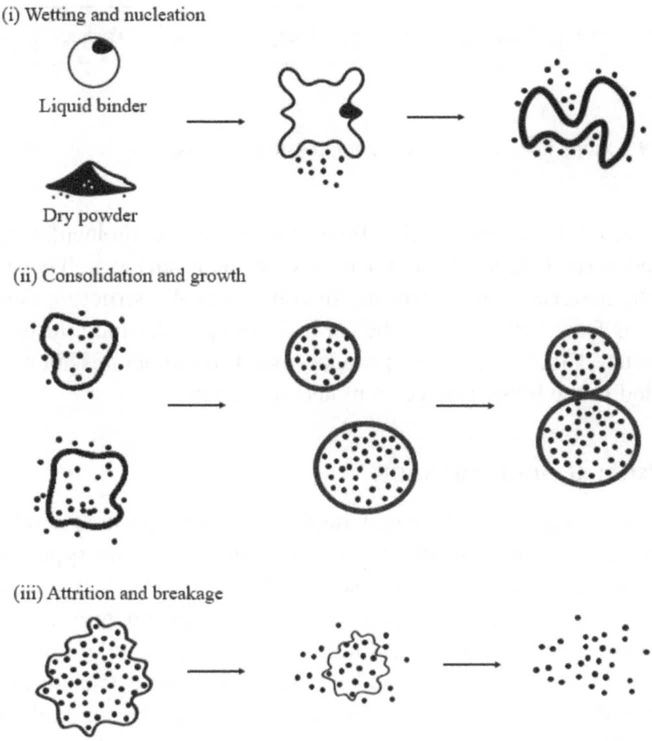

FIGURE 10.8 Mechanisms of wet granulation: (i) wetting and nucleation, (ii) consolidation and growth and (iii) attrition and breakage.

10.2.3 Direct Compression

Unlike dry and wet granulation, direct compression does not require a granulator. In this process, raw powdered material is compressed directly into tablets without any prior modification of the physical properties of tablet products through wet or dry granulation. This method involves fewer processing steps, leading to a reduction in unit operations and processing time. Besides, product stability is enhanced using this method since moisture and heat are not involved in the process. There are three common techniques available for direct compression, namely direct compression excipients, direct compression with induced dry binders and direct compression with induced die feeders. Figure 10.9 illustrates the typical process involved in the direct compression process. Some examples of pharmaceutical products produced using direct compression include chewable antacid tablets and vitamin C tablets.

10.2.4 Sintering

Sintering is defined as a thermal agglomeration process where particles are aggregated through heat treatment. It can be either a batch or continuous process. By

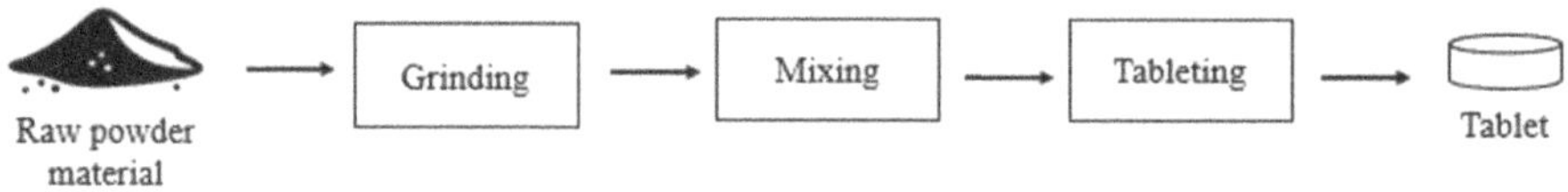

FIGURE 10.9 Steps involved in direct compression process.

heating the particles without melting them, the atoms and molecules diffuse within materials and form bridge-like connections between surfaces. These connections solidify as the material cools, resulting in cohesive, solid structure (Pietsch, 2008). This process is frequently used in the ceramic and powder metallurgy industries to solidify powdered materials or to produce metal agglomerates. For example, it is widely applied in the blast furnace ironmaking system.

10.2.5 Pressure Agglomeration

Pressure agglomeration, as its name implies, induces particle agglomeration by pressure. This process is typically carried out using two main types of equipment: dry confined-pressure devices and paste extrusion devices. Dry confined-pressure devices include briquetting, roll presses, tableting, piston presses and moulding presses, whereas screw extruders, cylinder pelletizers and pellet mills are examples of paste extrusion devices (Green and Perry, 2008). The extrusion process can be performed in either dry or wet modes to produce dense pellets or agglomerates. Most pressure agglomeration processes are conducted in batch mode. Additionally, pressure agglomeration techniques can be categorized based on the level of applied pressure and the configuration of the die or tool, as summarized in Table 10.2.

Depending on the equipment used, the products can take the form of continuous sheets, tablets or briquettes. The products produced using pressure agglomeration possess higher density compared to those produced through agitative agglomeration (Green and Perry, 2008). Pressure agglomeration can be selected if the process involves one or more of the following conditions or if there is a need to achieve any of the product characteristics as listed in Table 10.3.

In summary, various methods are available for particle agglomeration, each tailored to different applications and yielding distinct product attributes. The choice of method significantly influences the characteristics of the final product. Table 10.4 provides an overview of the different granulators and their respective applications. Despite the availability of numerous established techniques, there have been advancements in granulation technologies. These advancements include pneumatic dry granulation technology for dry granulation and several innovative technologies for wet granulation, such as reverse wet granulation, foam granulation, steam granulation, melt granulation, moisture-activated dry granulation, freeze granulation and thermal adhesion granulation.

10.3 PRINTING

In addition to coating and agglomeration, printing is another additive process that adds or applies components onto a substrate. To cater for different needs in various

TABLE 10.2
Summary of Pressure Agglomeration Techniques (Pietsch, 2008)

Level of Pressure Applied	Examples (With Its Die or Tool Configuration)	Product Density
Low (Extrusion)	• Gravity feed and roller(s) (flat die extruders) • Gravity feed and extrusion blades (screen and basket extruders) • Screw feeder(s) and extrusion blades (radial, axial and dome extruders)	Low
Medium (Pelleting)	• Hollow, perforated cylinder(s), feed from the inside • Hollow, perforated cylinder(s), feed from the outside • Gear-shaped press rollers, feed from the outside • Flat, perforated die plate with press roller(s) • Medium-pressure axial screw extruders	Medium
High	• High-pressure axial screw and ram extruders • Punch-and-die presses • Roller presses	High

TABLE 10.3
Key Considerations for Choosing Pressure Agglomeration: Process Conditions and Product Attributes (Pietsch, 2008)

Process Conditions	Product Attributes
Dry or elevated temperature processing	High density
Minimal or no binder usage	Low porosity
Feed with larger particle size	High final strength
No post-processing	Extended shelf life
Processing of elastic materials	Large piece
Automated operation	Defined shape
Simple clean-up	Specific mass
Rapid turn-over	Amendable to the production of near net-shape parts
High initial strength	

industries, there are many printing techniques available today, from conventional methods to advanced techniques. Some examples of conventional printing methods include screen printing, offset printing and inkjet printing. Screen printing can be used to produce radio-frequency identification (RFID) tags. Offset printing is mostly used for mass-production printing, such as printing newspapers. However, when there is a need to produce high-resolution colour prints, one should consider the use of inkjet printer. Inkjet printing has also found its application in fabricating conductive tracks for printed electronics. Note that inkjet printing can be used for both two-dimensional (2D) and three-dimensional (3D) printing.

TABLE 10.4
Comparison of Various Granulators and Their Applications (Green and Perry, 2008)

Method	Type	Granule Size (mm)	Granule Density	Production Scale	Application
Tumbling granulator	Drums	0.2–20	Moderate	0.5–800 ton/hr	Fertilizers, iron ore, non-ferrous ore, agricultural chemicals
	Disc				
Mixer granulators	Continuous high shear	0.1–0.5	Low	Up to 50 ton/hr	Chemicals, detergents, clays, carbon black
	Batch high shear	0.1–2	Moderate to high	Up to 500 kg per batch	Pharmaceuticals, ceramics, clay
Fluidized granulators	Fluidized beds	0.1–1	Low (agglomerated)	100 to 900 kg per batch	Batch: pharmaceuticals, agricultural chemicals, nuclear wastes
	Spouted beds		Moderate (layered)	50 tons/h if continuous	Continuous: fertilizers, inorganic salts, food, detergents
	Wurster coaters				
Pressure compaction	Extrusion	> 0.5	High to very high	Up to 5 ton/hr	Pharmaceuticals, catalysts, clays, minerals, ceramics, organic and inorganic chemicals
	Roll press	> 1		Up to 50 ton/hr	
	Tablet press	10		Up to 1 ton/hr	
Thermal processes	Sintering	2–50	High to very high	Up to 100 tons/h	Ferrous and non-ferrous ore, minerals, ceramics, cements

3D printing has recently emerged as a promising technology in various fields. In chemical engineering, it is used to fabricate components such as membranes, filters and adsorbents for water purification and gas separation. Its ability to rapidly prototype and produce complex structures enables the creation of innovative chemical processing equipment. With the use of 3D printing, geometries of packed columns and reactors can be customized to enhance the process performance. Thus, the following section will focus on 3D printing, providing an overview of its general working principle. The section also delves into the common 3D printing techniques along with their benefits and drawbacks.

10.3.1 Three-Dimensional (3D) Printing

3D printing is an additive manufacturing process that fabricates a wide range of physical models with complex geometries based on 3D digital data. Unlike screen printing or 2D inkjet printing, which produce text and images on paper or other flat surfaces, 3D printer prints real-life 3D objects, irrespective of the material used. Their applications can be found in various industries, such as healthcare, manufacturing, architecture and wastewater treatment. Some examples include the use of 3D printing to produce microfluidic devices for wastewater treatment and drug delivery. It is also applied to create porous structures for gas separation applications. 3D printing reduces the overall cost of product development as the process is rapid and uses lesser material as compared to traditional manufacturing methods.

Regardless of the 3D printing techniques used, all 3D printers work under the same fundamental principle, wherein they create objects by depositing layers of materials over one another. The key difference between these technologies usually stems from the type of raw material used and the manner in which the machine tries to deposit each layer to create the object. 3D printing technologies are generally divided into three main categories, including photopolymer laser stereolithography (SLA) technology, selective laser sintering (SLS) technology and fused deposition modelling (FDM), with FDM being the most commonly used technology. Though there are other techniques or categories that exist along the additive manufacturing process, the following sections will focus on discussing these common 3D printing methods in detail.

10.3.1.1 Stereolithography (SLA) Technology

Stereolithography (SLA) is a type of vat photopolymerization technique. It appears as one of the oldest 3D printing methods. It was the first commercially available 3D printing technique, introduced by Chuck Hull, the Co-Founder and Chief Technology Officer of 3D Systems, in the 1980s. This process uses a vat of liquid photosensitive resins as the printing material. The resin is selectively cured (solidified) by an ultraviolet (UV) laser beam into solid layers that form its desired shape. Note that the resin will remain in liquid form without UV laser irradiation. When the liquid resin is exposed to UV light, resin molecules undergo photochemical reactions. The molecules rapidly bind together like a chain to cure and build the 3D product.

The typical configuration of SLA comprises a container or vat for the liquid resin, a build platform, a UV laser and a control system equipped with Computer-Aided

Design (CAD) software. The build platform, where the 3D object is formed, is usually fitted with an elevator mechanism and partially submerged in a vat of liquid photosensitive resin. This allows the platform to move vertically after a layer is cured, creating space for the next layer to adhere to that previous layer. To initiate the printing process, a 3D model has to be first created using CAD software. The next step is to convert the model into a file format compatible with the SLA printer, usually in Standard Triangle Language (.STL) format. The .STL file, which now contains the details of the 3D model, is then transferred to the SLA printer. Based on the data given from the file, the printer then directs a UV laser beam to selectively cure the liquid photosensitive resin into a solid layer. The process is repeated to cure the resin layer by layer until the 3D object is fully formed according to its desired design. After the printing process is finished, the 3D product is rinsed to remove any residual resin and may go through additional post-curing and finishing steps to improve its appearance and attributes.

There are two types of SLA printers, namely top-down and bottom up. They can be distinguished based on the position of UV laser source, the direction of UV laser projection and the orientation of printing process. The UV laser source from a top-down SLA printer is positioned above the resin vat, which projects the UV light downwards to start the photopolymerization process. As the part is printed, the build platform moves down gradually. A bottom-up SLA printer, however, works in a similar manner but in the opposite direction. In the bottom-up configuration, the UV laser source is placed at the bottom of the vat. The UV light is projected upwards, and the build platform moves vertically upwards once a layer is formed. The top-down and bottom-up SLA printers are schematically shown in Figure 10.10.

3D printing with SLA technology offers several advantages. This is because the products produced by this technology are high-resolution products with a smooth surface finish. It is almost impossible to notice the layer lines with the naked eye. Besides, it takes a shorter amount of time to produce objects with complex geometries and highly detailed features. Nonetheless, there are few drawbacks associated with this technology. It is more costly compared to other existing technologies. The process also involves the use of toxic liquid resin and requires a longer time in post-processing.

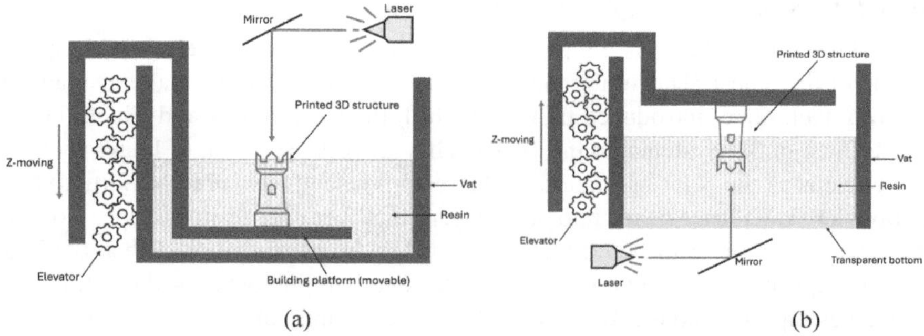

FIGURE 10.10 Schematic diagram of (a) top-down SLA printer (b) bottom-up SLA printer.

10.3.1.2 Fused Deposition Modelling (FDM) Technology

Fused Deposition Modelling (FDM), also known as Fused Filament Fabrication (FFF), is the most widely used 3D printing technology. It gains popularity because it is simple and most affordable for printing prototypes and/or functional parts. This technology falls under the material extrusion category of additive manufacturing. The process builds the object layer by layer by extruding a continuous filament of a thermoplastic polymer through a heated nozzle. Examples of the common thermoplastic polymers used in FDM include polylactic acid (PLA) and acrylonitrile butadiene styrene (ABS).

A typical FDM configuration includes a filament spool, a build platform, an extruder, an extrusion nozzle and a control system equipped with CAD software. Similar to SLA technology, a 3D CAD model must be developed and converted to .STL file format before the start of fabrication process. During printing, a spool of thermoplastic polymer is fed into the extruder, which is the print head. Note that the polymer filament is at room temperature and remains in its solid state. Next, the extruder is heated. When the nozzle reaches the targeted temperature, it partially melts the filament into a semi-molten state and extrudes it onto the build platform. The movement of both the nozzle and build platform is managed by software, which decodes the dimensions of object into *X, Y* and *Z* coordinates to guide the printing process (Singh and Davim, 2018). The nozzle moves horizontally and vertically (in the '*XY*' direction) to deposit the filament onto the build platform. Once a layer of material is deposited, the build platform moves vertically down (in the '*Z*' direction) to create space for the next layer. The filament layer that is newly extruded would then fuse with the previous layer. The process is repeated until the 3D object is fully built. Some extruders are also designed to include cooling fans that evenly cool the newly extruded filament layers to avoid them from warping. A schematic diagram of the FDM technology is illustrated in Figure 10.11.

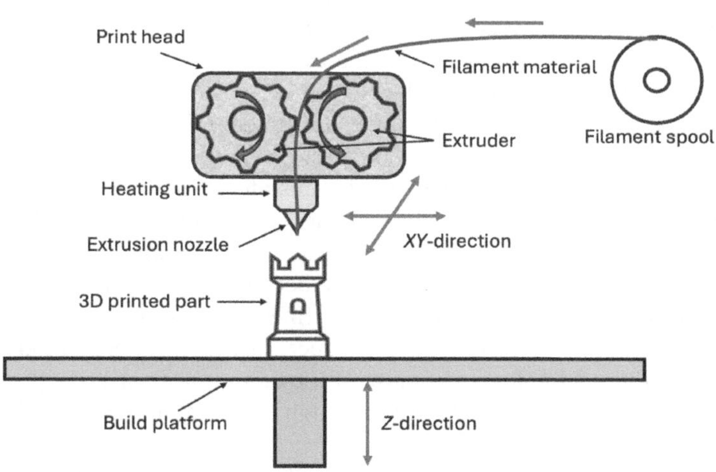

FIGURE 10.11 Schematic diagram of FDM technology.

FDM is relatively cheaper compared to other technologies. Moreover, it is easier to set up and makes it more accessible and user friendly to beginners. It supports a variety of thermoplastic polymers, allowing users to select materials with specific properties from a wide range for various applications. Nevertheless, 3D products made with FDM technology exhibit poorer surface quality, resulting in rougher surface and visible layer lines in the prints. The products often have lower mechanical properties and poor adhesion among layers, which makes them more susceptible to delamination. This issue is due to weaker mechanical performance of those commonly used thermoplastics in FDM. Because of this reason, new materials with enhanced properties are developed for use in FDM. The new materials include advanced polymer-based composites, nanocomposites and fibre-reinforced composites.

10.3.1.3 Selective Laser Sintering (SLS) Technology

Selective Laser Sintering (SLS) technology is another 3D printing technique available in the market. It is considered a type of powder bed fusion method in additive manufacturing that relies on a high-powered laser beam to selectively fuse fine powdered material, layer by layer, forming a 3D structure. The fine powder is sintered by the laser to create the 3D object. Note that the powders are not completely melted by the laser. The sintering process results in an atomic reaction that bonds the particles together and transforms the powdered material into a solid 3D structure. This technology can support a wide range of metals, polymers and alloy powders.

Like other 3D printing techniques, the SLS process starts with creating a 3D CAD model of the object, followed by converting it into an .STL file format. An SLS 3D printer normally includes a high-power laser beam (typically a carbon dioxide, CO_2 laser), a powder bed, build platform, roller and a control system equipped with CAD software. The powder, in the form of either metal, polymer or alloys, is evenly deposited across the build platform using the roller. The laser, which acts as a heating source, selectively scans a thin layer of powder, sinters and bonds the fine powdered particles together to form a solid layer. After one layer is formed, the build platform moves downward to make room for the subsequent layer of powder. The roller then deposits a new layer of powder on top of the previous one, and the laser again selectively sinters the layer, bonding it with the existing layer. The process is repeated layer by layer until the whole 3D object with the desired geometry is fully built. A schematic diagram of SLS technology is illustrated in Figure 10.12.

SLS technology offers several advantages. One key benefit is that the need for support structures can be eliminated. The unsintered powder naturally fills the empty spaces to support the part throughout the printing process. This sets SLS apart from technologies like FDM and SLA, which require support structures. Moreover, this process generates less waste because any unsintered material can be reused in future printing cycles. The layer lines are almost invisible in each SLS print. Nonetheless, the surface of parts produced using the SLS method is rougher compared to other technologies because of the nature of its printing process (Petzold et al., 2019). Another drawback is the high cost, due to the use of high-power lasers in the process.

In short, there is no single 3D printing method that works for all applications. Each 3D printing technology has its own strengths and limitations. As an example, FDM

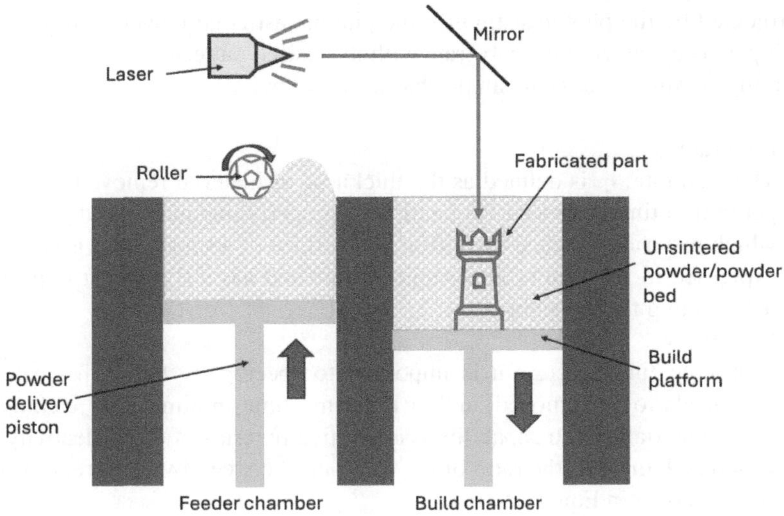

FIGURE 10.12 Schematic diagram of SLS technology.

technique may be suitable for fabricating simple parts rapidly and at low cost. But it falls short when product with complex geometries or high resolution is required. Besides, the choice of technology also affects the material selection, mechanical performance and the overall functionalities of the product. Hence, it is important to identify the design goals and their specific needs when selecting an appropriate printer for an application.

10.4 ETCHING

Etching is a surface removal process. There are two main types of etching, namely dry and wet etching. Dry etching uses gases or plasma as etchant, whereas wet etching uses a chemical solution. The surface layer of material is removed by either chemical reactions, physical interactions or both mechanisms, depending on the type of etching. Some examples of its applications are found in the manufacture of RFID tags, touch panels and microfluidic devices.

Etching is an important step in the photolithography process, which is used to transfer detailed patterns onto a substrate, typically a silicon wafer. Photolithography is a process normally found in semiconductor manufacturing and microfabrication industries for making integrated circuits and microchips. The process of photolithography begins with the coating of the substrate with a light-sensitive material known as photoresist. Once the substrate is coated with photoresist, it is exposed to ultraviolet (UV) light through a photomask, which defines the desired pattern. The pattern is developed on the substrate after UV light exposure because the areas exposed to UV light undergo changes in their chemical composition. The exposed photoresist is then developed to reveal the pattern that corresponds to the photomask design. Etching is then carried out to remove the topmost layer of the substrate in those areas which are not protected by the photoresist. Once the photoresist is no longer required, it must be stripped away, leaving the substrate with its desired pattern.

For any etching process, it can be characterized by:

i) Etch rate

The etch rate, r_e, is defined as the thickness of material removed or etched per unit of time. It is a basic quantitative property that indicates the speed at which material is etched away from the surface of a substrate. It is typically expressed in μm, nm or Å per unit of time and has a significant impact on determining the throughput.

ii) Selectivity

In any etching process, it is important to specifically target the material that needs to be removed, and at the same time minimize the etching of other materials. Such capability is quantified in terms of etch selectivity, S_e, which is defined as the ratio of the etch rates between two different materials, as shown in Equation 10.1.

$$Selectivity = \frac{r_{e1}}{r_{e2}} \tag{10.1}$$

where r_{e1} represents the etch rate of targeted material and r_{e2} denotes the etch rate of masking material or material below the film. High selectivity is desired for effective removal of the targeted material without significantly affecting the underneath or surrounding layers, thereby protecting the material beneath the mask.

iii) Profile

The etching profile refers to the geometry and shape of the material that results from the etching process. It describes how a material is etched from the substrate and can be characterized as either isotropic or anisotropic. Etching is said to be isotropic when it occurs simultaneously in both vertical and horizontal directions at equal rates. For this reason, it often results in an undercut profile, where some material is etched under the mask. On the other hand, anisotropic etching happens in a specific direction, typically vertically, that is, material is removed perpendicular to the surface. Such directional etching allows accurate transfer of the mask pattern as it creates well-defined and straight edges feature. Anisotropic etching is favoured in most applications that require high precision and control, such as microfabrication. The difference between isotropic and anisotropic structures is illustrated in Figure 10.13.

The degree of anisotropy, A_f, can be expressed by Equation 10.2.

$$A_f = \frac{vertical\, r_e - horizontal\, r_e}{vertical\, r_e} = 1 - \frac{horizontal\, r_e}{vertical\, r_e} \tag{10.2}$$

Thus, A_f equals zero for isotropic etching, whereas A_f equals one indicates perfect anisotropic etching.

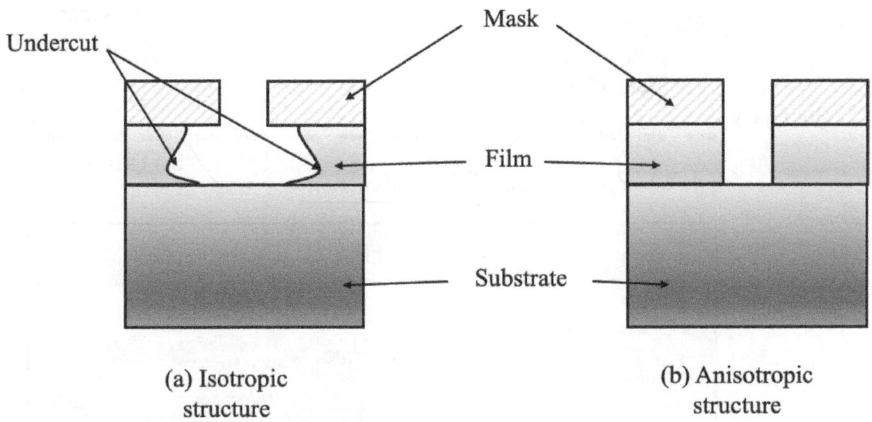

FIGURE 10.13 (a) Isotropic structure and (b) anisotropic structure.

iv) Uniformity

Uniformity measures the distribution of the etch rate, indicating how evenly material is removed across a wafer, multiple wafers or multiple batches of wafers. High etch rate uniformity is desirable in semiconductor production lines. Etch rate uniformity can be estimated using Equation 10.3 (May and Spanos, 2006).

$$Etch\,rate\,uniformity(\%) = \frac{maximum\,r_e - minimum\,r_e}{maximum\,r_e + minimum\,r_e} \times 100\% \quad (10.3)$$

The following sections will introduce both dry and wet etching techniques, along with a discussion of their respective advantages and disadvantages.

10.4.1 Wet Etching

Wet etching uses liquid chemicals, either acidic or basic, known as liquid etchants, to remove unwanted thin surface layers from a substrate. Examples of etchants include potassium hydroxide (KOH) for silicon etching, buffered oxide etchants (BOE) – a mixture of hydrofluoric acid (HF) and ammonium fluoride (NH_4F) for etching silicon oxide, as well as phosphoric acid (H_3PO_4) for silicon nitride removal. This process can be done by either immersing the substrate in the liquid etchants or spraying the etchants directly onto the substrate. During this process, redox reactions occur between the etchants and parts of the thin films, which then produce a soluble compound which can be dissolved and rinsed off. Lastly, the substrate is then dried to complete the process. The amount of time required for an etching process depends on the type and quantity of material to be removed.

There are three main steps in any wet etching process, as illustrated in Figure 10.14. It begins with the diffusion of reacting molecules from etchant solution through the boundary layer to the exposed film of the substrate surface. Chemical

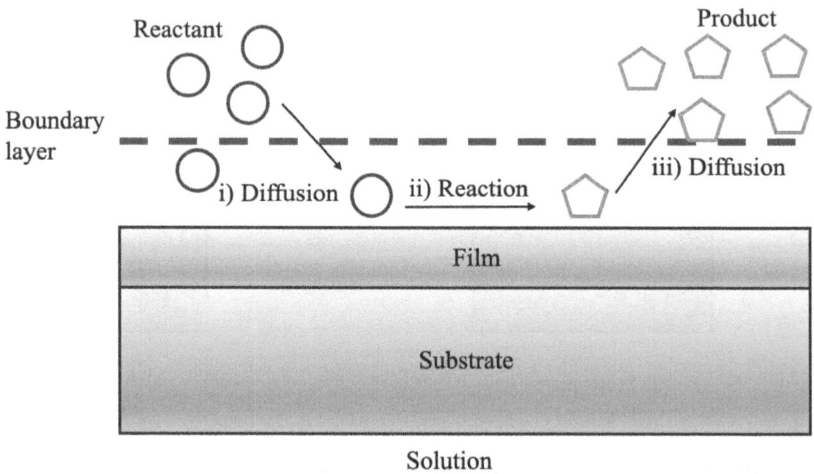

FIGURE 10.14 Main steps involved in wet etching process (May and Sze, 2003).

reactions then take place between the exposed film and the etchants. Finally, the reaction products diffuse away from the surface through boundary layer to the bulk solution. The etching rate is limited by the slowest step among these three processes.

Compared to dry etching, wet etching is relatively simpler to perform, as it involves less complex equipment. It also achieves a higher etching rate and offers good selectivity for most materials. However, the process is only suitable for products with feature size larger than 3 μ m due to its isotropic etching nature. For size smaller than 3 μ m, it should be subjected to dry etching, which can offer a better degree of control and precision.

10.4.2 Dry Etching

Dry etching involves either gases or plasma in removing materials from a surface. Unlike wet etching, it yields anisotropic structures, though the degree of anisotropy can vary depending on the specific dry etching method used. The common dry etching techniques include ion milling, plasma etching and reactive ion etching (RIE). These methods generally operate based on four basic mechanisms:

i) Chemical etching
Chemical etching occurs when neutral reactive atoms produced by the plasma react with the material's surface. This reaction forms volatile products that subsequently escape from the surface. The key advantages of this mechanism include high selectivity, high etch rate and minimal damage from ion bombardment. However, because there is little to no physical ion bombardment, this process often results in isotropic etching, where material is removed uniformly in all directions. The mechanism of chemical etching is used in classical plasma etching, a form of chemical dry etching

technology. Note that plasma is an ionized gas composed of equal numbers of positive and negative charges together with neutral particles.

ii) Sputtering

Sputtering is a physical process. In this process, positive ions are accelerated towards the surface of a substrate. When there is collision between ions and surface, large amounts of energy and momentum are transferred to the surface atoms. This energy transfer causes the surface atoms to be ejected from the surface, resulting in the removal of material. Although sputtering can produce anisotropic profiles, it is prone to substantial damage caused by ion bombardment and typically has low etch selectivity and low etching rate (May and Spanos, 2006). Sputtering is the key mechanism in ion milling technology, also referred to as ion etching, which relies entirely on a physical process. An ion milling system includes a vacuum chamber where an inert ion beam, usually made of argon ions, Ar^+, is accelerated to the surface of substrate, removing the material through ion bombardment.

iii) Ion-enhanced etching

Ion-enhanced etching integrates both physical and chemical etching mechanisms to remove a material. It is often applied in reactive ion etching (RIE) technology. Interestingly, despite its name, the etching primarily occurs due to the reactive neutrals, not the ions. In this process, a physical method is used to produce reactive ions to take part in the chemical etching. These energetic ions are accelerated towards the substrate. Sputtering occurs when they bombard the surface. At the same time, neutral reactive species from the plasma react with the substrate to form volatile product. The combined effects of physical sputtering and chemical reactions enhance the overall etching process. It provides reasonably good selectivity and produces anisotropic features. This mechanism is prevalent in many applications today.

iv) Sidewall inhibitor

In this case, a polymer-containing gas such as methane, CH_4 or octafluorocyclobutane, C_4F_8, is introduced to the plasma environment (Khanna et al., 2019). The addition of these inhibitor species leads to the formation of a polymeric layer that deposits onto all exposed surfaces of the substrate. This layer helps to inhibit etching in the horizontal direction, thereby preventing undercutting. This mechanism is able to produce highly anisotropic structures. This mechanism is used in a relatively new technology known as deep reactive ion etching (DRIE). Figure 10.15 provides a summary of the four various mechanisms of dry etching.

10.5 BREAKAGE

Breakage plays an important role in the size reduction process in the chemical product industry, especially in the food, cosmetic and pharmaceutical sectors. It is normally used to facilitate transport and handling, improve shelf life and product stability, to produce products with desired particle size and performance, etc. One can view breakage based on different states of matter.

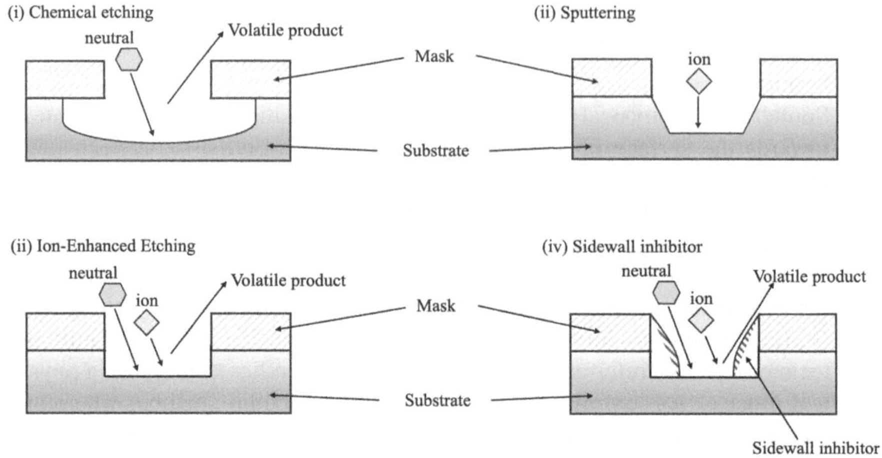

FIGURE 10.15 Four different mechanisms of dry etching: (i) chemical etching, (ii) sputtering, (iii) ion-enhanced etching and (iv) sidewall inhibitor (Shul and Fleming, 2000).

For solid materials, it is often referred to *particle breakage*. By reducing the particle size of solids, some unwanted substances can be separated using mechanical methods. Besides, it increases the surface area of solids, which in turn enhances their reactivity in chemical reactions. For liquid-liquid systems, it is referred to droplet breakup. The most common process used to reduce the size of liquid droplet is known as *homogenization*. In gas-liquid system, it is referred to *breakup of bubbles*. Though coalescence, where droplets or bubbles combine, is important in many industrial processes, one should not neglect the importance of droplet and/or bubble breakage. For example, homogenization is a critical process in producing hand lotion as it helps to reduce the particle size, which allows for more rapid absorption into the skin and easier spreadability. Another typical example found in the food industry is milk homogenization, where the process involves breaking up the fat particles and dispersing them evenly throughout the milk to improve its stability and shelf life. The following sections will discuss the type of breakage involved in solid, liquid and gas systems.

10.5.1 Solid Particle Breakage

In solid processing, particle breakage or *comminution* is the general term used for size reduction. The common terminology used in solid particle breakage is presented in Table 10.5.

Solids can be broken into smaller pieces using various mechanisms. The mechanisms include (i) cutting or shearing, (ii) compression or crushing, (iii) impact, (iv) attrition or rubbing and (v) the combination of impact and attrition (Smith et al., 2004). Cutting is normally used to produce particles with distinct geometric forms, such as slices and cubes. During the cutting process, solid substance is cut by a sharp blade. The cutting mechanism is illustrated in Figure 10.16.

TABLE 10.5
Terminology Used in Solid Particle Breakage (Rhodes, 2008)

Terminology	Size Range of Product (mm)
Coarse crushing	100–1000
Crushing	100
Fine crushing, coarse grinding	10
Intermediate grinding, milling	1
Fine grinding	0.1
Ultrafine grinding	0.01

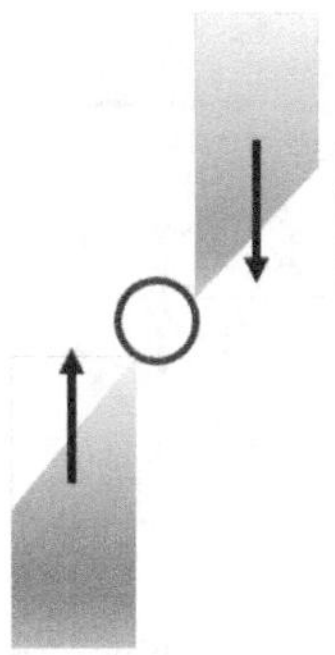

FIGURE 10.16 Mechanism of cutting or shearing (note that represents direction of forces).

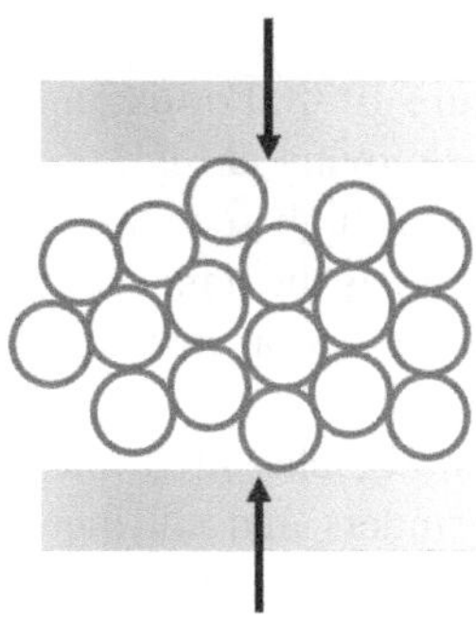

FIGURE 10.17 Mechanism of compression or crushing (note that represents direction of forces).

Compression is used when the feed material is hard, and the required final product is relatively coarse in size. By applying pressure, the solid substance is crushed between rollers, as illustrated in Figure 10.17.

Impact produces fine, medium or coarse products, whereas attrition is employed when very fine products are required from soft, nonabrasive material (Smith et al., 2004). Impact works by applying a sudden impact force to the particles for size

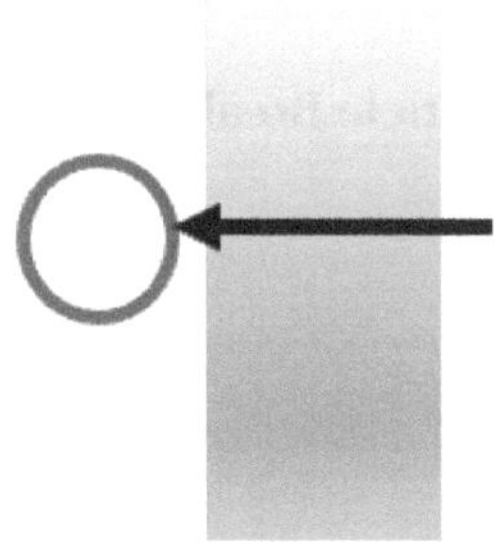

FIGURE 10.18 Mechanism of impact (note that represents direction of forces).

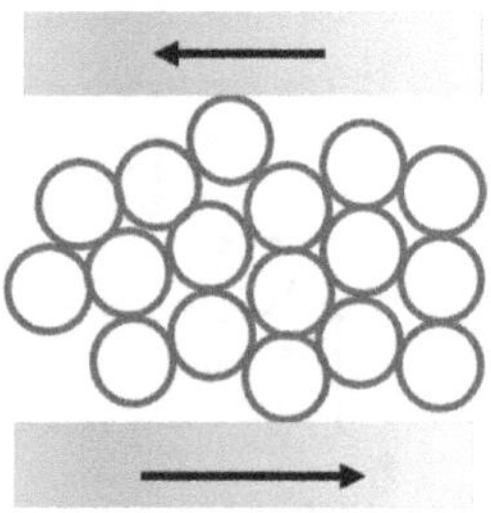

FIGURE 10.19 Mechanism of attrition or rubbing (note that represents direction of forces).

reduction, as illustrated in Figure 10.18. For example, a solid substance is hit by an object such as a hammer or bar moving at high speed. Alternatively, impact occurs when the particle, which moves at high speed, strikes a stationary surface. On the other hand, in attrition or rubbing, as the name suggests, solid substance breaks into smaller pieces due to the rubbing action between two surfaces. Its mechanism is illustrated in Figure 10.19.

Size reduction equipment can be classified into four main categories, including crushers, grinders, ultrafine grinders and cutting equipment (Smith et al., 2004). Crushers normally employ compression to break large chunks of solid substance into either coarse or fine products. Grinders often use impact, a combination of impact and attrition, or a combination of compression and attrition to achieve intermediate and fine particles. Ultrafine grinders mainly operate by attrition and sometimes are combined with impact or compression to achieve ultrafine products. Cutting equipment uses cutting action to produce a product with a definite shape and size. Table 10.6 shows the characteristics of the common types of size reduction equipment used in industries.

From Table 10.6, it is obvious that the choice of size reduction equipment depends on the stressing mechanism, the size of feed and product and its capacity. Besides, the selection of size reduction equipment is also affected by the following properties of material (Rhodes, 2008):

TABLE 10.6
Characteristics of Common Types of Size Reduction Equipment (Couper et al., 2005)

Equipment	Mechanism	Reduction Ratio	Feed Size (mm)	Product Size (mm)	Capacity (tonne/hr)	Power Consumption (kW)
Gyratory crushers	Compression	8	200–2000	25–250	100–500	100–700
Jaw crushers	Compression	8	100–1000	25–100	10–1000	5–200
Cone crusher	Compression	8	50–300	5–50	10–1000	20–250
Impact breaker	Impact	40	50–300	1–10	10–1000	100–2000
Hammer mill	Impact	400	5–30	0.01–0.1	0.1–5	1–100
Rod mill	Impact and attrition	10	5–20	0.5–2	20–500	100–4000
Ball mill	Impact and attrition	100	1–10	0.01–0.1	10–300	50–5000
Jet mill	Impact and attrition	300	1–10	0.003–0.05	0.1–2	2–100

a) Hardness
Hardness is a surface property of the material, which is measured using an arbitrary scale known as the Mohs' scale. Graphite (a very soft material) is rated as 1, whereas diamond (a very hard material) has a Mohs' scale of 10. Any material having a Mohs' scale rating below 3 is considered soft material, while material with a rating higher than 7 is considered hard material. It is often more difficult to reduce the size of a hard material.

b) Abrasiveness
When selecting a size reduction equipment, one should always consider the abrasiveness of a material as it often limits the choice of machine that can be used. This property is often associated with hardness. It indicates the tendency of a material to cause wear and tear of the machine parts that it comes into contact with. For example, mills operated at low speeds, such as ball mills, should be used to grind a very abrasive material to reduce the tendency of wear and tear.

c) Toughness
Toughness is a property that measures how well a material resists the propagation of a crack. One should not confuse toughness with hardness. A material can be hard but brittle. More often, a soft but tough material tends to cause more problem in size reduction compared to a hard but brittle material. It is harder to reduce the size of a soft but tough material.

d) Fibrous nature
If a material is fibrous, one should consider the size reduction equipment employing cutting action, such as shredders or cutters. It is not suitable to crush a fibrous material.

e) Cohesivity/adhesivity
Cohesivity and adhesivity refer to the 'stickiness' a material has between itself and other surfaces. A sticky material may have the tendency to clog the machine. This property is affected by the size and moisture content of a material. The cohesivity and adhesivity of a material can be decreased by reducing its moisture content or having a larger particle size. One may consider using wet grinding if the stickiness issue is caused by the particle size.

f) Melting point
Extra care is required when handling materials with a low melting point. This is mainly because heat is generated during the milling process, and such heat may cause the materials to melt, which in turn increases their stickiness or toughness. This issue may be overcome by having ambient or refrigerated air as the carrier medium.

There are three well-known empirical models used to predict energy requirements for solid size reduction. This includes Rittinger's law, Kick's law and Bond's law, which were all derived from the same differential equation:

$$\frac{dE}{dx_p} = -C\frac{1}{x_p^N} \tag{10.4}$$

where E represents the energy required per unit mass of material, x_p is the size of unit mass of material, C and N are the constants with $C = C_R$ and N = 2 for Rittinger, $C = C_K$ and N = 1 for Kick as well as $C = C_B$ and N = 1.5 for bond.

According to Rittinger's Law, the amount of energy needed to reduce a particle size is directly proportional to the area of new surface created (von Rittinger, 1867). The energy required per unit mass of material, E_R, can be estimated using Equation 10.5. It is common practice to assume that this law is applicable to ultrafine grinding (Rhodes, 2008).

$$E_R = C_R\left(\frac{1}{x_{p2}} - \frac{1}{x_{p1}}\right) \tag{10.5}$$

where x_{p1} and x_{p2} represent the initial and final size of the product.

On the other hand, Kick's law states that the amount of energy required for size reduction is directly proportional to the ratio of initial particle size to final particle size (Kick, 1885). The energy required per unit mass of material, E_K, can then be estimated using Equation 10.6.

$$E_K = C_K ln\left(\frac{x_{p1}}{x_{p2}}\right) \tag{10.6}$$

Based on Equation 10.6, it simply means that the energy required to reduce particle size from 1 cm to 1 mm is the same as that to reduce particle size from 10 μm to 1 μm. This is unrealistic in reality. Kick's law is useful for predicting energy required to reduce size of large particle (coarse crushing and crushing) (Rhodes, 2008).

Bond's law indicates that the amount of energy required for crack propagation is proportional to the new crack length produced (Bond, 1952), which can be expressed by Equation 10.7. It is most suitable for intermediate particle size (Rhodes, 2008).

$$E_B = C_B\left(\frac{1}{\sqrt{x_{p2}}} - \frac{1}{\sqrt{x_{p1}}}\right) \quad (10.7)$$

Example 10.1. Breakage Energy

A material made up of initial average size of 35 mm is crushed to an average size of 9 mm. This size reduction process consumes 30 kJ/kg of energy. Determine the energy consumption to crush the same material of average size 25 mm to an average size of 5 mm, assuming

a) Rittinger's law
b) Kick's law
c) Bond's law

Solution

a) Apply Rittinger's law to estimate C_R; Equation 10.5 becomes

$$30 = C_R\left(\frac{1}{9} - \frac{1}{35}\right)$$

$$C_R = 363.5$$

Energy consumption to reduce average size from 25 mm to 5 mm can then be estimated:

$$E_R = 363.5\left(\frac{1}{5} - \frac{1}{25}\right)$$

$$E_R = 58.2\,\text{kJ/kg}$$

b) Apply Kick's law to estimate C_K; Equation 10.6 becomes

$$30 = C_K\, ln\left(\frac{35}{9}\right)$$

$$C_K = 22.1$$

Energy consumption to reduce average size from 25 mm to 5 mm can then be estimated:

$$E_K = 22.1 ln\left(\frac{25}{5}\right)$$

$$E_K = 35.6\ \text{kJ/kg}$$

c) Apply Bond's law to estimate C_B; Equation 10.7 becomes

$$30 = C_B\left(\frac{1}{\sqrt{9}} - \frac{1}{\sqrt{35}}\right)$$

$$C_B = 182.6$$

Energy consumption to reduce average size from 25 mm to 5 mm can then be estimated:

$$E_B = 182.6\left(\frac{1}{\sqrt{5}} - \frac{1}{\sqrt{25}}\right)$$

$$E_B = 45\ \text{kJ/kg}$$

10.5.2 Homogenization

From a classical chemical engineering point of view, homogenization is often referred to as achieving uniform concentration in a reactor through intensive mixing. In the food, pharmaceutical and cosmetic sectors, homogenization is an important process used to reduce the size of particles in fluid to achieve a stable emulsion. For example, one of the most important industrial applications is milk homogenization. During this process, it breaks up the fat globules to a smaller size (less than 1 μm) to ensure that the particles are suspended evenly in milk. It helps to avoid agglomeration, segregation and settling of particles during storage, which then promotes improved shelf life, stability and taste of the product. Homogenized milk is whiter in colour and has better mouth feel.

Homogenization breaks up droplets through a mechanical process, often performed in a device known as homogenizer. This process occurs through two main mechanisms, including microturbulence and cavitation. Some examples of homogenizers include high-speed blender, colloid mill, high-pressure valve homogenizer and ultrasonic homogenizer. High-pressure valve homogenizer is normally used for milk homogenization. The operating pressure ranges between 120 to 600 bar (Saravacos and Maroulis, 2011). Smaller particle size can be achieved with an increase in pressure. Colloid mills are often used for baby soups, baby foods and mayonnaise production.

10.5.3 Breakup of Bubbles

Bubbles breakup happens in gas-liquid system. This is particularly important in bioprocessing as the size of bubbles affects the performance of aerobic bioreaction. The size of the bubble has a significant impact on the rate of oxygen transfer, OTR, from gas to liquid, which can be estimated by Equation 10.8.

$$OTR = k_L a\left(C_{AL}^{*} - C_{AL}\right) \tag{10.8}$$

where OTR represents the rate of oxygen transfer per unit volume of fluid (mol m^{-3} s^{-1}), k_L represents the mass transfer coefficient (m s^{-1}), a is the gas-liquid interfacial area per unit volume of liquid (m^2 m^{-3}), C_{AL}^{*} is the oxygen saturation concentration (mol m^{-3}) and C_{AL} is the oxygen concentration in the media (mol m^{-3}).

The bubble size has a significant impact on the magnitude of $k_L a$. One can increase a by having smaller bubbles, as smaller bubbles have a higher interfacial surface area. Besides, smaller bubbles also increase k_L as they do not rise rapidly due to low terminal velocity. There is more time for mass transfer to occur due to longer residence time. Thus, in aerobic fermentation, it is undesirable to have small bubbles coalesce into bigger bubbles as it reduces the oxygen transfer.

Though it is desirable to have small bubbles in aerobic bioreactor, there are some rules of thumb to follow. For example, bubbles with a size lesser than 1 mm diameter should be avoided, especially in viscous fermentation broths. This is because very small bubbles remain trapped in the fermentation medium for an extended period due to their low terminal velocity (Doran, 2013). In addition, a large amount of energy is released when small bubbles burst, which may in turn damage the microbial cells. For bubbles with a size less than 2 to 3 mm in diameter, k_L is reduced due to the dominant effect of surface tension (Doran, 2013). Ideally, it is recommended to have bubble size bigger than about 3 mm in fermentation broth to have a beneficial impact on magnitude of k_L (Doran, 2013).

The size of bubbles is influenced by the choice of sparger and agitation system. Air bubbles are normally produced at the sparger. Some examples of spargers include porous spargers made up of sintered metal, glass or ceramic, orifice spargers and point or nozzle spargers. Small bubbles are normally produced at sintered spargers, whereas large bubbles are formed at orifice and nozzle spargers. The choice of sparger is important when there is no other mechanism available to disperse the bubbles in reactors. However, when there is an agitator or impeller available, says in stirred reactors, the design of sparger is of lesser importance. This is because the dispersion of bubbles is influenced by the impeller. The higher the impeller speed, the smaller the bubbles, thus increasing the rate of oxygen transfer. However, higher impeller speed often implies that higher energy consumption is required, and it can also lead to increased shear stress.

REFERENCES

Bond, F. C. (1952). The Third Theory of Comminution. *Trans AIME Min Eng*, *193*, 484–494.

Couper, J. R., Penney, W. R., Fair, J. R., & Walas, S. M. B. T. (2005). Chapter 12 – Disintegration, Agglomeration, and Size Separation of Particulate Solids. In Couper, J. R., Penney, W. R., Fair, J. R., & Walas, S. M. B. T. (eds.), *Chemical Process Equipment* (2nd ed., pp. 359–395). Gulf Professional Publishing.

Doran, P. M. (2013). *Bioprocess Engineering Principles* (2nd ed.). Academic Press.

Green, D. W., & Perry, R. H. (2008). Section 21: Solid-Solid Operations and Processing. In *Perry's Chemical Engineers' Handbook* (8th ed., pp. 2226–2372). McGraw-Hill Education.

Holmberg, K., & Matthews, A. (2009). Deposition Processes and Coating Structures. In *Coatings Tribology: Properties, Mechanisms, Techniques and Applications in Surface Engineering* (pp. 7–39). Elsevier Science.

Khanna, R., Bevlin, K., Geerpuram, D., Yang, J., Ren, F., & Pearton, S. (2019). 12 – Dry Etching of Ga2O3. In Pearton, S., Ren, F., & Mastro, M. B. T.-G. O. (eds.), *Metal Oxides* (pp. 263–285). Elsevier.

Kick, F. (1885). *Das Gesetz der proportionalen Widerstände und seine Anwendungen: Nebst Versuchen über das Verhalten verschiedener Materialien bei gleichen Formänderungen sowohl unter der Presse als dem Schlagwerk.* A. Felix.

May, G. S., & Spanos, C. J. (2006). Technology Overview. In *Fundamentals of Semiconductor Manufacturing and Process Control* (pp. 25–81). https://doi.org/10.1002/0471790281.ch2

May, G. S., & Sze, S. M. (2003). *Fundamentals of Semiconductor Fabrication.* New York: Wiley.

Petzold, S., Klett, J., Schauer, A., & Osswald, T. A. (2019). Surface Roughness of Polyamide 12 Parts Manufactured Using Selective Laser Sintering. *Polymer Testing*, *80*, 106094.

Pietsch, W. (2008). *Agglomeration Processes: Phenomena, Technologies, Equipment.* Wiley.

Rhodes, M. J. (2008). Particle Size Reduction. In *Introduction to Particle Technology* (pp. 311–335). Hoboken, NJ: Wiley.

Saravacos, G. D., & Maroulis, Z. B. (2011). *Food Process Engineering Operations* (1st ed.). CRC Press.

Shul, R. J., & Fleming, J. G. (2000). *Bulk Si Micromachining for Integrated Microsystems and MEMS Processing BT – Handbook of Advanced Plasma Processing Techniques* (R. J. Shul & S. J. Pearton (eds.), pp. 419–458). Berlin, Heidelberg: Springer. https://doi.org/10.1007/978-3-642-56989-0_10

Singh, R., & Davim, J. P. (2018). Fused Deposition Modelling: Applications and Advancements. In *Additive Manufacturing: Applications and Innovations* (1st ed., pp. 128–184). CRC Press.

Smith, J., McCabe, W., & Peter Harriott, E. (2004). Operations Involving Particulate Solids. In *Unit Operations of Chemical Engineering.* McGraw-Hill Education.

von Rittinger, P. R. (1867). *Lehrbuch der Aufbereitungskunde: in ihrer neuesten Entwicklung und Ausbildung systematisch dargestellt.* Ernst & Korn.

Zhang, L., Fung, K. Y., Wibowo, C., & Gani, R. (2018). Advances in Chemical Product Design. *Reviews in Chemical Engineering*, *34*(3), 319–340.

10.6 PROBLEMS

1. A feed material having an average size of 0.5 mm is fed to a milling machine at the rate of 15 tonnes per hour to produce products having an average size of 0.2 mm. For this process, the crusher consumes 30 kW.
   a) If the capacity of the milling machine is reduced to 10 tonnes per hour, which law, Kick's or Rittinger's law, is more applicable for estimating the new power consumption? Justify the reason and calculate the new power consumption in kW.
   b) Using the same law that you have assumed in part (a), calculate the energy consumption in kJ/kg when the feed material is now further reduced to have an average size of 0.1 mm using the same milling machine.
   c) Assuming the capacity of milling machine remains at 10 tonnes per hour, what is the percentage increase in power consumption when reducing the product size to 0.1 mm (as in part b) compared to when the product size is 0.2 mm?

11 Case Studies

11.1 SKIN-FRIENDLY SHAVING PRODUCT

Shaving is an important grooming practice among men to promote hygiene and enhance professional appearance, which also fosters a sense of confidence and self-care.

11.1.1 Identification of Needs

One of the challenges in the design of common consumer products is that the main leads have been extensively explored by the companies. Therefore, the focus should be on identifying an unexplored sub-category of consumers who may have specialized needs for such products. A survey has been conducted among the current users of shaving creams to identify the needs from a consumer point of view. This step can also suggest whether any unfulfilled needs exist for this product. The needs identified through the survey are as follows:

i) Chemical free
ii) Cleansing ability along with shaving
iii) Cooling effect to provide a feeling of freshness
iv) Do not clog pores
v) Easy to carry
vi) Easy to rinse with water
vii) Foaming effect
viii) Moisturization
ix) Non-toxic
x) Pleasant smell
xi) Price of product
xii) Sensitive skin friendly
xiii) Skin care

In the next step, these needs have been classified based on the Kano model.

Must be

1. Non-toxic
2. Foaming effect
3. Easy to rinse with water
4. Do not clog pores
5. Sensitive skin friendly
6. Cooling effect

DOI: 10.1201/9781032708911-11

One dimensional

1. Price of the product
2. Easy to carry

Attractive

1. Moisturizing
2. Chemical free
3. Pleasant smell
4. Skin care
5. Cleansing ability along with shaving

It is important to note that the needs such as ‘sensitive skin friendly’ and ‘cooling effect’ typically come under attractive needs. However, in this case, these are the unfulfilled needs that are required to be addressed in a new product. In other words, if these needs are not addressed, there is no need to design a new shaving product, as the rest of the needs are currently addressed by the existing products.

In the next step, the underlying quantitative properties have been identified as shown (except the obvious ones) in Table 11.1. This step is used to define the specifications of the product.

The toxicity is usually expressed based on LC50 or LD50, which are the concentrations or doses at which 50% of the exposed population dies. These values are estimated via *in vivo* or *in vitro* experimental testing. To estimate the LC50 or LD50 of a mixture, acute toxicity estimate (ATE_{mix}) has been used. ATE_{mix} can be estimated using Equation 11.1 (Van Cott et al., 2018):

$$\frac{100}{ATE_{mix}} = \sum_{n} \frac{c_i}{ATE_i} \tag{11.1}$$

TABLE 11.1
Need Translation

Need	Attribute/s	Description
Non-toxic	Ingredient toxicity level, LD 50	Measure of toxicity
Foaming effect	Bubble stability, foaming rate	To determine how fast the formation of the foam to achieve the targeted foam volume
Easy to rinse with water	Hydrophilic-lipophilic Balance (HLB) Griffin's Scale	Higher HLB Scale valued surfactants are more hydrophilic and thus are more water soluble
Moisturization	Humectant	Based on AI
Do not clog pores	Comedogenic rating	Higher comedogenic rating indicates a higher tendency to clog pores
Sensitive skin friendly	AI selection	
Pleasant smell	Presence of synthetic or natural fragrance ingredients	Identify the ingredients and concentration
Bubble stability	Modulus G	Measurement of the elasticity of the foam

TABLE 11.2
Benchmark and Target Comparison

Attributes	Benchmark	Target
ATE_{mix}	14876	>5000
pH	8.2–8.5	7–8.5
Price (USD $ per use)	0.1	0.2–0.4
Mass (g)	175	1–5
Hydrophilic-lipophilic balance (HLB)	15	8–16
Comedogenic ratings	2–3	<4
Modulus G (Pa)	119.5	>119.5
Volume of foam produced in 3s (ml)	100	>100
Surface tension (mN/m)	20	25–35

where C_i is the concentration of ingredient in (% w/w or % v/v)
ATE_{mix} is the acute toxicity estimate (LD_{50}) of mixture,
ATE_i is the acute toxicity estimate (LD_{50}) of ingredient.

Measurement of the likelihood that an ingredient will clog pores can be estimated through *Comedogenic Scale*. This rates the ingredients from 0 to 5, where lower numbers indicate less likelihood to clog the pores. It is to be noted that the concentration of the ingredients also plays a major role in the clogging effect. Even a highly comedogenic ingredient can be harmless at low concentrations.

Bubble stability can be quantified using *Modulus G*, which is a measurement of the elasticity of the foam. It can be estimated using Equation 11.2

$$G = \frac{1.4\phi(\phi - 0.64)\gamma}{R} \tag{11.2}$$

where ϕ is the air fraction, γ is the surface tension and R is the bubble radius. The benchmark values are $\phi = 0.8$, $\gamma = 20\,\text{nm}$ and $R = 30\,\mu\text{m}$ (Pierre et al., 2013). From Equation 11.2, the specification for G can be estimated as 119.5 Pa.

The complete list of specifications and benchmarks is shown in Table 11.2.

11.1.2 Idea Generation and Classification

In this step, a number of ideas to meet the needs are generated. The partial list of raw ideas is shown in Table 11.3.

The following is a possible way to classify the needs after removing the redundant ideas:

Chemical Additive

A) Synthetic Additive
   1. Hormone Inhibition Shaving Cream

TABLE 11.3
Raw Ideas to Meet the Needs

1. 2 in 1 Shaving Product
2. Chemical-Free Shaving Product
3. Combination of Razor and Shaving Foam
4. Edible Shaving Product
5. Effervescent Shaving Tablet
6. Epilate the Breed
7. Face Mask with Breed Removal Effect
8. Fragrance-Free Shaving Product
9. Go to Barber Shop to Shave
10. Hair Removal Patch
11. Hormone Inhibition Shaving Cream
12. Just Use Razor
13. Laser Breed Removal
14. Multipurpose Razor
15. Multimode Electrical Razor
16. Nano Spray Shaving Product
17. Permanent Hair Removal
18. Cleansing Shaving Product
19. Razor with Solid Lubricant
20. Rechargeable Laser Razor
21. Refillable Shaving Foam
22. Shaving Product with Disinfectant
23. Skincare Shaving Product
24. Smart Razor with AI
25. Taking Hormone Inhibition Medicine
26. Use Substitution to Replace Shaving Product
27. UV Breed Treatment
28. UV Protection Shaving Product
29. Vegan Shaving Product
30. Wear Mask

   2. 2 in 1 Shaving Product
   3. Skincare Shaving Product
   4. UV Protection Shaving Product
   5. Shaving Product with Disinfectant

B) Natural Additive
   1. Chemical-Free Shaving Product
   2. Fragrance-Free Shaving Product

Medical

A) Medicine
   1. Taking Hormone Inhibition Medicine
B) Surgery
   1. Laser Breed Removal
   2. UV Breed Treatment

Mechanical

1. Combination of Razor and Shaving Foam
2. Rechargeable Laser Razor
3. Multipurpose Razor
4. Nano Spray Shaving Product
5. Multimode Electrical Razor
6. Razor with Solid Lubricant

Existing method

1. Epilate The Breed
2. Go to Barber Shop to Shave
3. Use Substitution to Replace Shaving Product
4. Just Use Razor
5. Wear Mask
6. Vegan Shaving Product

New direction

1. Refillable Shaving Foam
2. Hair Removal Patch
3. Face Mask with Breed Removal Effect
4. Smart Razor with AI
5. Edible Shaving Product
6. Effervescent Shaving Tablet

A screening matrix has been used to shortlist promising ideas. Since the main objectives of developing a new product are convenience and comfort, these targets are given a higher weightage. Cost and safety of the new product are also to be considered in the initial stage. Cost could be a main concern for this mature product, whereas convenience and comfort are also important. A sample of screening matrix is shown in Table 11.4.

Based on the results from the screening matrix, effervescent shaving tablet and 2 in 1 shaving product are promising options. However, this conclusion also depends on the chosen selection criteria, relative weightages and the scores given to each of the options in the selected criteria. In order to increase the robustness of this conclusion, a sensitivity analysis is to be conducted where the sensitivity of the criteria,

TABLE 11.4
Raw Ideas to Meet the Needs

Selection Criteria	Weightage	Shaving Foam (Benchmark)	Effervescent Shaving Tablet	Hair Removal Patch	2 in 1 Shaving Product	Combination of Razor and Shaving Foam
Cost	0.30	5	4	3	4	2
Convenient	0.25	5	7	6	6	5
Comfort	0.25	5	6	2	6	6
Safety	0.20	5	5	3	5	5
Total		**5**	**5.45**	**3.5**	**5.2**	**4.35**

TABLE 11.5
More Emphasis on Comfort

Selection Criteria	Weightage	Shaving Foam (Benchmark)	Effervescent Shaving Tablet	Hair Removal Patch	2 in 1 Shaving Product	Combination of Razor and Shaving Foam
Cost	0.15	5	4	3	4	2
Convenient	0.25	5	7	6	6	5
Comfort	0.4	5	6	2	6	6
Safety	0.20	5	5	3	5	5
Total		**5**	**5.75**	**3.35**	**5.5**	**4.95**

scores and weightages to the conclusion are analysed. In this example, we have considered the following situations:

i) Customers may consider comfort to be more important.
ii) The scores given to cost may be too high as the alternative options may be more expensive.
iii) Customers consider that quality of the shave is the most important attribute.

The analysis of these scenarios is shown in Tables 11.5–11.7.

From the sensitivity analysis, we can conclude that the effervescent shaving tablet and 2 in 1 shaving product are promising options and less sensitive to the minor variations in the criteria used for the product evaluation. However, the purpose of screening matrix is to shortlist promising options, and the emphasis here is the elimination of the ideas that are clearly inferior to the rest. In addition, the ranking of ideas at screening is not necessarily an indication of the superiority of the ideas. The intention here is only to choose ideas that need to be analysed during the selection stage.

11.1.3 Manufacture

The process flow diagram for the commercial production of shaving tablet is shown in Figure 11.1.

TABLE 11.6
Change of Weightage in Cost

Selection Criteria	Weightage	Shaving Foam (Benchmark)	Effervescent Shaving Tablet	Hair Removal Patch	2 in 1 Shaving Product	Combination of Razor and Shaving Foam
Cost	0.30	5	3	2	3	2
Convenient	0.25	5	7	6	6	5
Comfort	0.25	5	6	2	6	6
Safety	0.20	5	5	3	5	5
Total		**5**	**5.15**	**3.2**	**4.9**	**4.35**

TABLE 11.7
Change of Selection Criteria

Selection Criteria	Weightage	Shaving Foam (Benchmark)	Effervescent Shaving Tablet	Hair Removal Patch	2 in 1 Shaving Product	Combination of Razor and Shaving Foam
Cost	0.20	5	4	3	4	2
Convenient	0.25	5	7	6	6	5
Comfort	0.25	5	6	2	6	6
Quality	0.30	5	4	3	5	5
Total		**5**	**5.25**	**3.5**	**5.3**	**4.65**

11.2 SELF-CLEANING SURFACE HYGIENE PRODUCT

Surface hygiene refers to keeping surfaces clean and sanitary, especially those in contact with food, hands or objects that may harbour microorganisms. It is crucial for preventing the spread of infectious diseases and maintaining a healthy environment in both residential and commercial settings.

11.2.1 Identification of Needs

Identifying customer needs is crucial in the product design process to ensure that the product developed is relevant and useful to the target market. A survey has been conducted among the current users of surface hygiene products covering aspects such as current product usage, satisfaction levels, challenges faced and desired features. The survey gathered responses from 90 participants, providing a diverse set of perspectives and experiences. The results highlighted common challenges that customers face with existing surface hygiene products. Based on these challenges, the unmet needs were identified and listed as follows:

i) Able to provide a long-lasting cleaning effect
ii) Able to provide shine and lustre

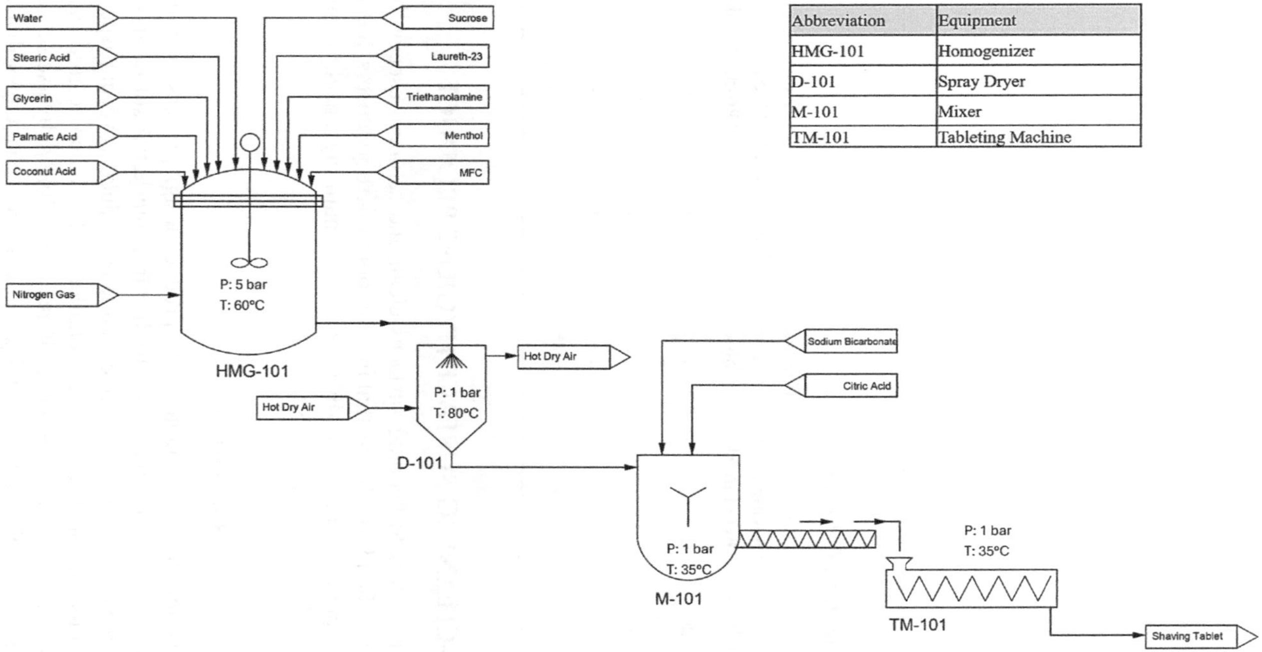

Abbreviation	Equipment
HMG-101	Homogenizer
D-101	Spray Dryer
M-101	Mixer
TM-101	Tableting Machine

FIGURE 11.1 PFD for the commercial production of shaving tablet.

iii) Able to produce fragrance
iv) Compatible with different surface materials
v) Drying time
vi) Easy to apply/use
vii) Effective in eliminating pathogens
viii) Effective in removing stains
ix) Economical
x) Environmentally sustainable
xi) Non-allergenic
xii) Non-corrosive
xiii) Non-flammable
xiv) Non-toxic
xv) No-rinse formula
xvi) Portable
xvii) Residue-free/stain-free upon application
xviii) Self-cleaning

Using the Kano model, the identified needs have been classified into three groups:

Must be

i) Compatible with different surface materials
ii) Effective in eliminating pathogens
iii) Effective in removing stains
iv) Non-allergenic
v) Non-corrosive
vi) Non-toxic
vii) Residue-free/stain-free upon application

One dimensional

i) Able to provide long-lasting cleaning effect
ii) Economical
iii) Drying Time

Attractive

i) Easy to apply/use
ii) Able to produce fragrance
iii) Environmentally sustainable
iv) Non-flammable
v) No-rinse formula
vi) Portable
vii) Able to provide shine and lustre
viii) Self-cleaning

TABLE 11.8
Need Translation

Need	Attribute/s	Description
Non-toxic	Ingredient toxicity level, LD50	Measure of toxicity
Effective in eliminating pathogens	Anti-microbial coverage	Extent of surface area treated by the active ingredients per spray application
Effective in removing stains	Stain removal coverage	Extent of surface area treated by the active ingredients per spray application
Drying time	Evaporation rate	Rate at which a material transitions from its liquid state to vapour
Able to produce fragrance	Presence of synthetic or natural fragrance ingredients	Concentration of active ingredients
Non-corrosive	pH	Measure of corrosiveness
Self-cleaning	Wettability	Measure of contact angle to determine wettability behaviour of a surface
Residue-free	Solubility	Measure of solubility of ingredients in solvents

In the next step, the underlying quantitative properties have been identified as shown (except the obvious ones) in Table 11.8. This step is used to define the specifications of the product.

Based on the results of the survey, we found that majority of the participants use disinfectant liquid sprays, and Dettol is the preferred brand. Hence, the Dettol Anti-Bacterial Surface Cleanser was set as our benchmark. The active ingredients, their roles and weight composition were identified.

The anti-microbial and stain removal coverages are calculated based on the extent of surface area treated by the active ingredients per spray application. In this calculation, the effects of gravity, gas flow and high-pressure operation are neglected to simplify the model. It is assumed that the spray is applied from 200 mm away from the surface, with a spray angle of 60°. Under these conditions, the spray coverage is determined to have a diameter of 231 mm (Spray Coverage, 2024). This setup provides a standardized way to estimate the effectiveness and reach of each spray application in treating surface areas. The spray coverage area is calculated using Equation 11.3. Then the volume fractions of the active ingredients that take part in either antimicrobial or stain removal action were calculated using their respective weight compositions and densities, as outlined in Equation 11.4. The volume fractions are then multiplied by the coverage area determined to give the active ingredient coverage areas, as per Equation 11.5.

$$Spray\ Coverage\ Area = \frac{\pi D^2}{4} \tag{11.3}$$

$$Volume\ of\ Active\ Ingredient\ (AI) = \sum \frac{Mass\ of\ AI}{Density\ of\ AI} \tag{11.4}$$

$$Coverage = Spray\ Coverage\ Area \times Volume\ Fraction\ of\ AI \tag{11.5}$$

The evaporation rate measures the rate at which a material transforms from the liquid to vapour state. This is relevant to the benchmark as the majority of the active ingredients are liquids at room temperature. The Stiver and Mackay equation (11.6) was used to calculate the evaporation rate of the benchmark product (Stiver and Mackay, 1984).

$$\phi_j = \frac{k_j P_j M_{wj}}{RT} \tag{11.6}$$

where $\phi_j (gm^{-2} s^{-1})$ is evaporation rate, $k_j (ms^{-1})$ is mass transfer coefficient, P_j is partial pressure (Pa), M_{wj} is molecular weight $(g\,mol^{-1})$, R is the gas constant $(Pa\,m^3\,mol^{-1}\,K^{-1})$ and $T(K)$ is the liquid temperature. k_j is given by Equation 11.7.

$$k_j = 0.002\,U_1 \tag{11.7}$$

U_1 represents the wind speed at a height of 1 m, and it was assumed to be 1.5 m/s.

From the wind speed, the mass transfer coefficient can be worked out, and with the use of the product vapour pressure and molecular weight, the evaporation rate was found to be 0.278 g/m^2s.

HLB is a numerical value that characterizes the balance between the hydrophilic and lipophilic properties of non-ionic surfactants. Higher HLB values indicate greater water solubility, while lower values indicate greater oil solubility. The wettability of surfactants dictates the cleaning efficiency of a surface cleaning product. The HLB of the benchmark was computed via the group contribution method (Mistry, 2021), shown in Equation 11.8, considering only the non-ionic surfactants in the formulation.

$$HLP = 7 + \Sigma\, Hydrophobic\, Groups + \Sigma\, Hydrophilic\, Groups \tag{11.8}$$

With a HLB value of 11, the surfactant wettability of the benchmark was found to be hydrophilic.

In this case, Dermal LD50 values were prioritized over oral or inhalation exposure routes due to the inherent nature of the surface hygiene product, which frequently encounters direct contact with the skin. To estimate the LD50 of a mixture, ATE_{mix} has been used outlined in Equation 11.1 (Van Cott et al., 2018).

The complete list of specifications and benchmarks is shown in Table 11.9.

11.2.2 Idea Generation and Classification

In this step, several ideas to meet the needs are generated. The partial list of raw ideas is shown in Table 11.10.

The following is a possible way to classify the needs after removing the redundant ideas:

Existing

1. Anti-Static Surface Cleaner
2. Antibacterial Surface Cleaner

TABLE 11.9
Benchmark and Target Comparison

Attributes	Benchmark	Target
pH	9.3–10.3	6.5–7.5
Wettability	Hydrophilic	Superhydrophobic/ superhydrophilic
$LD50 / ATE_{mix} [mg / kg]$	31393	> 2820
Evaporation rate $[g / m^2 s]$	0.28	< 0.28
Surface coverage (antimicrobial) $[cm^2]$	0.628	≥ 3
Surface coverage (stain removal) $[cm^2]$	28.4	≥ 40

TABLE 11.10
Raw Ideas to Meet the Needs

1. Anti-Static Surface Cleaner
2. Antibacterial Surface Cleaner
3. Antiviral Nanocoating
4. Aromatherapy Cleaning
5. Automatic Cleaning Robot
6. Baking Soda and Vinegar
7. Biofilm Disruption Technology
8. Bioluminescent Cleaner
9. Bleaching
10. Cleaning Cloth
11. Cleansing Foam
12. Coca-Cola
13. Colour-Changing Formula
14. Electrostatic Cleaner
15. Fogging
16. Hydrophobic Surface Coating
17. Hydrogel Surface Coating
18. Infrared Surface Sterilization
19. Multi-Surface Cleaner
20. Oxygen-Activated Cleaner
21. Photocatalytic Surface Coating
22. Plant-Based Enzymatic Cleaner
23. Powder Surface Cleaner
24. Probiotic Surface Cleaner
25. Self-Healing Coating
26. Smart Dispensing System
27. Tablet
28. UV-C Disinfection

3. Bleaching
4. Cleaning Cloth
5. Multi-Surface Cleaner

New Direction

1. Antiviral Nanocoating
2. Aromatherapy Cleaning
3. Bioluminescent Cleaner
4. Colour-Changing Formula
5. Electrostatic Cleaner
6. Hydrogel Surface Coating
7. Hydrophobic Surface Coating
8. Photocatalytic Surface Coating
9. Self-Healing Coating
10. Oxygen-Activated Cleaner

Natural Additives

1. Biofilm Disruption Technology
2. Plant-Based Enzymatic Cleaner
3. Probiotic Surface Cleaner

Alternative Delivery Methods

1. Automatic Cleaning Robot
2. Infrared Surface Sterilization
3. Smart Dispensing System
4. UV-C Disinfection

A screening matrix has been used to shortlist promising ideas for developing a new surface hygiene product. The primary objective of this new product is the ability of the product to effectively eliminate various types of stains from different surfaces and the product's capability to provide a sustained cleaning effect comfort, so these targets are given higher weightage in the evaluation process. Additionally, safety and preservation of surface integrity are critical considerations, particularly as it is important to ensure the product is safe for use around humans and pets, and free from harmful chemicals. The product should not damage or degrade the surfaces it is applied to, maintaining their original condition. Moreover, the product should be easy to use, with straightforward application methods that do not require special tools or extensive effort. A sample of the screening matrix is shown in Table 11.11.

Based on the results from the screening matrix, the ideas with the highest total scores are Hydrophobic Surface Coating (6.60), Photocatalytic Surface Coating (6.55) and Hydrogel Surface Coating (5.85), making them promising options. However, these conclusions depend heavily on the specific selection criteria, their assigned weightages and the scores given to each option. To ensure the robustness of these conclusions, a sensitivity analysis is necessary. This analysis will evaluate

TABLE 11.11
Raw Ideas to Meet the Needs

Selection Criteria	Weigh-tage	Benchmark	UV-C Disin-fection	Hydro-phobic Surface Coating	Photo-catalytic Surface Coating	Hydrogel Surface Coating	Plant-Based Enzymatic Cleaner	Oxygen-Activated Cleaner
Effectiveness in removing stains	0.30	5.00	7.00	7.00	8.00	6.00	7.00	7.00
Long-lasting cleaning effect	0.25	5.00	3.00	8.00	7.00	6.00	4.00	5.00
Safety	0.20	5.00	2.00	4.00	4.00	4.00	8.00	5.00
Preservation of surface integrity	0.15	5.00	3.00	8.00	8.00	9.00	3.00	6.00
Ease of application	0.10	5.00	2.00	5.00	4.00	4.00	5.00	5.00
Total		**5.00**	**3.90**	**6.60**	**6.55**	**5.85**	**5.65**	**5.75**

the impact of variations in criteria, scores and weightages on the overall rankings and the selection of the optimal product idea. Different scenarios will be considered to assess the stability and reliability of the chosen options under various conditions.

For the first case of sensitivity analysis, the weighting factors are adjusted in accordance with ratings obtained from the conducted survey, without altering the individual scores. The results showed that the top three ideas remained the same, as shown in Table 11.12.

For the second sensitivity analysis case (Table 11.13), the weighting factors are adjusted by taking price into consideration as consumers may be reluctant to buy products that are too costly. The consideration was based on 500 ml of product. In this case, the Hydrogel Surface Coating was replaced by the Plant-Based Enzymatic Cleaner, demonstrating its susceptibility to priorities or fluctuations.

From the sensitivity analysis, we can conclude that Hydrophobic Surface Coating and Photocatalytic Surface Coating are promising options and are less sensitive to minor variations in the criteria used for product evaluation. However, the primary purpose of the screening matrix is to shortlist promising options, emphasizing the elimination of ideas that are clearly inferior to the rest. The ranking of ideas at the screening stage does not necessarily indicate the superiority of the ideas. The intention here is to identify ideas that warrant further analysis during the selection stage.

TABLE 11.12
Adjustment of Weighting Factors in Accordance with Ratings Obtained from the Conducted Survey

Selection Criteria	Weightage	Benchmark	UV-C Disinfection	Hydrophobic Surface Coating	Photocatalytic Surface Coating	Hydrogel Surface Coating	Plant-Based Enzymatic Cleaner	Oxygen-Activated Cleaner
Effectiveness in removing stains	0.206	5.00	7.00	7.00	8.00	6.00	7.00	7.00
Long-lasting cleaning effect	0.201	5.00	3.00	8.00	7.00	6.00	4.00	5.00
Safety	0.199	5.00	2.00	4.00	4.00	4.00	8.00	5.00
Preservation of surface integrity	0.199	5.00	3.00	8.00	8.00	9.00	3.00	6.00
Ease of application	0.195	5.00	2.00	5.00	4.00	4.00	5.00	5.00
Total		**5.00**	**3.43**	**6.41**	**6.22**	**5.81**	**5.41**	**5.61**

TABLE 11.13
Inclusion of Price Considerations, Benchmarked against a Standardized Volume of 500 ml Per Product Unit

Selection Criteria	Weightage	Benchmark	UV-C Disinfection	Hydrophobic Surface Coating	Photocatalytic Surface Coating	Hydrogel Surface Coating	Plant-Based Enzymatic Cleaner	Oxygen-Activated Cleaner
Effectiveness in removing stains	0.30	5.00	7.00	7.00	8.00	6.00	7.00	7.00
Long-lasting cleaning effect	0.20	5.00	3.00	8.00	7.00	6.00	4.00	5.00
Safety	0.20	5.00	2.00	4.00	4.00	4.00	8.00	5.00
Preservation of surface integrity	0.1	5.00	3.00	8.00	8.00	9.00	3.00	6.00
Ease of application	0.10	5.00	2.00	5.00	4.00	4.00	5.00	5.00
Price	0.10	5.00	1.00	3.00	2.00	3.00	8.00	6.00
Total		**5.00**	**3.70**	**6.10**	**6.00**	**5.40**	**6.10**	**5.80**

11.2.3 Selection and Risk Assessment

To further refine and validate the promising options for surface hygiene products, the following steps were taken:

i) Comparison of the three ideas using quantitative scientific tools.
ii) Selection of the best idea using selection matrices.
iii) Risk assessment.

In the selection section, scientific tools are required to quantify the selected product ideas from the concept screening matrix. With detailed engineering calculations and literature review, the selected ideas could be further compared to determine the best/robust invention to tackle the customers' needs. The three scientific tools that were used for comparison with the product specifications are Contact Angle, Antimicrobial and Stain Removal Coverage and Evaporation Rate, to measure the effectiveness of cleaning properties of a product. The estimated product attributes are presented in Table 11.14.

11.2.3.1 Idea Selection and Sensitivity Analysis

In the process of selecting the best surface hygiene product, we used a selection matrix to compare different ideas based on attributes determined. The key attributes compared with the benchmark included wettability and surface coverage of active ingredients, which were prioritized from an engineering perspective. Based on Table 11.15, Photocatalytic Surface Coating is the best option in this scenario.

TABLE 11.14
Product Attributes (Khannyra et al., 2021, 2022; Laudon et al., 2006)

Attributes	Target	Hydrophobic Surface Coating	Photocatalytic Surface Coating	Hydrogel Surface Coating
pH	6.5–7.5	7–9	7	7
Contact Angle	-	132°	9.3°	103°
Wettability	Superhydrophobic/ superhydrophilic	Hydrophobic	Superhydrophilic	Hydrophobic
$LD50 / ATE_{mix} [mg / kg]$	> 2820	8745	5004	5420
Evaporation rate $[g / m^2 s]$	< 0.28	0.20	-	0.17
Surface coverage (antimicrobial) $[cm^2]$	≥ 3	229.51	490.83	0.76
Surface coverage (stain removal) $[cm^2]$	≥ 40	183.00	490.83	38.54

TABLE 11.15
Emphasis on Wettability and Surface Coverage of Active Ingredients

Selection Criteria	Weightage	Benchmark	Hydrophobic Surface Coating	Photocatalytic Surface Coating	Hydrogel Surface Coating
pH	0.10	5.00	6.00	7.00	7.00
Wettability	0.30	5.00	7.00	8.00	6.00
LD50/ATE_{mix}	0.15	5.00	4.00	3.00	3.00
Evaporation rate	0.15	5.00	7.00	9.00	8.00
Surface coverage of active ingredients	0.30	5.00	8.00	9.00	6.00
Total	**1.00**	**5.00**	**6.75**	**7.60**	**5.95**

TABLE 11.16
Emphasis on Wettability and Evaporation Rate

Selection Criteria	Weightage	Benchmark	Hydrophobic Surface Coating	Photocatalytic Surface Coating	Hydrogel Surface Coating
pH	0.10	5.00	6.00	7.00	7.00
Wettability	0.30	5.00	7.00	8.00	6.00
LD50/ATE_{mix}	0.15	5.00	4.00	3.00	3.00
Evaporation rate	0.30	5.00	7.00	9.00	8.00
Surface coverage of active ingredients	0.15	5.00	8.00	9.00	6.00
Total	**1.00**	**5.00**	**6.60**	**7.60**	**6.25**

In the first sensitivity analysis (Table 11.16), the weighting factors were adjusted to prioritize the long-lasting effect of the product, giving more emphasis to the evaporation rate. Despite this adjustment, the Photocatalytic Surface Coating remained the best option, confirming its effectiveness and durability.

In the second sensitivity analysis (Table 11.17), we adjusted the weighting factors to prioritize the toxicity of the product, focusing more on the LD50 values. Even with this shift in emphasis, the Photocatalytic Surface Coating still emerged as the best option. This demonstrates the robustness of Photocatalytic Surface Coating against various evaluation criteria and changing priorities.

Through both the initial selection matrix and subsequent sensitivity analyses, Photocatalytic Surface Coating consistently performed as the top choice, making it a reliable and versatile option for surface hygiene applications.

TABLE 11.17
Emphasis on Wettability and Safety

Selection Criteria	Weightage	Benchmark	Hydrophobic Surface Coating	Photocatalytic Surface Coating	Hydrogel Surface Coating
pH	0.10	5.00	6.00	7.00	7.00
Wettability	0.30	5.00	7.00	8.00	6.00
LD50/ATE_{mix}	0.30	5.00	4.00	3.00	3.00
Evaporation rate	0.15	5.00	7.00	9.00	8.00
Surface coverage of active ingredients	0.15	5.00	8.00	9.00	6.00
Total	**1.00**	**5.00**	**6.15**	**6.70**	**5.50**

TABLE 11.18
Risk Assessment of Hydrophobic Surface Coating

Criteria	Probability	Consequences	Risk Level
Consumer acceptability	0.20	0.50	0.10
Technology maturity	0.30	0.30	0.09
Surface compatibility	0.30	0.40	0.12
Environmental impact	0.30	0.60	0.18
Wrong assumption/estimation on calculations	0.30	0.50	0.15

11.2.3.2 Risk Assessment Criteria

There are many potential hazards in product development, mostly due to the limited information. The possible risks in this case were identified and listed next:

i) Consumer Acceptability: Product quality, convenience, overall satisfaction and others
ii) Technology Maturity: Technology development, stability and reliability
iii) Surface Compatibility: Compatibility of materials, interaction between surfaces/environment
iv) Environmental Impact: Pollution, health and safety
v) Estimation on Calculations: Underestimation/overestimation; the probability and impact of each risk for all three products were estimated using engineering techniques and is tabulated in Tables 11.18–11.20.

Photocatalytic Surface Coating is the final selection, based on risk assessment and selection matrices. The identified solution is the best, even if the priorities change. Hence, it is a robust decision.

TABLE 11.19
Risk Assessment of Photocatalytic Surface Coating

Criteria	Probability	Consequences	Risk Level
Consumer acceptability	0.20	0.50	0.10
Technology maturity	0.30	0.30	0.09
Surface compatibility	0.20	0.10	0.02
Environmental impact	0.20	0.40	0.08
Wrong assumption/estimation on calculations	0.30	0.50	0.15

TABLE 11.20
Risk Assessment of Hydrogel Surface Coating

Criteria	Probability	Consequences	Risk Level
Consumer acceptability	0.40	0.50	0.20
Technology maturity	0.50	0.60	**0.30**
Surface compatibility	0.20	0.20	0.04
Environmental impact	0.20	0.30	0.06
Wrong assumption/estimation on calculations	0.40	0.50	0.20

11.2.4 Manufacture

The process flow diagram for the commercial production of Photocatalytic Surface Coating is shown in Figure 11.2.

11.3 ACTIVATED CHARCOAL PACKAGING PRODUCT

Unpleasant odours in life can cause environmental pollution and negatively affect the mood of those around. Using activated charcoal products for odour removal is an effective practice to promote a fresh environment, enhance air quality and foster a sense of cleanliness and comfort.

11.3.1 Identification of Needs

Activated charcoal in fabric sachet form is common in odour removal. This case study focuses on the packaging material that meets user needs. A survey was conducted among current users of activated charcoal to identify their requirements from a consumer perspective. This step also helps determine if there are

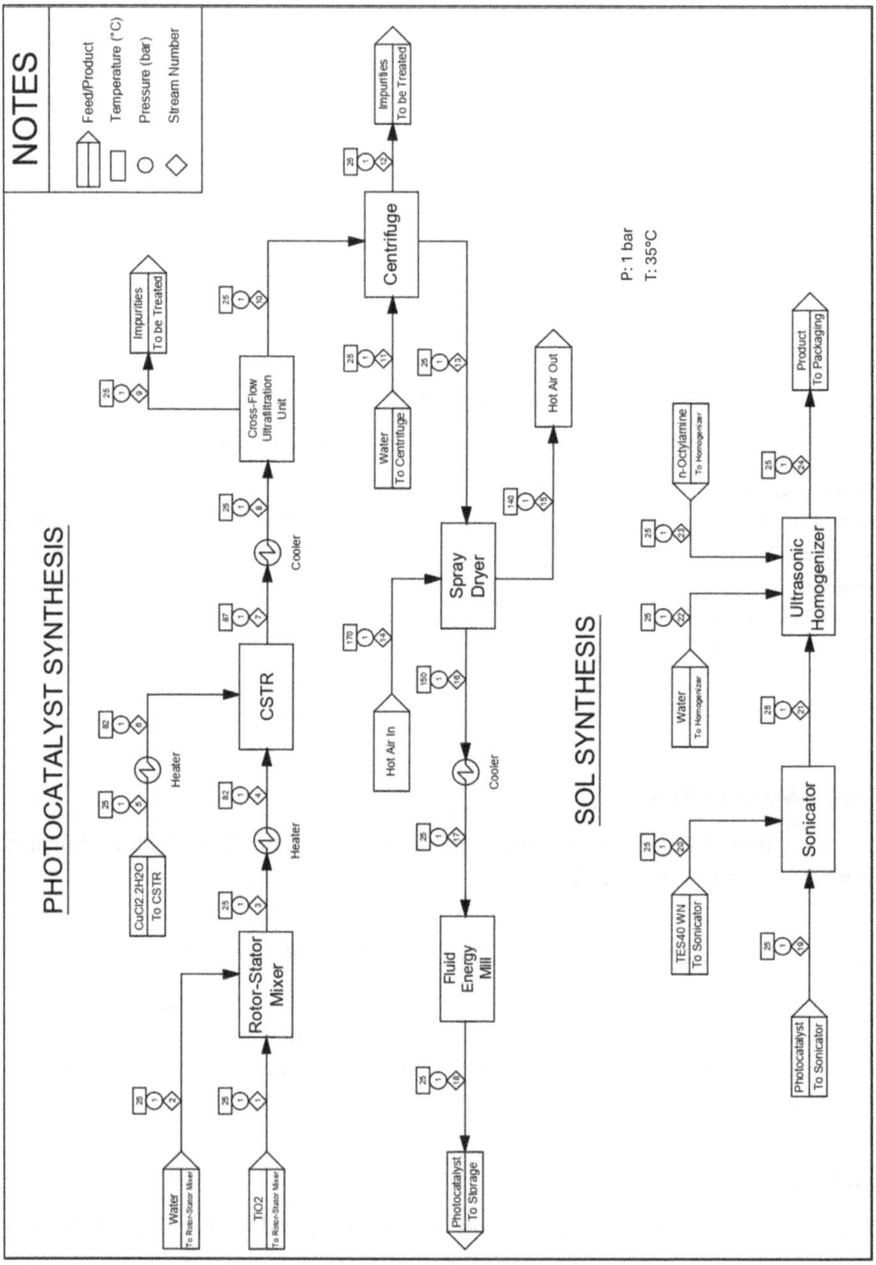

FIGURE 11.2 PFD for the commercial production of Photocatalytic Surface Coating.

any unfulfilled needs for this product. The needs identified through the survey are as follows:

i) Remove bad odour
ii) High efficiency
iii) Convenient to use
iv) Long shelf life
v) Waterproof
vi) Leakproof packaging
vii) Non-toxic
viii) Cost friendly
ix) Alcohol-free formulation
x) Versatile function
xi) Antimicrobial properties
xii) Allergen and irritant reduction

In the next step, these needs have been classified based on the Kano model.

Must be

1. Remove bad odour
2. High efficiency
3. Leakproof packaging
4. Non-toxic
5. Waterproof

One dimensional

1. Versatile function
2. Convenient to use
3. Cost friendly
4. Long shelf life

Attractive

1. Antimicrobial properties
2. Allergen and irritant reduction
3. Alcohol-free formulation

In the next step, the underlying quantitative properties have been identified as shown (except the obvious ones) in Table 11.21. This step is used to define the specifications of the product.

The waterproofing capability of a product can be estimated by calculating its water vapour permeability. Water vapour permeability (WVP) measures a material's ability to allow water vapour to pass through it. Materials with low water vapour permeability are more waterproof, as they prevent moisture from penetrating. The rate at which water vapour passes through the material is measured, and lower rates indicate better waterproofing.

TABLE 11.21
Need Translation

Need	Attribute	Description
Non-toxic	Ingredient toxicity level, LD 50	Measure of toxicity.
Long shelf life	Off-gassing characteristics	Estimate release of gases from a material over time.
Efficiency	Odour absorption rate	Measure the rate of odour particle absorption.
Waterproof	Water vapour permeability	Measure the permeability of water of a material.
Leakproof packaging	Tensile strength	Measure the amount of stress a packaging material can withstand before breaking.
Allergen and irritant reduction	Pore size estimation	Determine filtration efficiency for both odour particles and dust mites.
Antimicrobial properties	Time-kill kinetics	Evaluate the efficacy of antimicrobial agents by measuring the rate at which they kill microorganisms over time.
Cost friendly	Manufacturing cost	Determine cost of raw material and manufacturing cost.

Assumptions made are as follows:

- Temperature: 35°C
- Thickness of sachet: 0.00015m
- Relative humidity in packaging: 40%
- Relative humidity at room condition: 85%
- Saturation pressure at 35°C: 690Pa

The pressure difference between water vapour at different conditions is first calculated using Equation 11.9.

$$\Delta P = \left(\frac{RH\ at\ room\ temperature}{100} \times SWP\ at\ 35^{\circ}\mathrm{C}\right) - \left(\frac{RH\ in\ packaging}{100} \times SWP\ at\ 35^{\circ}\mathrm{C}\right) \tag{11.9}$$

where *RH* is the relative humidity (%), *SWP* is saturation water vapour pressure (Pa)

Permeance is a measure of the rate at which water vapour passes through a material under specific conditions. WVTR refers to the amount of water vapour that passes through a material per unit area per unit time. Thus, Equation 11.10 is used for the calculation of permeance.

$$Permeance = \frac{WVTR}{\Delta P}\left(\frac{g}{m^2.day.Pa}\right) \tag{11.10}$$

TABLE 11.22
Benchmark and Target Comparison

Attributes	Benchmark	Target
Tensile strength (MPa)	287 MPa	≥ 287 MPa
Cost (USD per unit)	1.75	≤ 1.75
Water vapour permeability (perm)	0.193 perm	≤ 0.193 perm
Dust mite resistance	90%	≥ 90%

where WVTR is water vapour transmission rate (g/m^2.day).

Finally, the water vapour permeability is calculated by multiplying the permeance by the thickness of the material, as shown in Equation 11.11. This is to indicate the overall resistance of a material to the water vapour transmission.

$$WVP = Permeance \times Thickness \tag{11.11}$$

where WVP is the water vapour permeability (perm).

Dust mite resistance can be performed by estimating the pore size, which is used to determine the filtration efficiency. The typical size of dust mite allergen ranges from 1.1 microns to 9.7 microns. Thus, by calculating the pore size of the material, the ability to block dust mite can be estimated.

Pore size refers to the diameter of the openings or voids within a material. Pores smaller than the particle size can trap particles, while larger pores allow particles to pass. Overall adsorption efficiency depends on the particle size, pore size distribution and overall surface area. Equation 11.12 is used to estimate the radius of the material.

$$r = \frac{4\sigma \cos\theta}{P} \tag{11.12}$$

where σ is the surface tension (N/m), P is the bubble point pressure (Pa).

The complete list of specifications and benchmarks is shown in Table 11.22.

11.3.2 Idea Generation and Classification

In this step, ideas to meet the needs are generated. The partial list of raw ideas is shown in Table 11.23.

The following is a possible way to classify the needs:

Redundant Ideas

1. Wear mask
2. Change room

Improvements on Existing Products

1. Charcoal-infused beads with fragrance particles
2. Charcoal-embedded textiles

TABLE 11.23
Raw Ideas to Meet the Needs

1. Wear mask
2. Charcoal-infused beads with fragrance particles
3. Charcoal-embedded textiles
4. Plastic pouch as packaging material
5. Replace activated charcoal with easily produced chemical product
6. Self-deodorizing garbage bin
7. Bioluminescent odour-absorbing plant
8. Ultrasonic toothbrush with odour-neutralizing frequencies
9. Sweat-activated deodorizing patches
10. Volcanic ash air filters
11. Odour-powered generators
12. Scent-cancelling headphones
13. Natural refrigerator deodorizer
14. Fragrance spray pen
15. Biodegradable odour-absorbing pouches
16. Natural odour-neutralizing pet wipes
17. Fragrance-infused plant watering bead
18. Activated charcoal-embedded night light
19. UV-C light air purifier
20. Rechargeable odour-eliminating shoe deodorizer
21. Portable personal air purifier necklace
22. Scented furniture
23. Odour-eliminating mattress encasement
24. Charcoal-infused paint additives
25. Scented face mask
26. Natural room spray with disinfectant
27. Smart kitchen vents with odour sensors
28. Odour camouflage perfume
29. Change room
30. Robotic odour patrol units

3. Plastic pouch as packaging material
4. Replace activated charcoal with easily produced chemical product
5. Scented furniture
6. Scented face mask
7. Natural room spray with disinfectant

Innovative Additions to Existing Products

1. Fragrance spray pen
2. Charcoal-infused paint additives
3. UV-C light air purifier

New direction

1. Self-deodorizing garbage bin
2. Bioluminescent odour-absorbing plant
3. Ultrasonic toothbrush with odour-neutralizing frequencies
4. Sweat-activated deodorizing patches
5. Volcanic ash air filters
6. Odour-powered generators
7. Scent-cancelling headphones
8. Natural refrigerator deodorizer
9. Biodegradable odour-absorbing pouches
10. Natural odour-neutralizing pet wipes
11. Fragrance-infused plant watering bead
12. Activated charcoal-embedded night light
13. Rechargeable odour-eliminating shoe deodorizer
14. Portable personal air purifier necklace
15. Odour-eliminating mattress encasement
16. Smart kitchen vents with odour sensors
17. Odour camouflage perfume
18. Robotic odour patrol units

11.3.2.1 Idea Selection and Sensitivity Analysis

A screening matrix has been used to shortlist promising ideas. The objective of designing this new product is to determine the best form of representing the activated charcoal to maximize its efficiency in odour removal. Thus, the efficiency is given the highest weightage. The versatility of new products is essential to fit in different scenarios so that it is convenient for users. As activated charcoal can cause messiness in surroundings if in contact with water, the cleanliness factor is considered. Finally, the cost of the product is an important factor to customers. A sample of screening matrix is shown in Table 11.24.

TABLE 11.24
Raw Ideas to Meet the Needs

Selection Criteria	Weightage	Fabric Sachets (Benchmark)	Charcoal-Embedded Textiles	Plastic Pouch	Activated Charcoal Sticks
Efficiency	0.4	5	6	7	5
Versatility	0.3	5	2	5	6
Cleanliness	0.15	5	3	8	1
Cost	0.15	5	1	4	6
Total		**5**	**3.6**	**6.1**	**4.85**

Based on the results from the screening matrix, plastic pouch and activated charcoal sticks are promising options. However, this conclusion also depends on the chosen selection criteria, relative weightages and the scores given to each of the options in the selected criteria. In order to increase the robustness of this conclusion, a sensitivity analysis is to be conducted where the sensitivity of the criteria, scores and weightages to the conclusion is analysed. In this example, we have considered the following situations:

i) Customers may not think that versatility of product is the most important attribute.
ii) The scores given to cleanliness may be too high as the alternative options may cause similar levels of messiness as benchmark.
iii) Customers consider cost of product as the most important attribute.

The sensitivity analysis of these scenarios is shown in Tables 11.25–11.27.

From the sensitivity analysis, we can conclude that the plastic pouch and activated charcoal stick are promising options and are less sensitive to the minor variations in the criteria used for the product evaluation. However, the purpose of screening matrix is to shortlist promising options, and the emphasis here is the elimination

TABLE 11.25
More Emphasis on Cost

Selection Criteria	Weightage	Fabric Sachets (Benchmark)	Charcoal-Embedded Textiles	Plastic Pouch	Activated Charcoal Sticks
Efficiency	0.35	5	6	7	5
Versatility	0.15	5	2	5	6
Cleanliness	0.1	5	3	8	1
Cost	0.4	5	1	4	6
Total		**5**	**3.1**	**5.6**	**5.15**

TABLE 11.26
Change of Weightage in Versatility

Selection Criteria	Weightage	Fabric Sachets (Benchmark)	Charcoal-Embedded Textiles	Plastic Pouch	Activated Charcoal Sticks
Efficiency	0.4	5	6	7	5
Versatility	0.3	5	7	8	4
Cleanliness	0.15	5	3	8	1
Cost	0.15	5	1	4	6
Total		**5**	**5.1**	**7**	**4.25**

TABLE 11.27
Change of Selection Criteria

Selection Criteria	Weightage	Fabric Sachets (Benchmark)	Charcoal-Embedded Textiles	Plastic Pouch	Activated Charcoal Sticks
Efficiency	0.4	5	6	7	5
Versatility	0.3	5	2	5	6
Cost	0.15	5	1	4	6
Environmental Impact	0.15	5	1	4	8
Total		**5**	**3.3**	**5.5**	**5.9**

TABLE 11.28
Quantitative Comparison

Attributes	Target	Charcoal-Embedded Textiles	Plastic Pouch	Activated Charcoal Sticks
Waterproof capability	≤ 0.193 perm	0.658	0.097	0.084
Dust mite resistance	≥ 90%	95%	98%	99%
Leakproof performance	≥ 287 MPa	40	170	70

of the ideas that are clearly inferior to the rest. In addition, the ranking of ideas at screening is not necessarily an indication of the superiority of the ideas. The intention here is only to choose ideas that need to be analysed during the selection stage.

11.3.3 Selection and Risk Assessment

To further refine and validate the promising packaging material options for activated charcoal, the three proposed ideas will be compared using quantitative scientific tools. A selection matrix will be utilized to identify the optimal choice. Additionally, a risk assessment will be conducted to evaluate potential challenges and ensure the chosen material meets all necessary criteria.

A selection matrix is implemented through comprehensive engineering calculations to conduct a quantitative analysis. The three attributes evaluated are waterproof capability, dust mite resistance and leakproof performance. By integrating detailed engineering calculations with an extensive literature review, the proposed ideas can be thoroughly compared to identify the most effective solution for addressing customer needs. The estimated product attributes are presented in Table 11.28.

Selection matrix is constructed as shown in Table 11.29 based on the quantitative comparison to decide the best product.

Plastic pouch has the highest total score of 7.8, making it the most suitable packaging material for activated charcoal among the three options. This decision is based

TABLE 11.29
Selection Matrix

Selection Criteria	Weightage	Charcoal-Embedded Textiles	Plastic Pouch	Activated Charcoal Sticks
Waterproof capability	0.5	1	8	9
Dust mite resistance	0.3	6	8	8
Leakage proof	0.2	2	7	3
Total		**2.7**	**7.8**	**7.5**

TABLE 11.30
Risk Assessment for Charcoal-Embedded Textiles

Criteria	Probability	Consequences	Risk
Consumer acceptability	0.5	0.4	0.2
Technology maturity	0.6	0.5	0.3
Market competition	0.3	0.1	0.03
Environmental impact	0.2	0.2	0.04

on its strong performance across all evaluated criteria, particularly in waterproof capability and leakproof performance. Thus, the final product is chosen to be activated charcoal in plastic pouch packaging.

11.3.3.1 Risk Assessment Criteria

Risk assessment covers consumer acceptance, technology maturity, market competition and environmental impact. Consumer acceptance gauges the likelihood of market adoption based on preferences and demand. Technology maturity assesses the development stage, reliability and scalability of the technologies involved. Market competition examines existing competitors and potential barriers to entry. Environmental impact evaluates the product's lifecycle effects, including resource use, emissions and waste. This ensures the product meets consumer needs, adheres to technological standards, remains competitive and aligns with sustainability goals. Tables 11.30–11.32 show the risk assessment for all three designs.

The plastic pouch generally shows a lower overall risk compared to the other products, with particularly low risks in consumer acceptability and market competition. Charcoal-embedded textiles exhibit moderate risks across most criteria, while activated charcoal sticks demonstrate the lowest risks in terms of environmental impact and consumer acceptability.

11.3.4 Manufacture

The manufacturing process for producing activated charcoal in a plastic pouch begins with the grinding of activated charcoal in a grinder to maximize the efficiency of

TABLE 11.31
Risk Assessment for Plastic Pouch

Criteria	Probability	Consequences	Risk
Consumer acceptability	0.1	0.1	0.01
Technology maturity	0.3	0.3	0.09
Market competition	0.6	0.1	0.06
Environmental impact	0.5	0.3	0.15

TABLE 11.32
Risk Assessment for Activated Charcoal Stick

Criteria	Probability	Consequences	Risk
Consumer acceptability	0.1	0.1	0.01
Technology maturity	0.3	0.2	0.06
Market competition	0.5	0.1	0.05
Environmental impact	0.2	0.1	0.02

end products. The ground charcoal is sent into a washer to be cleaned with stream of water. The washed charcoal is then transferred into a dryer to remove moisture. Simultaneously, the plastic pouches are prepared by cutting them into desired sizes using a cutting machine. The dried charcoal is then directed to a filling machine where it is filled into the plastic pouches. After filling, the pouches are sealed using a sealing machine. Finally, the filled and sealed pouches are packaged using a packaging machine. Throughout this process, hot air is utilized, and wastewater is managed accordingly. The final product is a packaged activated charcoal pouch ready for distribution. The process flow diagram for the commercial production of shaving tablet is shown in Figure 11.3.

11.4 ILLUSTRATION OF COMPUTER-AIDED MOLECULAR DESIGN: SOLVENT DESIGN FOR ORGANIC RANKINE CYCLE

11.4.1 Background

Due to the global needs in solving energy and environmental emission issues, there is a demand for recovering waste heat produced from industrial applications, mainly low-grade waste heat. Studies have shown that low-grade waste heat makes up at least 50% of the total waste generated in industry (Hung et al., 1997). The main issue is that low-grade waste heat is generally discarded, which causes thermal pollution. Many technologies have been proposed to capture the waste heat generated. Some of the proposed technologies are plant or district heating, power plant co-location, absorption cooling, organic Rankine cycle, piezoelectrics, thermoelectrics,

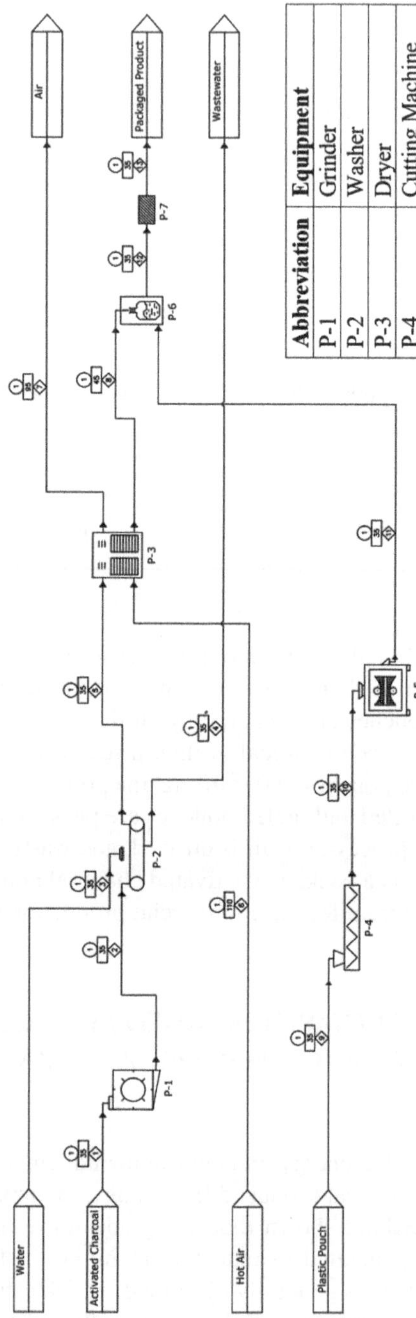

Abbreviation	Equipment
P-1	Grinder
P-2	Washer
P-3	Dryer
P-4	Cutting Machine
P-5	Sealing Machine
P-6	Filling Machine
P-7	Packaging Machine

FIGURE 11.3 PFD for the commercial production of a plastic pouch containing activated charcoal.

desalination, biomass co-location (Ebrahimi et al., 2014), chemical heat pumps, supercritical Rankine cycle, trilateral cycle, thermal energy storage (Chan et al., 2013) and condensing boiler (Walsh and Thornley, 2013).

There has been increasing interest in recovering low-grade waste heat using Organic Rankine Cycle (ORC) (Wang et al., 2013). A Rankine cycle is a heat engine with a vapour power cycle in which heat is converted into mechanical work. The conventional working fluid used is water. A general Rankine cycle is made up of pump, evaporator, turbine and condenser. The pressure of the working fluid is first increased through a pump, heat is then added to the working fluid to boiling through an evaporator, the high pressure and temperature working fluid is then expanded in the turbine to generate electricity. Lastly, heat is rejected from the working fluid through a condenser. The working fluid flows through the pump again, and the cycle continues. Meanwhile, for an ORC, the working principle is the same except that organic fluids are used as working fluid instead of water. ORC exhibits several advantages over conventional Rankine cycle. Rankine cycle operating using water does not provide efficient recovery of waste heat below 370°C. This can be resolved by utilizing ORC because organic compounds generally have a lower phase-change (boiling) temperature compared to water. Besides, ORC also offers advantages in terms of long service life, low maintenance costs, fully automatic and unmanned operation, excellent partial load behaviour and electric efficiency (Obernberger et al., 2002). ORC also demonstrates enhanced and economical use of energy in addition to reduction in CO, CO_2, NO_X and other atmospheric pollutants emissions (Mago et al., 2007; Dai et al., 2009). Another appealing ORC characteristic is that it can operate with a lower boiling temperature and higher vapour pressure, which enables the turbine exhaust to fall in the superheated region, hence preventing erosion of turbine caused by the presence of wet steam (Wang et al., 2013).

The organic fluid can be divided into three categories: dry fluid, wet fluid and isentropic fluid. The three categories are based on the saturation vapour curve on the temperature-entropy (*T-s*) diagram shown in Figure 11.4, in which a dry fluid exhibits positive slopes, a wet fluid shows negative slopes while an isentropic fluid with infinite slope. The turbine exhaust will contain lot of saturated liquid when wet liquid is used due to the negative slope of the saturation vapour curve. The turbine blades may be damaged and the isentropic efficiency of the turbine reduced because of the presence of liquid in the turbine. The recommended minimum dryness fraction at the turbine exhaust should be maintained above 85% (Desai and Bandyopadhyay, 2009). Meanwhile, using dry and isentropic fluids as working fluids can help to prevent

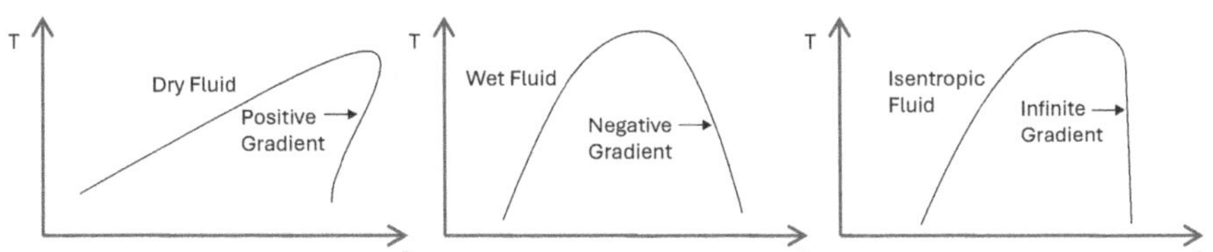

FIGURE 11.4 Saturation curves on T-s diagram.

damage in turbine as the turbine exhaust is always maintained at saturated or superheated vapour, hence no liquid formation in turbine (Hung, 2001).

The application of ORC in industries can be divided into three main categories (Wang et al., 2011). The first category is the utilization of geothermal energy. Liu et al. investigated the effects of turbine inlet temperature and condensation temperature on the ORC net power output by using butane, isobutane, pentane, isopentane and hexane as working fluids driven by geothermal water from 100°C to 150°C. They concluded that as the condensation temperature decreases, the ORC net power output increases, but the pump power input increases sharply. (Liu et al., 2013). Saleh et al. studied the thermodynamic screening of 31 pure component ORC working fluids operating between 30°C and 100°C for geothermal power plants by considering sub-critical and supercritical conditions. They deduced that high thermal efficiency can be achieved by utilizing dry fluids with high boiling temperature (butane) together with the use of internal heat exchanger (IHE) to cool down the superheated vapour discharged from turbine (Saleh et al., 2007).

The second category is mainly focused on solar energy. Marion et al. investigated a subcritical ORC combining with a solar thermal collector utilizing refrigerants R134a, R227ea and R365mfc operating with a fixed temperature sink to establish the optimum configuration by varying fluid mass flow rate (Marion et al., 2012). Rayegan and Tao established an algorithm to compare ORC working fluids based on their molecular components, temperature-entropy diagram and fluid effects on the thermal efficiency, net power generated, vapour expansion ratio and exergy efficiency of the ORC by considering 117 organic fluids (Rayegan and Tao, 2011). Tchanche et al. investigated the thermodynamic and environmental properties of 20 fluids by comparing the efficiencies, volume flow rate, mass flow rate, pressure ratio, toxicity, flammability, ozone depletion potential (ODP) and global warming potential (GWP). They concluded that fluids with high density, high latent heat of vaporization and high boiling point are preferable, and they identified that R134a is the most appropriate working fluid for small-scale solar applications (Tchanche et al., 2009).

The final category focuses on low-grade waste heat recovery. Hung et al. analysed the efficiency of ORC using benzene, ammonia, R11, R12, R134a and R113 as working medium and established a conclusion that isentropic fluids are most appropriate for recovering low-temperature waste heat (Hung et al., 1997). Hung also examined the working conditions of ORC and thermodynamic properties of dry working fluids such as benzene, toluene, p-xylene, R113 and R123 to compare the efficiency of ORC (Hung, 2001). Maizza and Maizza investigated the thermodynamic and physical properties of some unconventional working fluids to be utilized in ORC in which the heat source is supplied by waste energy sources in order to determine the thermodynamic efficiencies (Maizza and Maizza, 2001). Mago et al. investigated the usage of regenerative ORC from low-grade waste heat sources using dry fluids such as R113, R245ca, R123 and isobutene. The objective is to identify the configuration that provides the best thermal efficiency with minimum irreversibility performance by applying the combination of first law and second law analysis. They deduced that regenerative ORC enhances the efficiency and reduces the amount of waste needed to produce the same power with a reduced irreversibility (Mago et al., 2008).

In recent years, there has been a high demand to develop energy-efficient technologies for automobiles due to the increasing price of oils and fuels. The heat energy lost in exhaust gas can be retrieved by coupling it with an ORC so that the energy efficiency of the engine system can be improved. The selection of working fluid is crucial because it has a significant impact on the ORC system performance. However, most working fluids are selected from the conventional dataset; hence, the possibilities of discovering novel solutions or working fluids with enhanced properties are greatly reduced. Besides, the thermal efficiency of ORC using conventional working fluids is still considerably low. Hence, there is a need to develop an ORC that is able to recover low-grade waste heat from engine exhaust gas with better efficiency. In this paper, a framework that utilizes computer-aided molecular design (CAMD) algorithms is applied to develop an optimal working fluid for ORC with a wide range of optimization objectives. The objectives are specified in order to ensure that the working fluids selected provide optimal ORC system performance and possess favourable physical, chemical, economic, operating performance, safety and environmental properties.

11.4.2 CAMD in ORC Working Fluids Design

In order to achieve high ORC performance, one of the frequently used approaches to determine a suitable working fluid is by selecting organic fluid from the databank. Thermodynamic modelling is then carried out on each of the organic fluid to determine their respective efficiency. Quoilin et al. have proposed the operating map approach that concentrates on the interaction between working fluid and expansion machine (Quoilin et al., 2012). Shu et al. have chosen organic fluids from the alkane group as ORC working medium to recover waste heat of engine exhaust gas because they exhibit appropriate critical temperature and pressure. Besides, alkanes are environmentally friendly as they offer zero ozone depletion potential (ODP) and low global warming potential (GWP) (Shu et al., 2014). Wang et al. compared the optimized results of 13 working fluids that offer moderate vapour pressure in the evaporator, stability, appropriate critical temperature and ozone-safe (Wang et al., 2012). However, selecting the working fluids from the dataset often limits the opportunities for discovering novel and enhanced solutions from a massive amount of molecules that could be considered as potential ORC working medium candidates. This limitation can be avoided by applying computer-aided molecular design (CAMD) techniques.

CAMD techniques have already been proposed in literature to carry optimization on the selection of working fluids, which will offer a higher ORC efficiency in recovering waste heat from the low-grade heat sources (Papadopoulos et al., 2013; Lampe et al., 2012). The performance of ORC systems is highly dependent on the working fluids' properties. The properties are categorized into four groups and are shown next:

A. Thermodynamic Properties
   1) The fluid density should be high in both liquid and vapour phases as this leads to high mass flow rate (Papadopoulos et al., 2010). A low density will result to a higher volumetric flow rate, causing high pressure drop

in the heat exchangers and increased in equipment size (Bao and Zhao, 2013).

2) The fluid latent heat of vaporization should be high as this allows most of the heat to be added during phase change without the need to further undergo regenerative feed heating, which is used to regulate the superheating and expansion of vapour for the purpose of achieving higher efficiency. In addition, high latent heat is also related to fluids exhibiting a near-vertical saturated vapour line. This characteristic enables the formation of low moisture during expansion in turbine; hence, the need to condense a superheated fluid is avoided (Maizza and Maizza, 2001). However, some papers have proposed selecting a fluid with low latent heat of vaporization to avoid the formation of moisture during the expansion of saturated vapour (Yamamoto et al., 2001). In conclusion, a fluid with low latent heat of vaporization is preferable because these fluids are easier to be gasified; hence, more saturated vapours can be formed under an identical heat addition quantity condition during the evaporation process. The thermal efficiency can be improved because more amount of saturated vapours will enlarge the output power of the turbine (Wang et al., 2011).
3) The liquid fluid heat capacity should be low so that the work input needed to operate the pump is reduced and the work output from turbine can be increased (Chen et al., 2010).
4) The fluid viscosity should be low for both liquid and vapour phases for achieving high heat transfer coefficient in the heat exchangers with reduced power consumption (Mohanraj et al., 2009).
5) The fluid thermal conductivity should also be high to achieve high heat transfer coefficient in the heat exchangers (Bao and Zhao, 2013).
6) The fluid melting point must be lower than the lowest operating temperature of the ORC system (Bao and Zhao, 2013).
7) The fluid critical temperature must be higher than the highest operating temperature of the ORC system (Papadopoulos et al., 2010).
8) The fluid condensing gauge pressure should be operated above atmospheric pressure as vacuum operating condition may bring about air infiltration in ORC system (Quoilin et al., 2012).
9) The fluid thermal stability must be high because organic fluids will experience chemical deteriorations and decomposition at high temperature (Quoilin et al., 2012).
10) The fluid corrosiveness must be maintained at minimum level (Andersen and Bruno, 2005).

B. Environmental Properties (Quoilin et al., 2012)
1) The fluid ozone depletion potential (ODP) should be maintained at minimum level.
2) The fluid global warming potential (GWP) should be maintained at minimum level.

C. Safety Properties (Papadopoulos et al., 2010)
1) The fluid toxicity should be maintained at minimum level.

   2) The fluid flammability should be maintained at minimum level.

D. Process-Related Properties (Papadopoulos et al., 2010)
   1) The ORC process efficiency should be high in order to achieve high output power.
   2) The maximum operating pressure for ORC process should be low because extreme pressure operation requires the use of expensive equipment.
   3) The fluid mass flow rate should be kept low in order to reduce the operating costs.
   4) The fluid critical pressure should be higher than the maximum operating pressure, so that sub-critical operating condition is maintained.

Papadopoulos et al. performed the optimization using the thermodynamic properties as design objectives to come out with a set of molecules that will be included in the ORC process simulations for preliminary assessment of their performance in the process-related properties. The molecules generated will possess the favourable thermodynamic properties that are set as design objectives. The generated molecules will be further evaluated by determining their ORC system performance. However, some molecules generated may still give a low thermal efficiency because some target properties are not directly associated with the ORC thermal efficiency. The main purpose of almost all ORC working fluids studies is to identify the molecules that are able to offer enhanced ORC efficiency. Even though the generated set of molecules exhibit these favourable properties, this does not guarantee that the generated molecules are able to offer an improved ORC system efficiency.

In this paper, a new methodology is proposed to target the ORC thermal efficiency so that the molecules generated can offer an enhanced efficiency to the ORC system. Instead of targeting all the properties mentioned earlier, thermal efficiency will be the only property selected as design objective. Thermal efficiency can be approximately derived by incorporating the thermodynamic properties that have a direct impact on it. A simple correlation for thermal efficiency is developed, which incorporates all these specified thermodynamic properties. The effect of evaporating and condensing temperature on the thermal efficiency will also be studied. The temperature at which the optimum result is obtained is identified using two methods: (i) evaluating the performance of the molecules at a set of specified temperatures and (ii) applying disjunctive optimization to identify the temperature with the optimum result from a given case study temperature range.

11.4.3 Modelling Approach

The main objective of the project is to establish a systematic framework that applies CAMD for the development of optimal working fluid for ORC system. In this ORC working fluid design, the desired target properties will be defined as design objectives. The target properties identified need to be strongly associated with the ORC thermal efficiency as the main goal of this project is to develop an optimal working fluid which offers an enhanced ORC system performance. The model for each target property must then be developed. A few other properties are then identified as property constraints. Once the models are developed, the building blocks of the molecules must be identified. The molecules generated must be structurally feasible.

Lastly, optimization is carried out to optimize the specified design objectives, and a set of optimal molecules will then be generated. This proposed design methodology can be divided into three stages: problem formulation, model development and molecular design.

11.4.4 Problem Formulation

This stage begins with the definition of necessary design objectives. This can be done by identifying the desired target properties, such as thermodynamic, environmental, safety and process-related properties. One important note is that the working fluids need to be designed based on their ORC process performance. There are a number of common properties that have already been proposed, which are shown in Section 11.3.5, which can be employed to design and screen for the optimum working fluids. The main objective in this paper is to maximize ORC thermal efficiency. However, only a few of the suggested properties will directly affect thermal efficiency. In order to determine those properties, the equation for the thermal efficiency of an ORC system is first defined.

Figure 11.5 shows a general configuration of an ORC. The thermal efficiency of ORC can be calculated as shown in Equation 11.13:

$$\eta_{ORC} = \frac{W_{turbine} - W_{pump}}{Q_{source}} \times 100\% \tag{11.13}$$

where Q_{source} is the heat rejected from the heat source (kW), $W_{turbine}$ is the turbine work output (kW) and W_{pump} is the pump work load (kW). The equations for these three variables are defined.

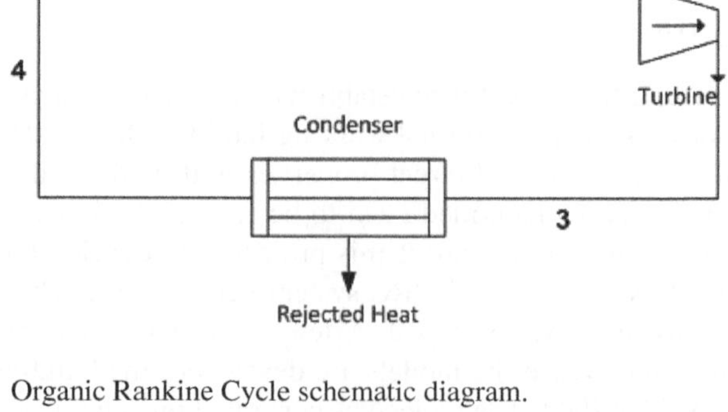

FIGURE 11.5 Organic Rankine Cycle schematic diagram.

In the evaporator stage, the assumptions made are that the working fluid experiences constant pressure (isobaric) heat addition and constant temperature (isothermal) phase change. The heat transferred is calculated as shown in Equation 11.14:

$$Q_{source} = M_f[c_{p,l}(T_{BP,U} - T_1) + H_v + c_{P,v}(T_F - T_{BP,U})] = M_f(H_2 - H_1) \quad (11.14)$$

where M_f is the molar flow rate of working fluid (kmol/s), $c_{P,l}$ is the liquid specific heat capacity (kJ/kmol.K), $c_{P,v}$ is the vapour specific heat capacity (kJ/kmol.K), $T_{BP,U}$ is the boiling point at upper operating pressure (K), T_1 is the subcooled temperature (K), T_F is the superheated temperature (K), H_v is the latent heat of vaporization (kJ/kmol), while H_1 and H_2 are the enthalpies at point 1 and point 2, respectively (kJ/kmol).

In the turbine stage, the assumptions made are that the working fluid experiences adiabatic and isentropic expansion, where the superheated or saturated vapour is converted into a state between superheated vapour and two-phase region. The turbine mechanical work output is calculated as shown in Equation 11.15:

$$W_{turbine} = M_f(H_2 - H_3) \quad (11.15)$$

where H_3 is the enthalpy at point 3 (kJ/kmol).

In the condenser stage, the assumptions made are that the working fluid experiences constant pressure (isobaric) heat rejection and constant temperature (isothermal) phase change, where the superheated vapour is first cooled to saturated vapour, and then it is condensed to saturated liquid. The heat transferred is calculated as shown in Equation 11.16.

$$Q_{\sin k} = M_f[C_{P},_v(T_3 - T_{BP,L}) + H_v] = M_f(H_3 - H_4) \quad (11.16)$$

where $Q_{\sin k}$ is the heat rejected by the working fluid (kW), T_3 is the temperature at point 3 (K), $T_{BP,L}$ S is the dew point at lower operating pressure (K), while H_4 is the enthalpy at point 4 (kJ/kmol).

In the pump stage, the assumptions made are that the working fluid experiences adiabatic and isentropic compression. The pump work load is calculated as shown in Equation 11.17:

$$w_{pump} = M_f v_l(P_{\max} - P_{\min}) = M_f(H_1 - H_4) \quad (11.17)$$

where v_1 is the liquid molar volume (m^3/kmol), P_{max} is the upper operating pressure (kPa) and P_{min}. is the lower operating pressure (kPa).

In order to achieve high ORC efficiency, the turbine work output, $W_{turbine}$, has to be high while both pump work load, W_{pump} and the heat rejected from the heat source, Q_{source}, have to be low. The properties involved in calculating Q_{source} are specific heat capacities ($c_{P,l}$ and $c_{P,v}$) and latent heat of vaporization (H_v), while the property involved in calculating W_{pump} is liquid molar volume (v_l). In order to keep both

Q_{source} and W_{pump} low, the value of $c_{P,l}$, $c_{P,v}$, H_v and v_l have to be minimum. Another property chosen to be included is the thermal conductivity of the organic liquid, k, which is considered crucial because a higher k can enhance heat transfer in the heat exchangers, eventually enhancing the thermal efficiency of ORC. Therefore, thermal conductivity of liquid (k) is to be high. However, $c_{P,v}$ will not be included as the fluid will only be heated to saturated vapour.

11.4.5 Model Development

In the second stage, all properties considered in the design objectives (k, $c_{P,l}$, H_v and v_l) can be calculated via different group contribution methods. In this work, only simple acyclic molecules are considered. Based on the group contribution methods mentioned in Chapter 5, only the first-order groups are applied. The group contribution model is shown in Equation 11.18.

$$f(X) = \sum_i N_i C_i \tag{11.18}$$

The group contribution models for some of the physical properties: normal boiling point (T_B), critical temperature (T_c) and critical pressure (P_c) are shown in Equations 11.19–11.21.

$$exp\left(\frac{T_B}{T_{B0}}\right) = \sum_i N_i T_{B1i} \tag{11.19}$$

$$exp\left(\frac{T_c}{T_{c0}}\right) = \sum_i N_i T_{c1i} \tag{11.20}$$

$$\left(P_c - P_{c1}\right)^{-0.5} - P_{c2} = \sum_i N_i P_{c1i} \tag{11.21}$$

There have been a lot of researches on the group contribution models for liquid thermal conductivity. The thermal conductivity of an organic liquid generally decreases with temperature, and it is a strong function of temperature. Sastri and Rao (1999) have proposed a new temperature-thermal conductivity relationship model for predicting liquid thermal conductivity, which is shown in Equations 11.22 and 11.23.

$$k = k_B a^{\left[1-\left[(1-T_r)/(1-T_{Br})^n\right]\right]} \tag{11.22}$$

$$k_B = \sum \Delta k_B + \sum \Delta k_{cor} \tag{11.23}$$

where k is the thermal conductivity of a saturated liquid at temperature T. (W/m.K), k_B is the thermal conductivity of a saturated liquid at the normal boiling point (W/m.K), T_r is the reduced temperature (T/T_c), T_{Br} is the reduced boiling point of the liquid (T_B/T_c), Δk_B is the contribution of the group/structure to thermal conductivity at the normal boiling point (W/m.K), Δk_{cor} is the correction factor to thermal conductivity at the normal boiling point (W/m.K). For alcohols and phenols, a and n take values of 0.856 and 1.23, respectively. For other liquids, a and n take values of 0.16 and 0.2, respectively (Sastri and Rao, 1999). Δk_{cor} takes different values for some specific cases, and it is relatively complex to apply that in the optimization model. In order to simplify the model, Δk_{cor} assumed to be negligible in this problem formulation.

Meanwhile, the ideal gas molar heat capacities, $c_{P,v}$, and liquid molar heat capacities, $c_{P,l}$ of the organic fluids can be calculated using the polynomial equations shown in Equations 11.24 and 11.25 (Buxton et al., 1999):

$$C_{P,v} = \left(\sum_j n_j \Delta_a^j - 37.93\right) + \left(\sum_j n_j \Delta_b^j + 0.21\right)T + \left(\sum_j n_j \Delta_c^j - 3.91\times10^{-4}\right) T^2 + \left(\sum_j n_j \Delta_d^j + 2.06\times10^{-7}\right)T^3 \quad (11.24)$$

$$\frac{c_{P,l} - c_{P,v}}{R} = 1.45 + 0.45\left(1-T_r\right)^{-1} + 0.25\omega\times\left[17.11 + 25.2\left(1-T_r\right)^{0.33} T_r^{-1} + 1.742\left(1-T_r\right)^{-1}\right] \quad (11.25)$$

where R is the universal gas constant, ω is the acentric factor, while Δ_a^j, Δ_b^j, Δ_c^j and Δ_d^j are group contributions parameters taken from Joback (Joback, 1989). The acentric factor, ω, can be determined using the proposed model as shown in Equation 11.26 (Constantinoua et al., 1995):

$$exp\left(\frac{\omega}{a}\right)^b - C = \sum_i N_i \omega_{1i} + D\sum_j M_j \omega_{2j} \quad (11.26)$$

where ω_{1i} is the contribution of the first-order group type i, which occurs N_i times, and ω_{2j} is the contribution of the second-order group type j, with M_j occurrences in a compound. The universal parameters a, b and C have the values of 0.4085, 0.5050 and 1.1507, respectively (Constantinoua et al., 1995). In this work, simple-structured molecules are considered; hence, only first-order group is applied, and D is assigned a zero value. Equation is thus reduced to:

$$exp\left(\frac{\omega}{a}\right)^b - C = \sum_i N_i \omega_{1i} \quad (11.27)$$

Constantinoua et al. also proposed a group contribution method model to estimate liquid molar volume, which is shown in Equation 11.28:

$$v_l - d = \sum_i N_i v_{1i} + F \sum_j M_j v_{2j} \tag{11.28}$$

where v_{1i} is the contribution of the first-order group type i, which occurs N_i times, and v_{2j} is the contribution of the second-order group type j, with M_j occurrences in a compound. The universal parameter d has the value of 0.01211 m^3kmol^{-1} (Constantinoua et al., 1995). As explained earlier, only first-order group is applied in this work.

Besides liquid thermal conductivity, the enthalpy of vaporization of organic components is also a strong function of temperature. Benkouider et al. have established a new group contribution method to estimate the enthalpy of vaporization of organic liquids as a function of temperature in which the model is shown in Equation 11.29.

$$H_v = \left(A - AT_r + AT_r^2\right)\left(c_1\left(1 - T_r\right)\right)^B \tag{11.29}$$

where c_1 is an adjustable parameter having a value of 0.003261; A and B are obtained by summing group contributions: $A = \sum N_i a_i$ and $B = \sum N_i b_i$; N_i is the number of groups of type i; a_i and b_i are the group contributions of group i (Benkouider et al., 2014).

11.4.6 Molecular Design

In the final stage, the developed model is employed to identify the optimum molecules that fulfil the design objectives or target properties. Firstly, the possible molecular groups acting as the potential building blocks are selected based on the property targets specified in the process design. Next, the structural constraints and target property range are specified and implemented in order to eliminate combination of infeasible solutions. In order to ensure that a molecule is generated, the summation for the number of occurrences for all selected groups must be greater than zero. Besides, in order to ensure that a molecule has no free bonds, the octet rule of structural feasibility is applied:

$$\sum_i N_i\left(2 - v_i\right) = 2g \tag{11.30}$$

In this work, the ORC is modelled to recover waste heat from the engine exhaust gas. In order to generate a set of optimal molecules, any optimization solver can be used. In this work, optimization is performed using LINGO 20.0. The general concept of optimization is to optimize (maximize or minimize) the value of the objective function, which is subjected to a number of constraints. The first step of optimization formulation is to determine the objective function and express the objective function as a mathematical function in terms of optimization variables (El-Halwagi, 2012). The objective function in this work is to maximize the thermal efficiency, and the four properties directly impacting the thermal efficiency are identified in Section 4.1: k, $c_{P,l}$, H_v and v_l. A simple correlation relating these four properties with thermal efficiency is developed as shown in Equation 11.31.

$$Z = \frac{k}{c_{P,l} \cdot H_v \cdot v_l} \tag{11.31}$$

By maximizing Z, k can be maximized while $c_{P,l}$, H_v and v_l are minimized simultaneously. Two hypotheses made on this are (i) the higher the Z-value, the higher the thermal efficiency of ORC is; and (ii) all four properties on the right-hand side of the Z-value equation contribute to the thermal efficiency of ORC with equal weightage. All group contribution models are included in the optimization model in order to determine the value of Z so that it can be optimization can be carried out.

The next step is to develop the constraints where all explicit relations, restrictions and limitations of the properties are identified. These can be expressed mathematically as equality and inequality constraints (El-Halwagi, 2012). The constraining conditions of the model are shown next:

(1) The amount of energy supplied by the heat source to the system is 1MW.
(2) The evaporating temperature studied ranges from 400 K to 600 K with 10 K intervals. 600 K is chosen as the upper limit as organic fluids would become chemically unstable at any temperature above this (Drescher and Brüggemann, 2007; Bao and Zhao, 2013).
(3) The working fluid is in the state of saturated vapour at the evaporator outlet, and the temperature at evaporator outlet is assumed to be similar to the evaporating temperature.
(4) The condensing temperature studied ranges from 310 K to 400 K with 10 K intervals.
(5) The working fluid is in the state of saturated liquid at the condenser outlet, and the temperature at condenser outlet is assumed to be similar to the condensing temperature.
(6) The critical temperature of the working fluid must be at least 20 K higher than the maximum operating temperature of the system for safety reasons.
(7) The critical pressure of the working fluid must be at least 5 bar higher than the maximum operating pressure of the system for safety reasons.
(8) The upper operating pressure is assumed to be equivalent to the saturated pressure of the working fluid at the evaporating temperature.
(9) The upper operating pressure must not exceed 25 bar (Tchanche et al., 2009; Bao and Zhao, 2013).
(10) The lower operating pressure is assumed to be equivalent to the saturated pressure of the working fluid at the condensing temperature.
(11) The lower operating pressure must not be lower than 0.05 bar (Drescher and Brüggemann, 2007).
(12) The isentropic efficiency of the pump is assumed to be at 0.85.
(13) The isentropic efficiency of the turbine is assumed to be at 0.8.

(14) The viscosity of the fluid generally increases as the fluid temperature decreases. Hence, the viscosity of the fluid at the lowest operating temperature must be less than 1.5 mPa.s.
(15) The flammability index of the working fluid should be less than 0.6 as any value above this is considered strongly flammable (Papadopoulos et al., 2010).
(16) The highest operating temperature should also be lower than the autoignition temperature.

After developing all the constraints, the group contribution methods model for some of the properties will be developed, which includes vapour (saturation) pressure [Constraints (8) and (10)], fluid viscosity [Constraint (14)], flammability index of the fluid [Constraint (15)] and autoignition temperature of the fluid [Constraint (16)].

Regarding Constraint 2, the evaporating temperature ranges from 400 K to 600 K with 10 K intervals. This is to observe the effect of evaporating temperature on thermal efficiency. Two different methods are proposed to evaluate this. In the first method, the temperature is fixed at a constant value, which ranges from 400 K to 600 K with 10 K intervals. A set of optimal molecules is generated at each temperature. However, this method is more time-consuming because the optimal results at each temperature must be identified first; only then can the overall optimal results be chosen among them. Another method is proposed, which involves the application of disjunctive optimization. The working principle is to first determine the optimum solution at the first temperature (e.g. 400 K) and the second temperature (e.g. 410 K). In the next step, select the best solution among these two optimum solutions. The selected solution will then 'square off' with the optimum solution at the next temperature (e.g. 420 K). The best solution is selected, and the same algorithm continues all the way until the final temperature (e.g. 600 K). This method is much more convenient as it is less time-consuming as compared to the first method.

In the final optimization formulation stage, the potential molecular building blocks must be identified. The generated set of molecules must also be structurally feasible. In this work, seven molecular groups or structures are chosen, in which Group 1, 2, 3, 4, 5, 6 and 7 represent CH_3, CH_2, CH, OH, C, CH_3O and CH_2O, respectively. These seven groups are selected because they are the most common building blocks used for the design of an ORC working fluid, which involves hydrocarbons, alcohols and ethers. Then, the number of occurrences of each group is limited to a specific number:

For Group 1 to 5: $N_i \leq 5$;
For Group 6 and 7: $N_i \leq 2$.

This is to prevent the optimization software to generate unbounded solutions. In this illustration, we only focus on acyclic molecules.

11.4.7 Results

The main objective of this optimization is to maximize the Z-constant. Based on the optimization results, seven feasible molecules with the highest Z-constant are selected and are shown in Table 11.33.

The ORC thermal efficiency for all seven molecules is then calculated and is shown in Table 11.34.

From the optimization results, the seven molecules with the maximum Z-constant value are identified and are to be further evaluated. Of the seven molecules, there are four molecules which are alcohols (ethanol, 1-propanol, 2-propanol and ethylene glycol), two molecules which are ethers (methylethyl ether and 1,1-dimethoxyethane), and one paraffin (butane). It can be observed that ethanol has the highest Z-constant

TABLE 11.33
Molecules with High *Z*-Constant

Molecule	Maximum Z-Constant $(\times 10^7)$	Evaporating Temperature at Maximum Z-Constant (K)
Ethanol	5.37	440
Methylethyl Ether	4.14	400
Ethylene Glycol	4.00	540
1-Propanol	3.75	490
2-Propanol	3.58	480
1,1-Dimethoxyethane	3.05	490
Butane	3.03	400

TABLE 11.34
Maximum Efficiency and Evaporating Temperature of the Selected Molecules

Molecule	Maximum Efficiency	Evaporating Temperature at Maximum Efficiency (K)
Ethanol	0.2346	470
Methylethyl Ether	0.1753	410
Ethylene Glycol	0.3142	600
1-Propanol	0.2448	500
2-Propanol	0.2278	480
1,1-Dimethoxyethane	0.2250	480
Butane	0.1628	400

value, which occurs when the evaporating temperature is at 440 K. It has the highest Z-constant value in the range of 400 K to 440 K. The molecule with the next highest Z-constant value is methylethyl ether at 400 K. Both methylethyl ether and butane can only operate at the lowest temperature of the studied temperature range as any temperature higher than that will violate the critical temperature constraint as the predicted critical temperature of both molecules using group contribution method (Marrero and Gani, 2001) are two of the lowest (433.6 K and 420.7 K respectively). Another alcohol, ethylene glycol, has the highest Z-constant value in the range of 440 K to 450 K and 490 K to 540 K, while 2-propanol and 1-propanol have the best results in the range of 450 K to 480 K and 480 K and 490 K, respectively.

The general trend for all molecules is that the Z-constant increases as the evaporating temperature increases. Hence, a higher evaporating temperature is preferable as it generates better results. However, a higher evaporating temperature is associated with higher evaporating pressure, which results in higher equipment cost. In order to limit the increased in evaporating pressure, the evaporating pressure must not exceed 30 bar. Besides, the critical temperature of the molecules also restricts the ORC from being operated below the critical temperature. In general, ethanol is the best working fluid to be operated at the lower temperature range (less than 440 K) as it has a lower critical temperature, both 1-propanol and 2-propanol can be considered for medium temperature range, while ethylene glycol can be used as working fluid for higher temperature range application (above 500 K) as it has the highest critical temperature. The working fluid at the evaporator outlet is in a saturated vapour state as studies have shown that superheating organic fluids will lead to an increase in system irreversibility based on second-law analysis. Superheating the fluids also does not affect the thermal efficiency of ORC significantly. Hence, the organic fluids are preferable to be operated at saturated conditions to minimize the total irreversibility of the system (Mago et al., 2007).

APPENDIX: LINGO CODING FOR COMPUTER-AIDED MOLECULAR DESIGN

```
model:
max = Z
Z = k/(Hv*Mv*Cpl);
!Manually input evaporating temperature (370K to 600K);
T=400;!unit=K;
!Manually input condensing temperature (310K to 400K); T_low =
310;!unit=K;
!Defining the group contribution n1=CH3, n2=CH2, N3=CH,
n4=OH, n5=C, n6=CH3O, n7=CH2O;

@GIN(n1); @GIN(n2); @GIN(n3); @GIN(n4);
@GIN(n5); @GIN(n6); @GIN(n7);

!Group contribution for critical temperature [Marrero and
Gani, 2001]; tc1 = 1.7506;
```

```
tc2 = 1.3327;
tc3 = 0.596;
tc4 = 5.2188;
tc5 = 0.0306;
tc6 = 3.4393;
tc7 = 2.4217;

!Group contribution for liquid molar volume [Constantinoua
et al., 1995]; v1 = 0.02614; @free(v1);
v2 = 0.01641; @free(v2); v3 = 0.00711; @free(v3); v4 =
0.00551; @free(v4); v5 = -0.0038; @free(v5); v6 = 0.03274;
@free(v6); v7 = 0.02311; @free(v7);

!Group contribution for enthalpy of vaporization (Term A)
[Benkouider et al., 2014];
ha1 = 360.27; @free(ha1); ha2 = 157.16; @free(ha2); ha3 =
-160.21; @free(ha3); ha4 = 1488.6; @free(ha4); ha5 = -370.48;
@free(ha5); ha6 = 959.45; @free(ha6); ha7 = 658.58;
@free(ha7);

!Group contribution for enthalpy of vaporization (Term B)
[Benkouider et al., 2014];
hb1 = 0.26051; @free(hb1); hb2 = 0.000698; @free(hb2);
hb3 = -0.26407; @free(hb3); hb4 = 0.28877; @free(hb4);
hb5 = -0.51473; @free(hb5); hb6 = 0.30676; @free(hb6);
hb7 = 0.03881; @free(hb7);

!Group contribution for normal boiling point [Marrero and
Gani, 2001]; tb1 = 0.8491; @free(tb1);
tb2 = 0.7141; @free(tb2); tb3 = 0.2925; @free(tb3);
tb4 = 2.567; @free(tb4); tb5 = -0.0671; @free(tb5);
tb6 = 1.7703; @free(tb6); tb7 = 1.3368; @free(tb7);

!Group contribution for thermal conductivity [Sastri and Rao,
1999]; k1 = 0.0545; @free(k1);
k2 = -0.0008; @free(k2); k3 = -0.06; @free(k3); k4 = 0.083;
@free(k4); k5 = -0.123; @free(k5); k6 = 0.0645; @free(k6);
k7 = 0.0092; @free(k7);

!Group contribution for ideal gas molar heat capacity [Joback,
1989];
cpa1 = 19.5; @free(cpa1); cpa2 = -0.909; @free(cpa2);
cpa3 = -23; @free(cpa3); cpa4 = 25.7; @free(cpa4);
cpa5 = -66.2; @free(cpa5); cpa6 = 45; @free(cpa6);
cpa7 = 24.591; @free(cpa7);

cpb1 = -0.00808; @free(cpb1); cpb2 = 0.095; @free(cpb2);
cpb3 = 0.204; @free(cpb3); cpb4 = -0.0691; @free(cpb4);
cpb5 = 0.427; @free(cpb5); cpb6 = -0.07128; @free(cpb6);
cpb7 = 0.0318; @free(cpb7);
```

```
!cpc and cpd values are not considered because these values
are relatively small compared to cpa and cpb,
it may causing some optimising scaling issues;
!cpc1 = 0.000153;!@free(cpc1);
!cpc2 = -0.0000544;!@free(cpc2);
!cpc3 = -0.000265;!@free(cpc3);
!cpc4 = 0.000177;!@free(cpc4);
!cpc5 = -0.000641;!@free(cpc5);
!cpc6 = 0.000264;!@free(cpc6);
!cpc7 = 0.0000566;!@free(cpc7);
!cpd1 = -0.0000000967;!@free(cpd1);
!cpd2 = 0.0000000119;!@free(cpd2);
!cpd3 = 0.00000012;!@free(cpd3);
!cpd4 = -0.0000000988;!@free(cpd4);
!cpd5 = 0.000000301;!@free(cpd5);
!cpd6 = -0.0000001955;!@free(cpd6);
!cpd7 = -0.0000000429;!@free(cpd7);
!cpd8 = 0.0000000652;!@free(cpd8);

!Group contribution for accentric factor [Constantinoua
et al., 1995]; w1 = 0.29602; @free(w1);
w2 = 0.14691; @free(w2); w3 = -0.07063; @free(w3);
w4 = 1.5237; @free(w4); w5 = -0.35125; @free(w5);
w6 = 0.52646; @free(w6); w7 = 0.44184; @free(w7);

!Group contribution for critical pressure [Marrero and Gani,
2001]; pc1 = 0.018615; @free(pc1);
pc2 = 0.013547; @free(pc2); pc3 = 0.007259; @free(pc3);
pc4 = -0.005401; @free(pc4); pc5 = 0.001219; @free(pc5);
pc6 = 0.020084; @free(pc6); pc7 = 0.017954; @free(pc7);

!Group contribution for lower flammability limit (LFL)
[Pintar, 1996]; LFL1 = 17.275;
LFL2 = 13.7022;
LFL3 = 10;
LFL4 = 2.1797;
LFL5 = 5.5291;
LFL6 = 14.2594;
LFL7 = 10.6866;

!Group contribution for upper flammability limit (LFL)
[Pintar, 1996]; HFL1 = 3.8461; @free(HFL1);
HFL2 = 1.4959; @free(HFL2); HFL3 = 0.2183; @free(HFL3);
HFL4 = 0.5856; @free(HFL4); HFL5 = -0.8422; @free(HFL5);
HFL6 = 1.6034; @free(HFL6); HFL7 = -0.7468; @free(HFL7);

!Group contribution for autoignition temperature [Pintar,
1996]; at1 = 301.91; @free(at1);
at2 = -10.86; @free(at2); at3 = -275.17; @free(at3);
at4 = 324.1; @free(at4); at5 = -570.43; @free(at5);
at6 = 283.31; @free(at6); at7 = -29.46; @free(at7);
```

```
!Equation for critical temperature (Tc) [Marrero and Gani,
2001];!unit=K; tc_RHS = n1*tc1+n2*tc2+n3*tc3+n4*tc4+n5*tc5+n6*
tc6+n7*tc7;
Tc = 231.239*@log(tc_RHS);!unit=K;

Tc >= T+20;!critical temperature must be at least 20K above
the evaporating temperature;

!Equation for liquid molar volume (Mv) [Constantinoua et al.,
1995];
!unit=m^3/kmol;
mv_RHS = n1*v1+n2*v2+n3*v3+n4*v4+n5*v5+n6*v6+n7*v7;
Mv = mv_RHS + 0.01211;!unit=m^3/kmol;
!Equation for enthalpy of vaporization (Hv)
[Benkouider et al., 2014];
!unit=kJ/kmol;
A = n1*ha1+n2*ha2+n3*ha3+n4*ha4+n5*ha5+n6*ha6+n7*ha7;
!summation of A terms;
B = n1*hb1+n2*hb2+n3*hb3+n4*hb4+n5*hb5+n6*hb6+n7*hb7;
!summation of B terms;
Tr = T/Tc;
c1 = 0.003261;
Hv = 1000*(A-A*Tr+A*Tr^2)*(c1*(1-Tr))^B;!unit=kJ/kmol;
!Equation for normal boiling point (NBP) [Marrero and Gani,
2001];!unit=K; tb_RHS =
n1*tb1+n2*tb2+n3*tb3+n4*tb4+n5*tb5+n6*tb6+n7*tb7;
NBP = 222.543*@log(tb_RHS);!unit=K;

!Equation for thermal conductivity (k) [Sastri and Rao, 1999];
!unit=W/m.K; TBr = NBP/Tc;
const_a1 = 0.16; const_n1 = 0.2;!constants for organic liquids
other than alcohols;
const_a2 = 0.856; const_n2 = 1.23;!constants for alcohols;
const_a = const_a1*(1-const_I1)+const_a2*const_I1;
const_n = const_n1*(1-const_I1)+const_n2*const_I1;
n4 <= 100*const_I1;!const_I1 is set to one when OH group is
present; @bin(const_I1);!const_I1 can only be binary integers;
kb = n1*k1+n2*k2+n3*k3+n4*k4+n5*k5+n6*k6+n7*k7;
k = kb*const_a^(1-((1-Tr)/(1-TBr))^const_n);!unit=W/m.K;

!Equation for ideal gas molar heat capacity (Cpv)
[Buxton et al., 1999];
!unit=kJ/kmol.K;
cpa = n1*cpa1+n2*cpa2+n3*cpa3+n4*cpa4+n5*cpa5+n6*cpa6+n7*cpa7;
cpb = n1*cpb1+n2*cpb2+n3*cpb3+n4*cpb4+n5*cpb5+n6*cpb6+n7*cpb7;
!cpc = n1*cpc1+n2*cpc2+n3*cpc3+n4*cpc4+n5*cpc5+n6*cpc6+n7*
cpc7+n8*cpc8;
!cpc and cpd are not considered;
!cpd = n1*cpd1+n2*cpd2+n3*cpd3+n4*cpd4+n5*cpd5+n6*cpd6+n7*
cpd7+n8*cpd8; T_avg = (T+T_low)/2;
Cpv = (cpa-37.93)+(cpb+0.21)*T_avg;!unit=kJ/kmol.K;
```

```
!Equation for accentric factor [Constantinoua et al., 1995];
w_RHS = n1*w1+n2*w2+n3*w3+n4*w4+n5*w5+n6*w6+n7*w7+1.1507;
w = 0.4085*(@log(w_RHS))^(1/0.505);

!Equation for liquid heat capacity (Cpl)
[Buxton et al., 1999];
!unit=kJ/kmol.K; Tr_avg = T_avg/Tc;
cpl_RHS = 1.45+0.45/(1-Tr_avg)+0.25*w*(17.11+25.2*((1-
Tr_avg)^0.33)/Tr_avg+1.742/(1-Tr_avg)); Cpl = cpl_
RHS*8.3145+Cpv;!unit=kJ/kmol.K;
!Equation for critical pressure (Pc) [Marrero and Gani, 2001];
!unit=bar; pc_RHS =
n1*pc1+n2*pc2+n3*pc3+n4*pc4+n5*pc5+n6*pc6+n7*pc7+0.108998;
Pc = 5.9827+(1/(pc_RHS^2));!unit=bar;

!Equation for vapour pressure of upper operating temperature
(VP_max) [Joback, 1989];
h_vp = TBr*(@log(Pc/1.01325))/(1-TBr); G_vp = 0.4835+0.4605*h_
vp;
k_vp = ((h_vp/G_vp)-(1+TBr))/(3+TBr)/((1-TBr)^2);
vpmax_RHS = (-G_vp/Tr)*(1-(Tr^2)+(k_vp*(3+Tr)*((1-Tr)^3)));
@free(vpmax_RHS);
VP_max = Pc*@exp(vpmax_RHS);!unit=bar;
Pc >= VP_max+5;!critical pressure must be at least 5 bar above
the evaporating pressure;
VP_max <= 25;!evaporating pressure cannot be greater
than 25 bar;

!Equation for vapour pressure of lower operating temperature
(VP_min) [Joback, 1989];
Tr_low = T_low/Tc;
vpmin_RHS = (-G_vp/Tr_low)*(1-(Tr_low^2)+(k_vp*(3+Tr_low)*
((1-Tr_low)^3))); @free(vpmin_RHS);
VP_min = Pc*@exp(vpmin_RHS);!unit=bar;
VP_min >= 0.05;!condensing pressure must not be lower than
0.05 bar;

!Equation for lower flammability limit (LFL) [Pintar, 1996];
LFL_const =
n1*LFL1+n2*LFL2+n3*LFL3+n4*LFL4+n5*LFL5+n6*LFL6+n7*LFL7;
LFL = 100/LFL_const;!unit=volume%;

!Equation for upper flammability limit (HFL) [Pintar, 1996];
HFL_const =
n1*HFL1+n2*HFL2+n3*HFL3+n4*HFL4+n5*HFL5+n6*HFL6+n7*HFL7;
HFL = 100/HFL_const;!unit=volume%;

!Equation for flammability (F-number) [Kondo et al., 2001];
F_const = LFL/HFL;
F_number = 1-F_const^0.5;
```

```
F_number < 0.65;!F_number must be lower than 0.65 as molecules
with any value higher than that is very flammable;

!Equation for autoignition temperature [Pintar, 1996];
Auto_T = n1*at1+n2*at2+n3*at3+n4*at4+n5*at5+n6*at6+n7*at7;
Auto_T > T+20;!autoignition temperature must be 20K higher
than evaporating temperature;
!Molecular structure constraints; n1+n2+n3+n4+n5+n6+n7>0;
!a molecule must be formed;
n1<=3; n2<=3; n3<=3; n4<=3; n5<=3;!the upper limit number of
each group; n6<=1; n7<=1;
!Free bonds for each group; val1 = 1;
val2 = 2;
val3 = 3;
val4 = 1;
val5 = 4;
val6 = 1;
val7 = 2;
!Structural constraint, the molecule generated must not
contain free bonds; (n1*val1+n2*val2+n3*val3+n4*val4+n5*val5+n
6*val6+n7*val7) - (2*(n1+n2+n3+n4+n5+n6+n7-1)) = 0;

end
```

REFERENCES

Andersen, W. C., & Bruno, T. J. (2005). Rapid Screening of Fluids for Chemical Stability in Organic. *Ind Eng Chem Res, 44*, 5560–5566.

Bao, J., & Zhao, L. (2013). A Review of Working Fluid and Expander Selections for Organic Rankine Cycle. *Renewable and Sustainable Energy Reviews, 24*, 325–342.

Benkouider, A. M. et al. (2014). Estimation of the Enthalpy of Vaporization of Organic Components as a Function of Temperature Using a New Group Contribution Method. *Journal of Molecular Liquids, 194*, 48–56.

Buxton, A., Livingston, A. G., & Pistikopoulos, E. N. (1999). Optimal Design of Solvent Blends for Environmental Impact Minimization. *AIChE Journal, 45*(4), 817–843.

Chan, C. W., Ling-Chin, J., & Roskilly, A. P. (2013). A Review of Chemical Heat Pumps, Thermodynamic Cycles and Thermal Energy Storage Technologies for Low Grade Heat Utilisation. *Applied Thermal Engineering, 50*(1), 1257–1273.

Chen, H., Goswami, D. Y., & Stefanakos, E. K. (2010). A Review of Thermodynamic Cycles and Working Fluids for the Conversion of Low-Grade Heat. *Renewable and Sustainable Energy Reviews, 14*(9), 3059–3067.

Constantinoua, L., Gania, R., & O'Connell, J. P. (1995). Estimation of the Acentric Factor and the Liquid Molar Volume at 298 K Using a New Group Contribution Method. *Fluid Phase Equilibria, 103*(1), 11–22.

Dai, Y., Wang, J., & Gao, L. (2009). Parametric Optimization and Comparative Study of Organic Rankine Cycle (ORC) for Low Grade Waste Heat Recovery. *Energy Conversion and Management, 50*(3), 576–582.

Desai, N. B., & Bandyopadhyay, S. (2009). Process Integration of Organic Rankine Cycle. *Energy, 34*(10), 1674–1686.

Drescher, U., & Brüggemann, D. (2007). Fluid Selection for the Organic Rankine Cycle (ORC) in Biomass Power and Heat Plants. *Applied Thermal Engineering, 27*(1), 223–228.

Ebrahimi, K., Jones, G. F., & Fleischer, A. S. (2014). A Review of Data Center Cooling Technology, Operating Conditions and the Corresponding Low-Grade Waste Heat Recovery Opportunities. *Renewable and Sustainable Energy Reviews, 31*, 622–638.

El-Halwagi, M. M. (2012). *Sustainable Design Through Process Integration: Fundamentals and Applications to Industrail Pollution Prevention, Resouce Conservation, and Profitability Enhancement.* Oxford: Elsevier.

Hung, T. C. (2001). Waste Heat Recovery of Organic Rankine Cycle Using Dry Fluids. *Energy Conversion and Management, 42*(5), 539–553.

Hung, T. C., Shai, T. Y., & Wang, S. K. (1997). A Review of Organic Rankine Cycles (ORCs) for the Recovery of Low-Grade Waste Heat. *Energy, 22*(7), 661–667.

Joback, K. G. (1989). *Designing Molecules Possessing Desired Physical Property Values,* Cambridge, MA: Massachusetts Institute of Technology.

Khannyra, S., Luna, M., Gil, M. L. A., Addou, M., & Mosquera, M. J. (2022). Self-Cleaning Durability Assessment of TiO2/SiO2 Photocatalysts Coated Concrete: Effect of Indoor and Outdoor Conditions on the Photocatalytic Activity. *Building and Environment, 211.* https://doi.org/10.1016/j.buildenv.2021.108743

Khannyra, S., Mosquera, M. J., Addou, M., & Gil, M. L. A. (2021). Cu-TiO2/SiO2 Photocatalysts for Concrete-Based Building Materials: Self-Cleaning and Air De-Pollution Performance. *Construction and Building Materials, 313.* https://doi.org/10.1016/j.conbuildmat.2021.125419

Kondo, S., Urano, Y., Tokuhashi, K., Takahashi, A., & Tanaka, K. (2001). Prediction of Flammability of Gases by Using F-number Analysis. *Journal of Hazardous Materials, 82*(2), 113–128.

Lampe, M., Groß, J., & Bardow, A. (2012). Simultaneous Process and Working Fluid Optimisation for Organic Rankine Cycles (ORC) Using PC-SAFT. *Computer Aided Chemical Engineering, 30*, 572–576.

Laudon, Matthew., Romanowicz, B. F., & Nano Science and Technology Institute. (2006). *NSTI Nanotech 2006: 2006 NSTI Nanotechnology Conference and Trade Show, Boston, May 7–11, 2006.* Nano Science and Technology Institute.

Liu, Q., Duan, Y., & Yang, Z. (2013). Performance Analyses of Geothermal Organic Rankine Cycles with Selected Hydrocarbon Working Fluids. *Energy, 63*, 123–132.

Mago, P. J., Chamra, L. M., & Somayaji, C. (2007). Performance Analysis of Different Working Fluids for Use in Organic Rankine Cycles. *Proceedings of the Institution of Mechanical Engineers, Part A: Journal of Power and Energy, 221*(3), 255–264.

Mago, P. J., Chamra, L. M., Srinivasan, K., & Somayaji, C. (2008). An Examination of Regenerative Organic Rankine Cycles Using Dry Fluids. *Applied Thermal Engineering, 28*(8–9), 998–1007.

Maizza, V., & Maizza, A. (2001). Unconventional Working Fluids in Organic Rankine-Cycles for Waste Energy Recovery Systems. *Applied Thermal Engineering, 21*(3), 381–390.

Marion, M., Voicu, I., & Tiffonnet, A.-L. (2012). Study and Optimization of a Solar Subcritical Organic Rankine Cycle. *Renewable Energy, 48*, 100–109.

Marrero, J., & Gani, R. (2001). Group-Contribution Based Estimation of Pure Component Properties. *Fluid Phase Equilibria, 183–184*, 183–208.

Mistry, S. (2021). *HLB Scale (Hydrophilic Lipophilic Balance) – Solution Pharmacy.* https://solutionpharmacy.in/hlb-scale/

Mohanraj, M., Jayaraj, S., & Muraleedharan, C. (2009). Environment Friendly Alternatives to Halogenated Refrigerants – A Review. *International Journal of Greenhouse Gas Control, 3*(1), 108–119.

Obernberger, I., Thonhofer, P., & Reisenhofer, E. (2002). Description and Evaluation of the New 1,000 kWel Organic Rankine Cycle Process Integrated in the Biomass CHP Plant in Lienz, Austria. *Euroheat & Power, 10*, 1–17.

Papadopoulos, A. I., Stijepovic, M., & Linke, P. (2010). On the Systematic Design and Selection of Optimal Working Fluids for Organic Rankine Cycles. *Applied Thermal Engineering, 30*(6–7), 760–769.

Papadopoulos, A. I. et al. (2013). Molecular Design of Working Fluid Mixtures for Organic Rankine Cycles. *Computer Aided Chemical Engineering, 32*, 289–294.

Pierre, J., Leroy, V., Saint-Jalmes, A., Dollet, B., Ben Salem, I., Crassous, J., Guillermic, R.-M., Drenckhan, W., & Elias, F. (2013). Shaving Foam: A Complex System for Acoustic Wave Propagation. *The Journal of the Acoustical Society of America, 133*(5), 3374–3374.

Pintar, A. J. (1996). *Technical Support Document DIPPR Project 912*. Houghton: Michican Technological University.

Quoilin, S. et al. (2012). *Working Fluid Selection and Operating Maps for Organic Rankine Cycle Expansion Machines*. Purdue University; University of Liège.

Rayegan, R., & Tao, Y. X. (2011). A Procedure to Select Working Fluids for Solar Organic Rankine Cycles (ORCs). *Renewable Energy, 36*(2), 659–670.

Saleh, B., Koglbauer, G., Wendland, M., & Fischer, J. (2007). Working Fluids for Low-Temperature Organic Rankine Cycles. *Energy, 32*(7), 1210–1221.

Sastri, S., & Rao, K. (1999). A New Temperature-Thermal Conductivity Relationship for Predicting Saturated Liquid Thermal Conductivity. *Chemical Engineering Journal, 74*, 161–169.

Shu, G. et al. (2014). Alkanes as Working Fluids for High-Temperature Exhaust Heat Recovery of Diesel Engine Using Organic Rankine Cycle. *Applied Energy, 119*, 204–217.

Stiver, W., & Mackay, D. (1984). Evaporation Rate of Spills of Hydrocarbons and Petroleum Mixtures. *Environmental Science & Technology, 18*, 834–840.

Tchanche, B. F., Papadakis, G., Lambrinos, G., & Frangoudakis, A. (2009). Fluid Selection for a Low-Temperature Solar Organic Rankine Cycle. *Applied Thermal Engineering, 29*(11–12), 2468–2476.

Van Cott, A., Hastings, C. E., Landsiedel, R., Kolle, S., & Stinchcombe, S. (2018). GHS Additivity Formula: Can It Predict the Acute Systemic Toxicity of Agrochemical Formulations that Contain Acutely Toxic Ingredients? *Regulatory Toxicology and Pharmacology, 92*, 407–419.

Walsh, C., & Thornley, P. (2013). A Comparison of Two Low Grade Heat Recovery Options. *Applied Thermal Engineering, 53*(2), 210–216.

Wang, E. H. et al. (2011). Study of Working Fluid Selection of Organic Rankine Cycle (ORC) for Engine Waste Heat Recovery. *Energy, 36*(5), 3406–3418.

Wang, J. et al. (2013). Thermodynamic Analysis and Optimization of an (Organic Rankine Cycle) ORC Using Low Grade Heat Source. *Energy, 49*, 356–365.

Wang, Z. Q., Zhou, N. J., Guo, J., & Wang, X. Y. (2012). Fluid Selection and Parametric Optimization of Organic Rankine Cycle Using Low Temperature Waste Heat. *Energy, 40*(1), 107–115.

Yamamoto, T., Furuhata, T., Arai, N., & Mori, K. (2001). Design and Testing of the Organic Rankine Cycle. *Energy, 26*(3), 239–251.

Index

G

H

I

J

K

L

M

N

O

P

Q

R

S

T

U

V

W

For Product Safety Concerns and Information please contact our
EU representative GPSR@taylorandfrancis.com Taylor & Francis
Verlag GmbH, Kaufingerstraße 24, 80331 München, Germany